# HVAC Water Chillers and Cooling Towers

Fundamentals, Application, and Operation

Second Edition

**MECHANICAL ENGINEERING**
A Series of Textbooks and Reference Books



**RECENTLY PUBLISHED TITLES**

*HVAC Water Chillers and Cooling Towers: Fundamentals, Application, and Operation, Second Edition,*
Herbert W. Stanford III

*Ultrasonics: Fundamentals, Technologies, and Applications, Third Edition,*
Dale Ensminger and Leonard J. Bond

*Mechanical Tolerance Stackup and Analysis, Second Edition,*
Bryan R. Fischer

*Asset Management Excellence,*
John D. Campbell, Andrew K. S. Jardine, and Joel McGlynn

*Solid Fuels Combustion and Gasification: Modeling, Simulation, and Equipment Operations, Second Edition, Third Edition,*
Marcio L. de Souza-Santos

*Mechanical Vibration Analysis, Uncertainties, and Control, Third Edition,*
Haym Benaroya and Mark L. Nagurka

*Principles of Biomechanics,*
Ronald L. Huston

*Practical Stress Analysis in Engineering Design, Third Edition,*
Ronald L. Huston and Harold Josephs

*Practical Guide to the Packaging of Electronics, Second Edition: Thermal and Mechanical Design and Analysis,*
Ali Jamnia

*Friction Science and Technology: From Concepts to Applications, Second Edition,*
Peter J. Blau

*Design and Optimization of Thermal Systems, Second Edition,*
Yogesh Jaluria

*Analytical and Approximate Methods in Transport Phenomena,*
Marcio L. de Souza-Santos

*Introduction to the Design and Behavior of Bolted Joints, Fourth Edition: Non-Gasketed Joints,*
John H. Bickford

# HVAC Water Chillers and Cooling Towers

## Fundamentals, Application, and Operation

## Second Edition

Herbert W. Stanford III

CRC Press
Taylor & Francis Group
Boca Raton London New York

CRC Press is an imprint of the
Taylor & Francis Group, an **informa** business

CRC Press
Taylor & Francis Group
6000 Broken Sound Parkway NW, Suite 300
Boca Raton, FL 33487-2742

First issued in paperback 2017



Version Date: 20111017

ISBN 13: 978-1-4398-6202-5 (hbk)
ISBN 13: 978-1-138-07171-1 (pbk)



**Visit the Taylor & Francis Web site at**
**http://www.taylorandfrancis.com**

**and the CRC Press Web site at**
**http://www.crcpress.com**

# Contents



## *SECTION A Water Chillers: Fundamentals, Application, and Operation*

## *PART I Chiller Fundamentals*

# PART III Chiller Operations and Maintenance



# SECTION B Cooling Towers: Fundamentals, Application, and Operation

# PART I Cooling Tower Fundamentals

# List of Figures

# Preface

This is the second edition of *HVAC Water Chillers and Cooling Towers*, which was first published in 2003. In the past 8 years, there have been major improvements to many chiller and cooling tower elements resulting in both improved performance and lower operating costs. Climate change and a new focus on "green" design have significantly impacted the selection of refrigerants and the application of chilled water systems. And, finally, the expanded use of digital controls and variable frequency drives, along with reapplication of some older technologies, especially ammonia-based absorption cooling, has necessitated updating of this text in a new, second edition.

There are two fundamental types of HVAC systems designed to satisfy building cooling requirements: *direct expansion (DX)* systems, where there is direct heat exchange between the building air and a primary refrigerant, and *secondary refrigerant* systems that utilize *chilled water* as an intermediate heat exchange media to transfer heat from the building air to a refrigerant.

Chilled water systems are the heart of central HVAC cooling, providing cooling throughout a building or a group of buildings from one source. Centralized cooling offers numerous operating, reliability, and efficiency advantages over individual DX systems and, on a life cycle basis, can have significantly lower total cost. And, chilled water systems, especially with water-cooled chillers, represent a much more "green" design option.

Every central HVAC cooling system is made up of one or more refrigeration machines or *water chillers* designed to collect excess heat from buildings and reject that heat to the outdoor air. The water chiller may use the *vapor compression refrigeration cycle* or an *absorption refrigeration cycle* (utilizing either lithium bromide or ammonia solutions). Vapor compression refrigeration compressors may be *reciprocating*, *scroll*, *helical screw*, or *centrifugal* type with electric or gas-fired engine prime movers.

The heat collected by any water chiller must be rejected to the atmosphere. This waste heat can be rejected by *air-cooling* in a process that transfers heat directly from the refrigerant to the ambient air or by *water-cooling*, a process that uses water to collect the heat from the refrigerant and then to reject that heat to the atmosphere. Water-cooled systems offer advantages over air-cooled systems, including smaller physical size, longer life, and higher operating efficiency (in turn resulting in reduced greenhouse gas contribution and atmospheric warming). The success of their operation depends, however, on the proper sizing, selection, application, operation, and maintenance of one or more *cooling towers* that act as heat rejecters.

The goal of this book is to provide the HVAC designer, the building owner and his or her operating and maintenance staff, the architect, and the mechanical contractor with definitive and practical information and guidance relative to the application, design, purchase, operation, and maintenance of water chillers and cooling towers. The first half of the book discusses water chillers, while the second half addresses cooling towers.

Each of these two topics is treated in separate sections, each of which is divided into three basic parts:

1. Under "Fundamentals," the basic information about systems and equipment is presented. How they work and their various components are presented and discussed.
2. Under "Design and Application," equipment sizing, selection, and application are discussed. In addition, the details of piping, control, and water treatment are presented. Finally, special considerations such as noise control, electrical service, fire protection, and energy efficiency are presented.
3. Finally, the "Operations and Maintenance" section takes components and systems from commissioning through programmed maintenance. Chapters on purchasing equipment include guidelines and recommended specifications for procurement.

*This is not an academic textbook, but a book designed to be useful on a day-to-day basis*, providing answers about water chiller and cooling tower use, application, and problems. Extensive checklists, design and/or troubleshooting guidelines, and reference data are provided.

**Herbert W. Stanford III, PE**
2011

# Author

**Herbert W. Stanford III, PE,** is a North Carolina native and a 1966 graduate of North Carolina State University with a BS in mechanical engineering. He is a registered professional engineer in North Carolina, South Carolina, and Maryland.

In 1977, he founded Stanford White, Inc., an engineering consulting firm located in Raleigh, North Carolina and semiretired in 1998.

Currently, Mr. Stanford is actively engaged in investigative and forensic engineering, teaching, and writing within a broad range of topics relative to buildings, especially heating, ventilating, and air-conditioning (HVAC) systems; indoor environmental quality; and building operations and maintenance.

Mr. Stanford developed the Facilities Condition Assessment Program for the State of North Carolina that is used for allocation of annual repair/replacement funding and the Life Cycle Cost Methodology used by North Carolina to evaluate the cost effectiveness of building design decisions.

Since his "semiretirement," he has taught a series of short courses on current building topics at the University of Toledo (Ohio), North Carolina State University, and the University of North Carolina at Charlotte.

He is a life member of the American Society of Heating, Refrigerating, and Air-Conditioning Engineers (ASHRAE).

Mr. Stanford is the author of *Analysis and Design of Heating, Ventilating, and Air-Conditioning Systems* (Prentice-Hall, 1988), a text on the evaluation, analysis, and design of building HVAC systems; *Water Chillers and Cooling Towers: Fundamentals, Application, and Operation* (Marcel Dekker, 2003); *The Health Care HVAC Technician* (MGI Systems, Inc., 2008), a program and training manual for hospital HVAC maintenance personnel; and *Guide to Effective Building Maintenance* (Fairmont Press, 2010).

# Section A

## Water Chillers: Fundamentals, Application, and Operation

# Part I

## Chiller Fundamentals

# 1 Refrigeration Machines

Heat flows from hot regions to cold regions, driven by temperature difference and nature's desire to "level" energy differences (the Third Law of Thermodynamics). To reverse this process in a system and move heat from a lower-temperature region to a higher-temperature region requires that "work" be done on the system. Thus, we use *refrigeration machines* to provide work to move heat from a cooled area and reject it to a hot area. The performance of these machines is usually characterized by a quantity known as the *coefficient of performance* (COP), defined as

$$\mathrm{COP} = \frac{\text{removed heat}}{\text{input work}}$$

Therefore, COP is a dimensionless ratio of how much heat is transferred out of the cooled space to the amount of work that is used to accomplish this task. Unlike typical definitions of "efficiency," the COP can be larger than unity. Higher values are better, indicating that more heat is removed for a given amount of work. COP is usually dependent on operating conditions, such as the temperatures of the cooled space and the hot space to which heat is to be rejected, and the type of refrigeration cycle utilized.

All refrigeration cycles hinge on one common physical characteristic: if a chemical compound (which we can call a *refrigerant*) changes phase from a liquid to a gas, which is called *evaporation*, the compound must absorb heat to do so. Likewise, if the refrigerant changes phase back from a gas to a liquid, which is called *condensation*, the absorbed heat must be rejected. Thus, all refrigeration cycles depend on circulating a refrigerant between a heat "source" (with heat to be removed, thus resulting in cooling) and a heat "sink" (where the collected heat can be rejected).

Overall, there are two basic refrigeration cycles in common use: the *vapor compression cycle* and the *absorption cycle*. Each of these cycles can be used to cool a secondary refrigerant, usually water, which is then used to cool the spaces in a building. The refrigeration machine utilized in this application is typically called a *water chiller* or simply a *chiller*.

## VAPOR COMPRESSION REFRIGERATION

### Refrigeration Cycle

The *vapor compression cycle*, wherein a chemical substance alternately changes from liquid to gas and from gas to liquid, actually consists of four distinct steps:

1. *Compression.* Low-pressure refrigerant gas is compressed, thus raising its pressure by expending mechanical energy. There is a corresponding increase in temperature along with the increased pressure.
2. *Condensation.* The high-pressure, high-temperature gas is cooled by outdoor air or water that serves as a "heat sink" and condenses to a liquid form at high pressure.
3. *Expansion.* The high-pressure liquid flows through an orifice in the expansion valve, thus reducing the pressure. A small portion of the liquid "flashes" to gas due to the pressure reduction.
4. *Evaporation.* The low-pressure liquid absorbs heat from indoor air or water and evaporates to a gas or vapor form. The low-pressure vapor flows to the compressor and the process repeats.

As shown in Figure 1.1, the vapor compression refrigeration system consists of four components that perform the four steps of the refrigeration cycle. The *compressor* raises the pressure of the initially low-pressure refrigerant gas. The

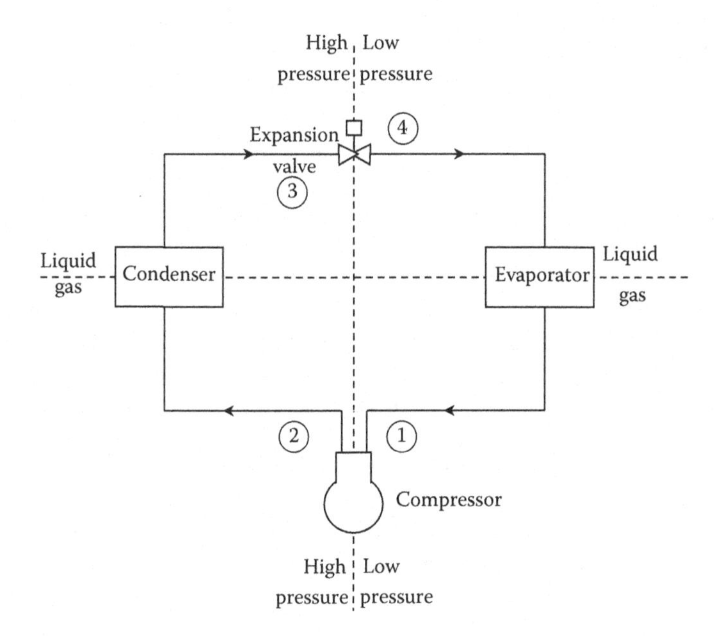


**FIGURE 1.1** Basic components of the vapor compression refrigeration system. Condition point numbers correspond to points on pressure–enthalpy chart (Figure 1.3).

*condenser* is a heat exchanger that cools the high-pressure gas so that it changes phase to liquid. The *expansion valve* controls the pressure ratio, and thus flow rate, between the high- and low-pressure regions of the system. The *evaporator* is a heat exchanger that heats the low-pressure liquid, causing it to change phase from liquid to vapor (gas).

Thermodynamically, the most common representation of the basic refrigeration cycle is made utilizing a *pressure–enthalpy chart*, as shown in Figure 1.2. For each refrigerant, the phase-change line represents the conditions of pressure and total heat content (enthalpy) at which it changes from liquid to gas and vice versa. Thus, each of the steps of the vapor compression cycle can easily be plotted to demonstrate the actual thermodynamic processes at work, as shown in Figure 1.3.

Point 1 represents the conditions *entering* the compressor. Compression of the gas raises its pressure from $P_1$ to $P_2$. Thus, the "work" that is done by the compressor adds heat to the refrigerant, raising its temperature and slightly increasing its heat content. Thus, point 2 represents the condition of the refrigerant *leaving* the compressor and *entering* the condenser. In the condenser, the gas is cooled, reducing its enthalpy from $h_2$ to $h_3$.

The portion between points 3 and 4 represents the pressure reduction that occurs in the expansion process. Due to a small percentage of the liquid evaporating as a result of the pressure reduction, the temperature and enthalpy of the remaining liquid are also reduced slightly. Point 4 then represents the condition *entering* the evaporator. The portion between points 4 and 1 represents the heat gain by the liquid, increasing its enthalpy from $h_4$ to $h_1$, completed by the phase change from liquid to gas at point 1.

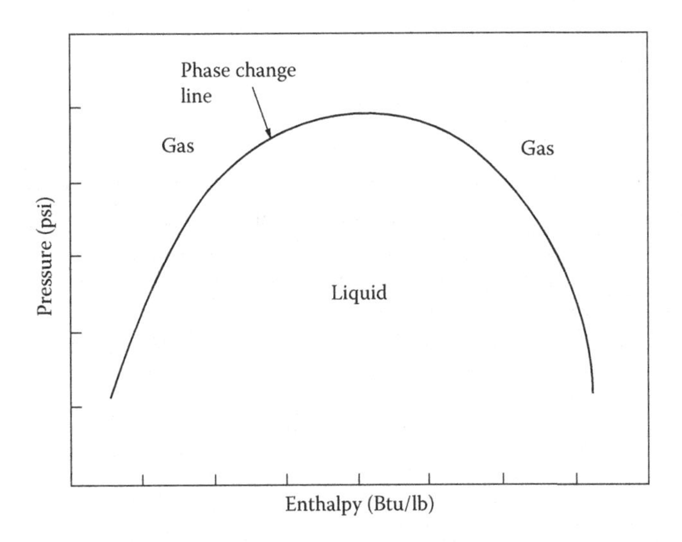


**FIGURE 1.2** Basic refrigerant pressure–enthalpy relationship.

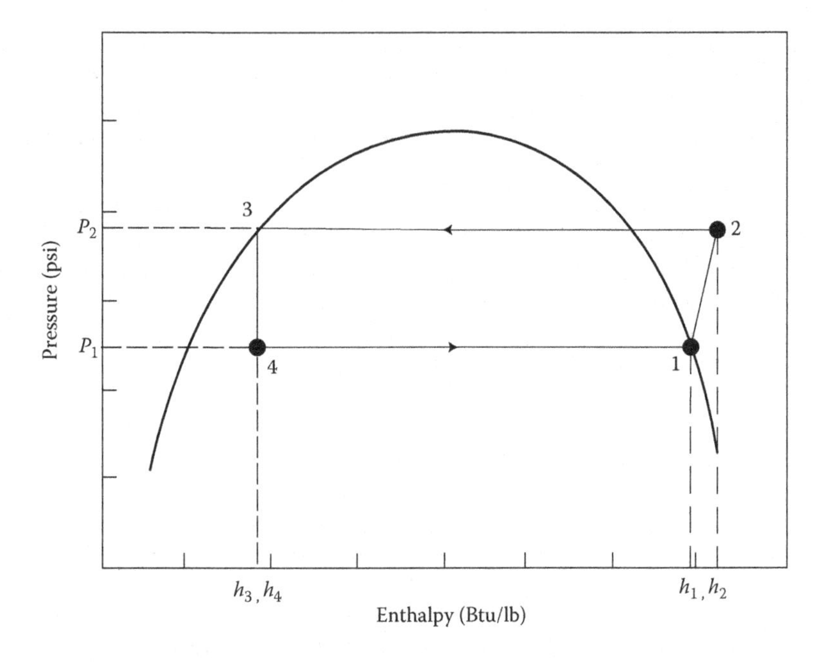


**FIGURE 1.3** Ideal refrigeration cycle imposed over a pressure–enthalpy chart.

For any refrigerant whose properties are known, a pressure–enthalpy chart can be constructed and the performance of a vapor compression cycle analyzed by establishing the high and low pressures for the system. (Note that Figure 1.3 represents an "ideal" cycle and in actual practice there are various departures dictated by second-law inefficiencies.)

## Refrigerants

Any substance that absorbs heat may be termed as a *refrigerant.* Secondary refrigerants, such as water or brine, absorb heat but do not undergo a phase change in the process. Primary refrigerants, then, are those substances that possess the chemical, physical, and thermodynamic properties that permit their efficient use in the typical vapor compression cycle.

In the vapor compression cycle, a refrigerant must satisfy several (sometimes conflicting) requirements:

1. The refrigerant must be chemically stable in both the liquid and vapor states.
2. Refrigerants must be nonflammable and have low toxicity.
3. The thermodynamic properties of the refrigerant must meet the temperature and pressure ranges required for the application.

Early refrigerants, developed in the 1920s and 1930s, used in HVAC applications were predominately chemical compounds made up of *chlorofluorocarbons* (CFCs) such as R-11, R-12, and R-503. While stable and efficient in the range of temperatures and pressures required within contained HVAC cooling systems, these CFC refrigerants were also used as aerosol propellants and cleaning agents in a wide range of industrial and commercial products. Once released into the air, these refrigerants had significantly adverse effects on the atmosphere.

CFC refrigerant gas was found to be long lived in the atmosphere. In the lower atmosphere, the CFC molecules absorb infrared radiation and, thus, contribute to atmospheric warming. Then, once it is in the upper atmosphere, the CFC molecule breaks down to release chlorine that destroys ozone and, consequently, damages the atmospheric ozone layer that protects the earth from excess UV radiation. These, and all other refrigerants, are now assigned an *Ozone Depletion Potential* (ODP) and/or *Global Warming Potential* (GWP), defined as follows:

- ODP of a chemical compound is the relative amount of degradation it can cause to the ozone layer.
- GWP is a measure of how much a given mass of a gas contributes to global warming. GWP is a relative scale that compares the greenhouse gas to carbon dioxide with a GWP, by definition, of 1.

Table 1.1 summarizes the ODP and GWP for a number of refrigerants commonly used in HVAC chiller systems.

The manufacture of CFC refrigerants in the United States and most other industrialized nations was eliminated by international agreement in 1996. While refrigeration equipment that utilizes CFC refrigerants is still in use, no new equipment using these refrigerants is now available in the United States or Europe.

Earlier on, to replace CFCs, researchers found that by modifying the chemical compound of CFCs by substituting a hydrogen atom for one or more of the chlorine or fluorine atoms resulted in a significant reduction in the life of the molecule and, thus, almost eliminated ODP and significantly reduced GWP. Some of these compounds, called *hydrochlorofluorocarbons* (HCFCs), are currently used in HVAC water chillers, especially R-22 and R-123.

While HCFCs have reduced the potential environmental damage from refrigerants released into the atmosphere, the potential for damage has not been totally eliminated. Again, under international agreement, this class of refrigerants is slated for phaseout for new equipment installations in 2010–2020, with total halt to manufacturing and importing into the United States mandated by 2030, as summarized in Table 1.2.

To replace HCFCs, a third class of refrigerants, called *hydrofluorocarbons* (HFCs), has been developed since about 1990, including R-134a, R-410A, and R-407C, all of which are commonly applied in HVAC equipment, although the latter two are primarily used only in smaller-packaged DX systems. This class of refrigerants has essentially no ODP and GWP levels that are 50–70% lower than CFCs.

**TABLE 1.1**
**ODP and CWP for Common Refrigerants**

| Refrigerant | Ozone Depletion Potential | Global Warming Potential |
|---|---|---|
| R-11 Trichlorofluoromethane | 1.0 | 4000 |
| R-12 Dichlorodifluoromethane | 1.0 | 2400 |
| R-13 B1 Bromotrifluoromethane | 10 | |
| R-22 Chlorodifluoromethane | 0.05 | 1700 |
| R-32 Difluoromethane | 0 | 650 |
| R-113 Trichlorotrifluoroethane | 0.8 | 4800 |
| R-114 Dichlorotetrafluoroethane | 1.0 | 3.9 |
| R-123 Dichlorotrifluoroethane | 0.02 | 0.02 |
| R-124 Chlorotetrafluoroethane | 0.02 | 620 |
| R-125 Pentafluoroethane | 0 | 3400 |
| R-134a Tetrafluoroethane | 0 | 1300 |
| R-143a Trifluoroethane | 0 | 4300 |
| R-152a Difluoroethane | 0 | 120 |
| R-290 Propane | 0 | 3 |
| R-401A (53% R-22, 34% R-124, 13% R-152a) | 0.37 | 1100 |
| R-401B (61% R-22, 28% R-124, 11% R-152a) | 0.04 | 1200 |
| R-402A (38% R-22, 60% R-125, 2% R-290) | 0.02 | 2600 |
| R-404A (44% R-125, 52% R-143a, R-134a) | 0 | 3300 |
| R-407A (20% R-32, 40% R-125, 40% R-134a) | 0 | 2000 |
| R-407C (23% R-32, 25% R-125, 52% R-134a) | 0 | 1600 |
| R-502 (48.8% R-22, 51.2% R-115) | 0.283 | 4.1 |
| R-717 Ammonia ($NH_3$) | 0 | 0 |

R-134a is utilized in *positive-pressure* rotary compressor water chillers offered by the vast majority of manufacturers. However, at least one manufacturer continues to offer *negative-pressure* centrifugal compressor water chillers using R-123 (an HCFC), creating a design dilemma for engineers and owners. *At this time, R-123 chiller has only 8–9 years of guaranteed refrigerant supply and new installations of chillers using this refrigerant should be avoided.*

Unfortunately, HFCs still have fairly high GWPs. Already, in Europe, there is legislation requiring that these refrigerants be eliminated. The European Union has issued a directive to phase out refrigerants with a GWP greater (ultimately) than 150 in autos beginning in 2011. This limitation is expected to be applied to HVAC systems beginning in 2015.

While no GWP-limiting legislation exists for the United States, there is a search underway for new low-GWP refrigerants to replace HFCs worldwide. This search is currently centering around older refrigerants such as R-245a (*propane*), especially in Europe, and R-717 (*ammonia*). Both refrigerants are highly flammable and ammonia has specific safety concerns and requirements that limit its potential application.

**TABLE 1.2**
**Implementation of HCFC Refrigerant Phaseout in the United States**

| Year Implemented | Clean Air Act Regulations |
|---|---|
| 2010 | No production and no importing of HCFC R-22 except for use in equipment manufactured prior to January 1, 2010. (Consequently there will be no production or importing of new refrigeration equipment using R-22. Existing equipment must depend on stockpiles or recycling for refrigerant supplies.) |
| 2015 | No production and no importing of any HCFC refrigerants except for use in equipment manufactured before January 1, 2020. |
| 2020 | No production or importing of HCFC R-22. (This is also the cutoff date for the manufacture of new equipment using HCFC refrigerants other than R-22 and should end the installation of new chillers using R-123.) |
| 2030 | No production or importing of any HCFC refrigerant. (While it is anticipated that the vast majority of equipment using R-22 will have been replaced by this date, there will still be a significant number of water chillers using R-123 still in operation. These chillers must depend on stockpiles or recycling for refrigerant supplies.) |

However, a new class of refrigerants called *hydrofluoro olefins* (HFOs) is now becoming available. The first of these refrigerants is Dupont's R-1234yf, designed to be a direct replacement for R-134a. While all research to date has focused on applying R-1234yf to automobile compressors, Dupont anticipates that this refrigerant, or blends of this refrigerant with R-744 (*carbon dioxide*), can reduce GWP by 50% over R-134a. Already, retrofits of existing R-134a compressors with R-1234yf have been implemented and the results are promising, with no oil problems or capacity loss and even small efficiency improvements.

*ASHRAE Standard 34-2010* classifies refrigerants according to their toxicity (A = nontoxic and B = evidence of toxicity identified) and flammability (1 = no flame propagation, 2 = low flammability, and 3 = high flammability). Thus, all refrigerants fall within one of the "safety groups," A1, A2, A2L, A3, B1, B2, B2L, or B3. The "L" designation indicates that a refrigerant has lower flammability than the range established for the "2" rating, but is not a "1."

Table 1.3 lists the safety group classifications for common refrigerants.

## ABSORPTION REFRIGERATION

### Absorption Refrigeration Cycle

The absorption refrigeration cycle is a relatively old technology. The concept dates back to the late 1700s and the first absorption refrigeration machine was built in the 1850s. However, by World War I, the technology and use of reciprocating compressors had advanced to the point where interest in and development of

**TABLE 1.3**
**HVAC Refrigerant Safety Groups**

| Refrigerant | Refrigerant Safety Classification |
|---|---|
| R-11 Trichlorofluoromethane | A1 |
| R-12 Dichlorodifluoromethane | A1 |
| R-13 B1 Bromotrifluoromethane | A1 |
| R-22 Chlorodifluoromethane | A1 |
| R-32 Difluoromethane | A2 |
| R-113 Trichlorotrifluoroethane | A1 |
| R-114 Dichlorotetrafluoroethane | A1 |
| R-123 Dichlorotrifluoroethane | B1 |
| R-124 Chlorotetrafluoroethane | A1 |
| R-125 Pentafluoroethane | A1 |
| R-134a Tetrafluoroethane | A1 |
| R-143a Trifluoroethane | A2 |
| R-152a Difluoroethane | A2 |
| R-290 Propane | A3 |
| R-401A (53% R-22, 34% R-124, 13% R-152a) | A1 |
| R-401B (61% R-22, 28% R-124, 11% R-152a) | A1 |
| R-402A (38% R-22, 60% R-125, 2% R-290) | A1 |
| R-404A (44% R-125, 52% R-143a, R-134a) | A1 |
| R-407A (20% R-32, 40% R-125, 40% R-134a) | A1 |
| R-407C (23% R-32, 25% R-125, 52% R-134a) | A1 |
| R-502 (48.8% R-22, 51.2% R-115) | B2 |
| R-717 Ammonia ($NH_3$) | A1 |
| R-744 Carbon dioxide ($CO_2$) | A1 |
| R-1234yf Hydrofluoride olefin | A2L |

absorption cooling essentially stagnated until the 1950s. During this period, the two-stage, indirect-fired absorption refrigeration machine was developed in the United States, while the direct-fired, two-stage concept was perfected in Japan and other Pacific-rim countries. The direct-fired option was developed primarily in response to government energy policies around the Pacific rim.

·The fundamental "single-stage" absorption cycle is represented in Figure 1.4. The absorption chiller has no compressor; heat, directly or indirectly, provides the motive force for refrigerant phase change. The *evaporator* consists of a heat exchanger, held at low pressure, with a separate refrigerant (typically, water) pump. The pump sprays the refrigerant over the tubes containing the chilled water, absorbs heat from the water, and evaporates as a low-pressure gas. The low-pressure gas flows to the *absorber*, due to the pressure differential. The absorber is at a pressure lower than the evaporator because the concentrated absorbent solution exerts a molecular attraction for the refrigerant. The absorbent solution is sprayed into contact with the refrigerant vapor. Condensing of the refrigerant

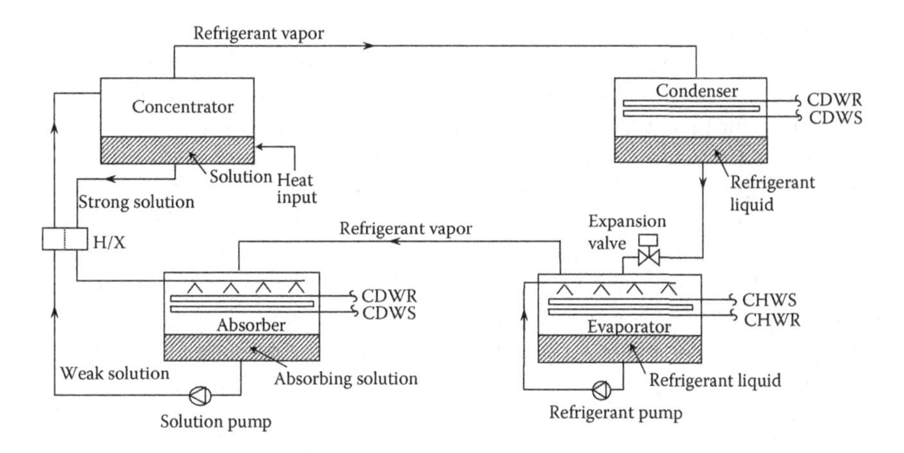


**FIGURE 1.4** Single-stage steam absorption chiller schematic.

occurs because the heat is absorbed by the absorbent. The absorbent, then, is cooled by condenser water.

The absorbent now consists of a dilute solution, due to its having absorbed water vapor refrigerant. The dilute solution is pumped to the *concentrator*, where heat is applied to reevaporate the refrigerant. The concentrated solution of the absorbent is then returned to the absorber. The refrigerant vapor goes to the condenser, where it is condensed by the condenser water. To improve efficiency, a heat exchanger is used to preheat the dilute solution, with the heat contained in the concentrated solution of the absorbent.

Leaks allow air to enter the refrigerant system, introducing noncondensable gases. These gases must be removed, or *purged*, to prevent pressure in the absorber increasing to the point where refrigerant flow from the evaporator will stop. The solution in the bottom of the absorber is relatively quiet and these gases tend to get collected at this point. They can be removed through the use of a vacuum pump, typically called a *purge pump*.

## Refrigerants

Today, there are two basic refrigerants used in absorption refrigeration chillers: water/*lithium bromide* and water/*ammonia*. Larger absorption units utilize water/lithium bromide solutions, while small units more commonly utilize water/ammonia solutions.

Lithium bromide is a corrosive, inorganic compound that has a very high absorption rate for water (hydroscopic). Thus, it makes an ideal "carrier" for the water refrigerant in absorption cycle chillers. However, the corrosion issues make lithium bromide solutions, especially at the higher temperatures associated with direct-fired chillers, which are difficult to address.

Since the water/lithium bromide solution in an absorption chiller is a corrosive salt solution, the primary potential for corrosion in these chillers is the ferrous metals used in them. The generator (or "concentrator") is the most critical location for potential ferrous corrosion since the highest salt concentration and highest temperatures are present in this heat exchanger, along with the potential impact of erosion corrosion as the refrigerant vapor is driven off by surface boiling.

Basic ferrous corrosion occurs when iron reacts with water to produce an iron oxide called *magnetite* and hydrogen. Under acidic conditions, this reaction is greatly accelerated. Also, in an absorption chiller, hydrogen is a "noncondensable" gas, that is, it does not act as a refrigerant, and the performance of the chiller is negatively impacted on.

There are basically two approaches to corrosion protection: (1) choose a metal compatible with the chemical environment in which it has to survive, or (2) modify the chemical environment so that it is less corrosive to carbon steel. The first approach, which requires the use of high-quality stainless steel for heat exchanger components, can add significantly to the cost of an absorption chiller and most manufacturers and owners have been unwilling to pay the premium involved. *Since chemical modification is much cheaper, this is the approach most commonly taken by chiller manufacturers.*

Two types of chemical modifications are generally made:

1. To reduce the acidity of the water/lithium bromide solution, a compatible alkaline, *lithium hydroxide*, is added. To reduce ferrous corrosion, it is desirable to maintain the solution as alkaline. But, since copper corrodes readily at higher alkaline levels, it is necessary to maintain the level high enough to help protect the steel, without being too high and accelerating copper corrosion.
2. To additionally protect the steel in a corrosive environment, a compatible *corrosion inhibitor* is added. In this case, this inhibitor is *lithium chromate*. Lithium chromate "passivates" the iron with which it comes into contact (i.e., makes it less reactive) by forming a protective molecular film on the surface. Lithium chromate has the advantage of working well at low-alkalinity levels, allowing the solution to be maintained at levels more suitable to the copper in the chiller.

This chemistry balance is complex and requires routine adjustments to maintain correctly. If the alkalinity is not adjusted properly, the solution will become acidic and accelerate ferrous metal corrosion. If alkalinity is too high, the copper in the chiller will corrode. If the lithium chromate level is too low, it offers poor protection to the ferrous metals in the chiller. But, if the level is too high, it can initiate pitting in the ferrous metals due to scaling and the resulting localized corrosion, while also increasing copper corrosion rates.

Water/ammonia solutions have much lower corrosion issues, but suffer from the safety concerns associated with the use of ammonia. Thus, the use of ammonia as the refrigerant in absorption cooling systems is typically limited to industrial applications.

## VAPOR COMPRESSION CYCLE WATER CHILLERS

As introduced in the section "Vapor Compression Refrigeration," a secondary refrigerant is a substance that does not change phase as it absorbs heat. The most common secondary refrigerant is water and *chilled water* is used extensively in larger commercial, institutional, and industrial facilities to make cooling available over a large area without introducing a plethora of individual compressor systems. Chilled water has the advantage that fully modulating control can be applied and, thus, closer temperature tolerances can be maintained under almost any load condition.

For very low-temperature applications, such as ice rinks, an antifreeze component, most often ethylene glycol or propylene glycol, is mixed with the water and the term *brine* (left over from the days when salt was used as antifreeze) is used to describe the secondary refrigerant.

In the HVAC industry, a chiller using the vapor compression cycle consists of one or more compressor(s), evaporator(s), and condenser(s), all packaged as a single unit. Where multiple compressors are used, it is typical to provide multiple, separate refrigeration circuits so that the failure of one compressor will not impact on the operation of the remaining compressors. The condensing medium may be water or outdoor air.

The evaporator, often called the *cooler*, consists of a shell-and-tube heat exchanger with refrigerant in the shell and water in the tubes. Coolers are designed for 3–11 fps water velocities when the chilled water flow rate is selected for a range of 10–20°F.

For air-cooled chillers, the condenser consists of an air-to-refrigerant heat exchanger and fans to provide the proper flow rate of outdoor air to transfer the heat rejected by the refrigerant.

For water-cooled chillers, the condenser is a second shell-and-tube heat exchanger with refrigerant in the shell and condenser water in the tubes. Condenser water is typically supplied at 70–85°F and the flow rate is selected for a range of 10–15°F. A cooling tower is typically utilized to provide condenser water cooling, but other cool water sources, such as wells, ponds, and so on, can also be used.

### SCROLL COMPRESSORS

Scroll compressors are positive-displacement orbital motion compressors that use two spiral-shaped scroll members, one that is fixed and the other that rotates, to compress refrigerant gas.

Scroll members are typically a geometrically identical pair, assembled 180° out of phase. Each scroll member is open on one end and bound by a base plate on the other. The two are fitted to form pockets between their respective base plates and various lines of contact between their walls. The flanks of the scrolls remain in contact, but the contact point moves progressively inward, compressing the refrigerant gas, as one scroll moves. Compression occurs by sealing gas in pockets of a given volume at the other periphery of the scrolls and progressively reducing the size of the pockets as the scroll relative motion moves them inward toward the discharge port.

Two different capacity control mechanisms are available. The most common approach to capacity control is *variable-speed control*, utilizing a variable frequency drive to control the rotational speed of the moving scroll. The cooling capacity, then, varies directly as a function of its speed. Another control method is called *variable displacement*, which incorporates "porting" holes in the fixed scroll. Capacity control is provided by disconnecting or connecting compression chambers on the suction side by closing or opening these porting holes.

Scroll compressors are available in capacities from 1.5 tons to about 40 tons and are applied in both single and multiple compressor configurations. The maximum chiller size typically applied is 80–160 tons, using a multiple compressor and usually with air-cooled condensing.

## Rotary Screw and Centrifugal Compressors

For larger chillers (150 tons to over 10,000 tons), rotary compressor water chillers are utilized. There are two types of rotary compressors applied: positive-displacement rotary screw compressors and centrifugal compressors.

Figure 1.5 illustrates the rotary *screw compressor* operation. Screw compressors utilize double-mating helically grooved rotors with "male" lobes and "female" flutes or gullies within a stationary housing. Compression is obtained by direct volume reduction through rotary motion. As the rotors begin to unmesh, a void is created on both the male and the female sides, allowing refrigerant gas to flow into the compressor. Further rotation starts the meshing of another male lobe with a female flute, reducing the occupied volume, and compressing the trapped gas. At a point determined by the design volume ratio, the discharge port is uncovered and the gas is released to the condenser.

Capacity control of screw compressors is typically accomplished by opening and closing a *slide valve* on the compressor suction to throttle the flow rate of refrigerant gas into the compressor. Variable-speed control can also be used to control the compressor capacity.

The design of a *centrifugal compressor* for refrigeration duty originated with Dr. Willis Carrier just after World War I. The centrifugal compressor raises the pressure of the gas by increasing its kinetic energy. This kinetic energy is then converted into static pressure when the refrigerant gas leaves the compressor and expands into the condenser. Figure 1.6 illustrates a typical centrifugal water chiller configuration. The compressor and motor are sealed within a single casing and a refrigerant gas is utilized to cool the motor windings during operation.

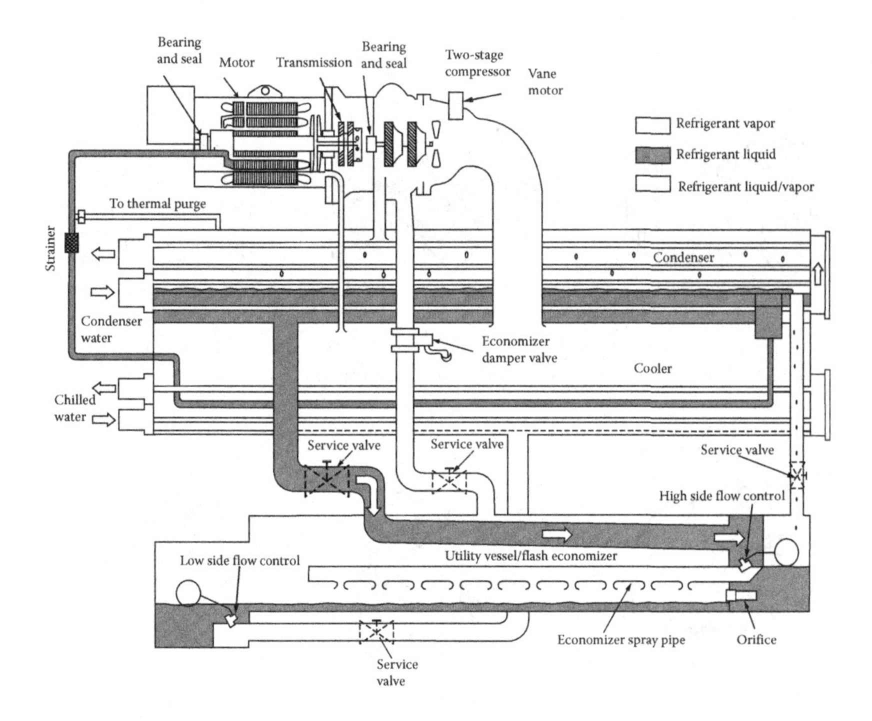


**FIGURE 1.5** Cutaway of a typical centrifugal water chiller.

Low-pressure gas flows from the cooler to the compressor. The gas flow rate is controlled by a set of *preswirl inlet vanes* and/or a variable frequency speed controller that regulates the refrigerant gas flow rate to the compressor in response to the cooling load imposed on the chiller.

Normally, the output of the chiller is fully variable within the range 15–100% of full-load capacity. The high-pressure gas is released into the condenser, where water absorbs the heat and the gas changes phase to liquid. The liquid, in turn, flows into the cooler, where it is evaporated, thus cooling the chilled water.

Centrifugal compressor chillers using R-134a are referred to as positive-pressure machines, while those using R-123 are considered to be negative-pressure machines, as defined by the evaporator pressure condition. At standard Air-Conditioning and Refrigeration Institute (ARI) rating conditions and using R-134a, the evaporator pressure is 36.6 psig and the condenser pressure is 118.3 psig, yielding a total pressure increase or *lift* provided by the compressor of 81.7 psig. However, for R-123, these pressure conditions are –5.81 psig in the evaporator and 6.10 psig in the condenser, yielding a total lift of 11.91 psig.

Mass flow rates for refrigerants in both positive- and negative-pressure chillers are essentially the same at ~3 lb/min ton. However, due to the significantly higher density of R-134a, its volumetric flow rate (cfm/ton), which defines impeller size, is over five times smaller than R-123 volumetric flow rate.

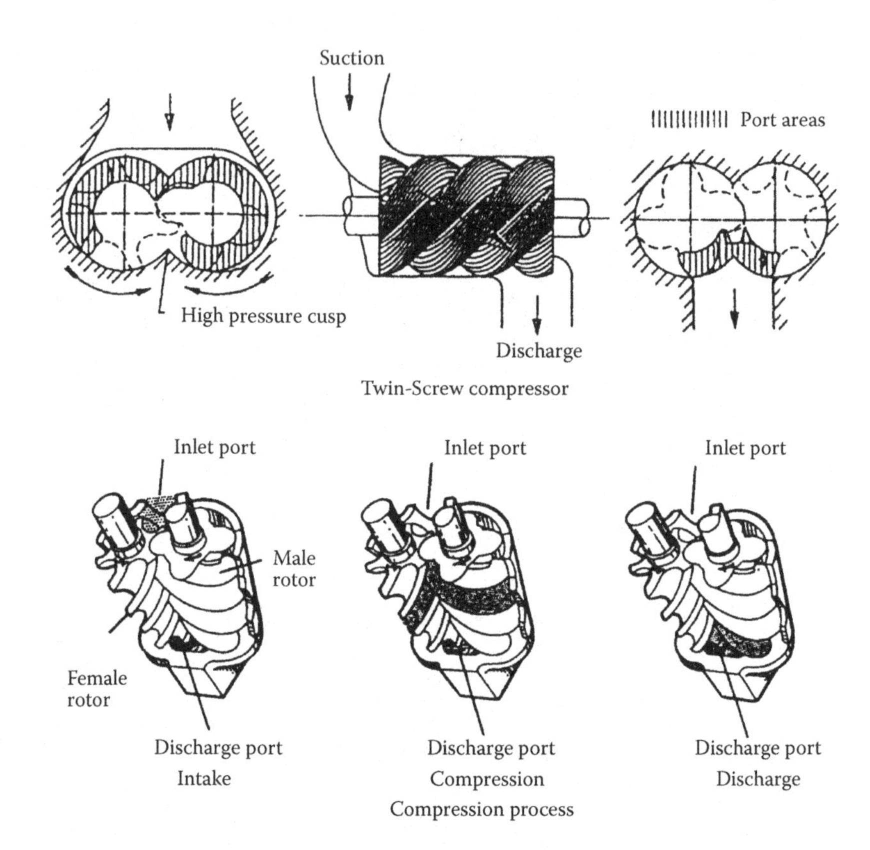


**FIGURE 1.6** Rotary screw compressor operation. (Courtesy of the American Society of Heating, Refrigerating, and Air-Conditioning Engineer's, Inc., Atlanta, GA.)

Early compressors using R-123 typically used large-diameter impellers (~40″ diameter) and direct-coupled motors that (at 60 Hz) turn at 3600 rpm. These large wheel diameters required by R-123 put a design constraint on the compressor and, to reduce the diameter, current designs typically utilize two or three impellers in series or *stages* to produce an equivalent pressure increase. In practice, the flow paths from the outlet of one stage to the inlet of the next introduce pressure losses that reduce efficiency to some degree.

Compressors using R-134a typically use much smaller impellers (about 5″ diameter) that are coupled to the motor through a gearbox or *speed increaser* and can operate at speeds approaching 30,000 rpm.

Since the evaporator in positive-pressure chillers is maintained at a pressure well above atmospheric, any leaks in the refrigeration system will result in a loss of refrigerant and the effect of any leaks is quickly evidenced by low refrigerant levels in the chiller. However, any leaks associated with a negative-pressure machine result in the introduction of atmospheric air (composed of *noncondensable gases* and water vapor) into the chiller.

Noncondensable gases create two problems:

1. The compressor does work when compressing the noncondensable gases, but they offer no refrigerating effect.
2. Noncondensable gases can "blanket" evaporator and condenser tubes, lowering heat exchanger effectiveness.

Noncondensable gases can lower the efficiency of the chiller by as much as 14% at full load.

Moisture introduced with atmospheric air is a contaminant that can allow the formation of acids within the chiller that can cause serious damage to motor windings (of hermetic motors) and bearings.

To remove potential noncondensable gases and moisture from negative-pressure chillers, these chillers are furnished with *purge units*. While purge units are very efficient at separating and venting noncondensable gases and moisture from the refrigerant, it is not 100% efficient and some refrigerant is vented to the atmosphere each time the purge unit operates. Additionally, to reduce the potential for leaks when chillers are off, the evaporator should be provided with an external heater to raise the refrigerant pressure to above atmospheric.

The energy requirement for a water-cooled rotary compressor chiller at peak load is a function of (1) the required leaving chilled water temperature, and (2) the temperature of the available condenser water. As the leaving chilled water temperature is reduced, the energy requirement to the compressor increases, as summarized in Table 1.4. Similarly, as the condenser water temperature increases, the compressor requires more energy (see Chapter 10). Thus, the designer and owner can minimize the cooling energy input by utilizing a rotary compressor chiller

**TABLE 1.4**
**Rotary Chiller Input Power Change as a Function of Chilled Water Supply Temperature**

| Leaving Chilled Water Temperature (°F) | Compressor Input Power (Approximate % Change) |
|---|---|
| 41 | +7 |
| 42 | +5 |
| 43 | +2 |
| 44 | 0 |
| 45 | –2 |
| 46 | –6 |
| 47 | –8 |
| 48 | –12 |
| 49 | –16 |

selected to operate with the highest possible leaving chilled water temperature and the lowest possible condenser water temperature.

## Electric-Drive Chillers

Electric-drive chillers may be configured as *hermetic* or *open*-drive machines. With open-drive chillers, the compressor and motor are separated, with their shafts being connected via a flexible coupling. The advantage of this concept is that in the event of motor failure, it does not contaminate the refrigerant and the motor can be readily replaced. The disadvantage is that the chiller motor is cooled by ambient air and these large motors may impose a high heat gain in the mechanical equipment room.

The alternative, and by far more popular design, contains both the motor and the compressor within a common, sealed enclosure. In this configuration, the compressor is rigidly connected directly to the motor shaft, eliminating the need for a flexible coupling. The motor is cooled by the refrigerant flow and thus imposes no heat gains that must be separately addressed. The only disadvantage is that in the event of motor failure, the refrigerant system is often contaminated, requiring a difficult and expensive cleaning in addition to replacing the motor.

The energy consumption by a rotary compressor chiller decreases as the imposed cooling load is reduced, as shown in Figure 1.7. These chillers operate efficiently at between ~30% and 100% load and *most efficiently* between 40% and 80% load. Within this capacity range, the refrigerant gas flow rate is reduced, yet the full heat exchange surface of the cooler and the condenser is still available, resulting in higher heat transfer efficiency.

Below about 30% load, the refrigerant gas flow rate is reduced to the point where heat pickup from the motor and mechanical inefficiencies have stabilized input energy requirements.

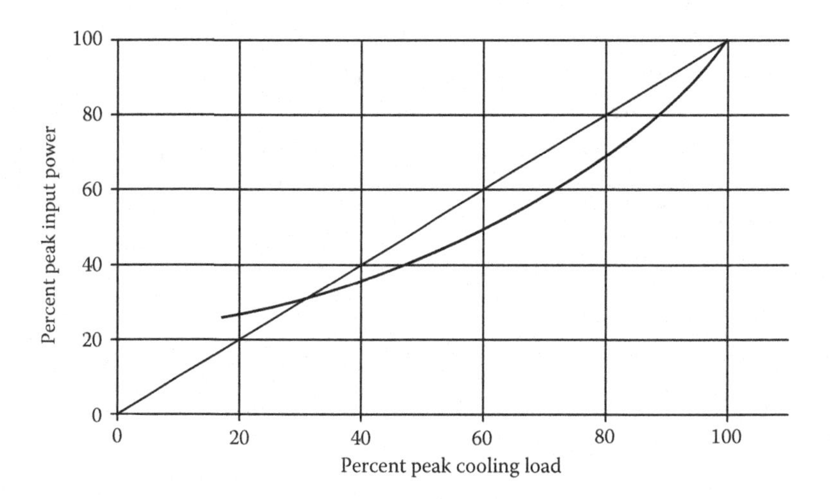


**FIGURE 1.7** Typical rotary compressor part load performance.

The vast majority of electric-drive rotary compressor water chillers utilize a single compressor. However, if the imposed cooling load profile indicates that there will be significant chiller usage at or below 30% of peak load, it may be advantageous to use a dual-compressor chiller or multiple single-compressor chillers.

The dual-compressor chiller typically uses two compressors, each sized for 50% of the peak load. At 50–100% of design load, both compressors operate. But, if the imposed load drops below 50% of the design value, one compressor is stopped and the remaining compressor is used to satisfy the imposed load. This configuration has the advantage of reducing the inefficient operating point to 15% of full load (50% of 30%), reducing significantly the operating energy penalties that would result from a single-compressor operation.

Negative-pressure chillers are typically somewhat more efficient than positive-pressure chillers. A peak load rating of 0.5 kW/ton or less is available for negative-pressure chillers, while positive-pressure chiller ratings below 0.55 kW/ton are difficult to obtain.

Positive-pressure chillers tend to be smaller and lighter than negative-pressure chillers, which can result in smaller chiller rooms and lighter structures. Negative-pressure chillers generally have a higher first cost than positive-pressure machines.

Driven by the evermore stringent requirements of each new edition of ASHRAE Standard 90.1, manufacturers are constantly trying to improve the efficiency of electric-drive rotary compressor water chillers. In the past few years, a *magnetic bearing* compressor has been offered that reduces compressor/motor friction losses, thus reducing power input requirements, by eliminating bearings. In this design, a magnetic field holds the compressor/motor shaft in alignment.

This technology was developed by Daiken Corporation of Japan and this manufacturer is currently the sole supplier of magnetic bearing motors that are used with centrifugal compressors. Daiken (and Donfoss, under license) sells the magnetic centrifugal compressor/motor for retrofit applications in the range of 70–270 tons.

McQuay/Daiken and JCI/York offer water chillers using magnetic bearing compressors. McQuay offers a dual-compressor machine in the capacities of 140–550 tons, while JCI/York offers their dual-compressor chiller in the range of 210–400 tons.

Packaged electric-drive chiller "modules" are available from several manufacturers (Multistack, Tandem, etc.) in 20-, 30-, 50-, and 70-ton packages, each with dual independent scroll compressors utilizing R-410A. Each chiller module is designed to be mix–matched to form water-cooled chillers featured by a total capacity of 20 to 600 tons.

Each chiller module is constructed with compressors, an evaporator, a condenser, and a control cabinet mounted on a common frame with a very small footprint, typically about 28″ wide and 48″ deep, allowing each module to pass through a normal 30″ doorway for installation. Chilled water and condenser water piping headers are included with quick-connect couplings for piping modules together. Modules are wired to provide for single-point electrical connection to serve each chiller configured from multiple modules.

The advantages of modular chillers over more typical factory-fabricated chillers are flexibility, especially in retrofit installations where space and accessibility are limited, and ease of future expansion. Their disadvantages include higher maintenance cost and lower energy efficiency (0.65 kW/ton or higher).

## Engine-Drive Chillers

In an open-drive configuration, natural gas- and propane-fueled spark ignition engines have been applied to rotary compressor systems. The full-load cooling COPs for engine-drive chillers are ~1.3–1.9 for helical screw compressors and 1.9 for centrifugal compressors. These low COPs can be improved if the engine water jacket heat and exhaust heat can be recovered to heat service hot water or for other uses.

Engine-drive chillers have been around for many years, but their application, most typically utilizing natural gas for fuel, has been limited by a number of factors:

1. Higher first cost
2. Air quality regulations
3. Much higher maintenance requirements
4. Short engine life
5. Noise
6. Larger physical size
7. Lack of integration between engine and refrigeration subsystems

Since the mid-1980s, manufacturers have worked very hard to reduce these negatives with more compact designs, emissions control systems, noise abatement measures, basic engine improvements, and development of overall systems controls using microprocessors.

However, *the maintenance requirements for engine-drive chillers remain high, adding about $0.03–$0.04/ton-hour to the chiller operating cost.* Currently, the engines used for chillers are either spark ignition engines based on automotive blocks, heads, and moving components (below about 400 tons capacity) or spark ignition engines using diesel blocks and moving components (for larger chillers). While the automotive-derivative engines are advertised to have a 20,000 h useful life, the real life may be much shorter, requiring an engine replacement every 2 years or so. The diesel-derivative engines require an overhaul every 10–12,000 h (equivalent to a diesel truck traveling 500,000 miles at 50 mph).

Newer engines use *lean burn* technology to improve combustion and reduce CO and $NO_x$ emissions. By adding catalytic converters to the exhaust and additional emissions controls, natural gas-fired engine-drive chillers can meet stringent California air quality regulations.

Gas engine-drive chillers remain more expensive than electric-drive units, and including maintenance costs, have higher overall operating costs (see Table 1.5). However, engine-drive chillers may be used during peak cooling load periods to reduce seasonal peak electrical demand charges (see the section "System Peak Cooling Load and Load Profile" in Chapter 2).

## Condensing Medium

The heat collected by the water chiller, along with the excess compressor heat, must be rejected to a heat sink. Directly or indirectly, ambient atmospheric air is typically used as this heat sink.

For *air-cooled* chillers, the condenser consists of a refrigerant-to-air coil and one or more fans to circulate outdoor air over the coil. The performance of the condenser is dependent on the airflow rate and the air's dry bulb temperature.

Air-cooled condenser airflow rates range from 600 to 1200 cfm/ton with a 10–30°F approach between the ambient dry bulb temperature and the refrigerant condensing temperature. For typical HVAC applications, the condensing temperature is about 105–120°F. Thus, the ambient air temperature must be no greater than 95°F with a loss of efficiency. As the ambient air temperature increases, the condensing temperature increases and net cooling capacity decreases by about 2% for each 5°F increase in condensing temperature.

Water-cooled chillers typically use a *cooling tower* (or some large water source like a river, lake, or ocean) to reject condenser heat to the atmosphere and Chapters 9 through 17 of this book address this topic in detail. At the chiller, with 85°F (or lower) condenser water temperature supplied from the cooling tower, condensing temperatures are reduced to 94–98°F, reducing the "lift" required of the compressor and significantly improving the chiller COP when compared with air-cooled machines.

**TABLE 1.5**
**Chiller Efficiency and Estimated Energy Cost**

| Electrical Input (kW/ton) | Heat Input (Mbh/ton) | Cost ($/ton-hour[a]) | COP | Driver/Compressor Type |
|---|---|---|---|---|
| | 7.5 | 0.097 | 1.6 | Engine-drive screw |
| | 6.3 | 0.056 | 1.9 | Engine-drive centrifugal |
| 1.2 | — | 0.112 | 3.25 | Electric-drive air-cooled scroll |
| 1.0 | — | 0.093 | 3.5 | Electric-drive air-cooled rotary |
| 0.55 | — | 0.051 | 5.8 | Electric-drive water-cooled rotary |

[a] Based on 2009 (the last year for which data are available) U.S. average commercial energy costs.

Table 1.5 illustrates the relative efficiency and operating cost for the various types of chillers with both air- and water-cooled condensing.

## ABSORPTION CHILLERS

### Lithium Bromide Absorption Chillers

The vast majority of water chiller systems utilized for HVAC applications in the United States are vapor compression cycle systems. However, in some applications, principally in large cities and at large universities and hospital complexes, steam distribution systems are available. In the past years, the cost of steam was often cheaper than the cost of electricity and was used to provide cooling, utilizing the *absorption refrigeration cycle*. This is generally no longer the case and little or no new absorption cooling is utilized except where a waste heat source is available, such as with cogeneration or some industrial processes, or where the use of absorption cooling during peak cooling load periods may allow a reduction in seasonal electric demand charges (see the section "System Peak Cooling Load and Load Profile" in Chapter 2).

Absorption chillers using lithium bromide are defined as *indirect-fired* or *direct-fired* and may be *single stage* or *two stage*, as follows:

1. The indirect-fired single-stage machine uses low- to medium-pressure steam (5–40 psig) to provide the heat for the absorption process. This type of chiller requires ~18,500 Btu/h per ton of cooling effect, resulting in a chiller COP of about 0.67.
2. The indirect-fired two-stage chiller utilizes high-pressure steam (at least 100 psig) or high-temperature hot water (400°F or higher) and requires ~12,000 Btu/h per ton of cooling effect, resulting in a chiller COP of 1.0.
3. The direct-fired chiller, as its name implies, does not use steam but utilizes a natural gas and/or fuel oil burner system to provide heat. These chillers are two-stage machines resulting overall COP of 1.0–1.1.

For the indirect-fired units, the overall COP must be reduced to account for the losses in the steam production in the boilers. With a typical boiler firing efficiency of 80–85%, this reduces the overall COP for the single-stage system to ~0.54 and to ~0.80 for the two-stage system.

As absorption cooling has a COP of only 0.54–1.1, it competes poorly with electric-drive rotary compressor chillers, as shown in Table 1.5. (A two-stage, direct-fired absorption chiller will cost ~$0.107/ton-hour of cooling to operate, almost twice the cost of an equivalent electric-drive chiller.)

Other factors that must be considered for absorption chillers include the following:

1. Absorption chillers require ~50% more floor area than the equivalent electric-drive (vapor compression cycle) chiller. Additionally, due to

their height, mechanical equipment rooms must be 6–10 ft taller than the rooms housing electric-drive chillers. Finally, because the liquid solution is contained in long, shallow trays within an absorption chiller, the floor must be closer to an absolute level.
2. Absorption chillers will weigh at least twice as much the equivalent electric-drive chiller.
3. Due to their size, absorption chillers are sometimes shipped in several sections, requiring field welding for final assembly.
4. While most electric chillers are shipped from the factory with their refrigerant charge installed, the refrigerant and absorbent (including additives) must be field installed in absorption chillers.
5. While noise and vibration are real concerns for electric-drive chillers (see the section "Noise and Vibration" in Chapter 6), absorption chillers (unless direct-fired) are quiet and essentially vibration free.
6. Due to the potential for crystallization of the lithium bromide in the chiller if it becomes too cool, the condenser water temperature must be kept above 75–80°F.
7. An emergency power source may be required if lengthy power outages are common. Without power and heat input, the chiller begins to cool and the lithium bromide solution may crystallize. However, because an absorption chiller has a very small electrical load requirement (usually less than 10 kW), a dedicated backup generator is not a major element.
8. The heat rejection rate from the condenser is 20–50% greater than for the equivalent electric-drive chiller, requiring higher condenser water flow rates and larger cooling towers and condenser water pumps.
9. Finally, an indirect-fired absorption chiller will be at least 50% more expensive to purchase than the equivalent electric-drive chiller. Direct-fired absorption chillers will cost almost twice as much as an electric machine, and have the added costs associated with providing combustion air and venting (stack).

Direct-fired absorption cycle chillers should be carefully evaluated as an alterative anytime an engine-drive vapor compression cycle chiller is being considered. Even though the energy costs for the absorption chiller may be higher, the increased maintenance costs associated with engine-drive systems may make the absorption chiller more cost effective.

### Ammonia Absorption Chillers

Small-capacity (3–5 tons) ammonia absorption chillers and heat pumps are marketed for commercial and industrial applications. These units are direct-fired (typically natural gas) air-cooled chillers and generally have a COP of about 0.5.

## CHILLED WATER FOR HVAC APPLICATIONS

Typical applications for chilled water systems include large buildings (offices, laboratories, hospitals, universities, etc.) or any multibuilding campuses where it is desirable to provide cooling from a central facility.

As shown in Figure 1.8, the typical water-cooled HVAC system has three heat transfer loops:

*Loop 1*. Cold air is distributed by one or more air-handling units to the spaces within the building. Sensible heat gains, including heat from temperature-driven transmission through the building envelope; direct solar radiation through windows; infiltration; and internal heat from people, lights, and equipment, are "absorbed" by the cold air, raising its temperature. Latent heat gains, moisture added to the space by air infiltration, people, and equipment are also absorbed by the cold air, raising its specific humidity. *The resulting space temperature and humidity condition is an exact balance between the sensible and latent heat gains and capability of the entering cold air to absorb those heat gains.*

The distributed air is returned to the air handling unit, mixed with the required quantity of outdoor air for ventilation, and then directed over the cooling coil where chilled water is used to extract heat from the air, reducing both its temperature and moisture content so that it can be distributed once again to the space.

As the chilled water passes through the cooling coil in counterflow to the air, the heat extraction process results in increased water temperature. The chilled water temperature leaving the cooling coil (chilled water return) will be 8–16°F warmer than the entering water temperature (chilled water supply) at design load. This temperature difference (*range*) establishes the flow requirement via the following relationship:

$$F_{chw} = \frac{Q}{500 \times \text{range}}$$

where

$F_{chw}$ = chilled water flow rate (gpm)

$Q$ = total cooling system load (Btu/h)

Range = chilled water temperature rise (°F)

500 = conversion factor (Btu min/gal °F h) (1 Btu/lb °F × 8.34 lb/gal × 60 min/h)

*Loop 2*. The warmer-returned chilled water enters the water chiller where it is cooled to the desired chilled water supply temperature by transferring the heat extracted from the building spaces to a primary refrigerant. This process, obviously, is not "free" since the compressor must do work on the refrigerant for cooling to occur and, thus, must consume energy in the process. Since most chillers are *refrigerant-cooled*, the compressor

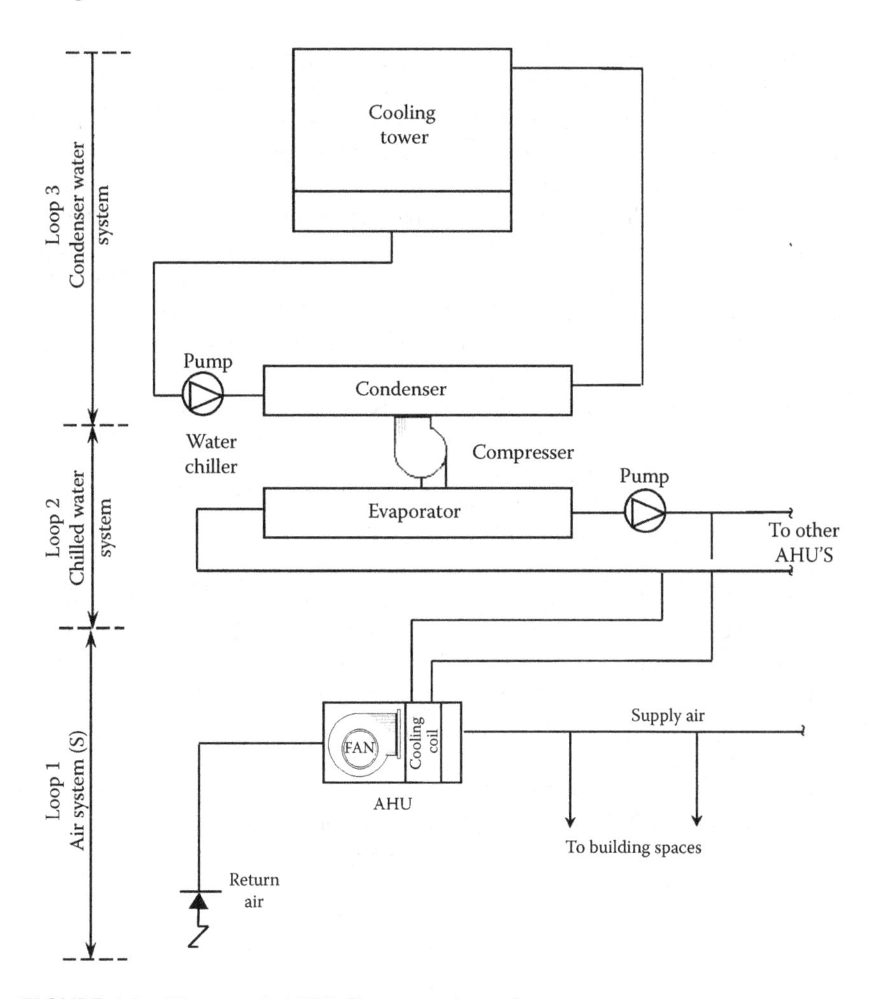


**FIGURE 1.8** Water-cooled HVAC system schematic.

energy, in the form of heat, is added to the building heat and both must be rejected through the condenser.

*Loop 3*. The amount of heat that is added by the compressor depends on the efficiency of the compressor. This *heat of compression* must then be added to the heat load on the chilled water loop to establish the amount of heat that must be rejected by the condenser to a *heat sink*, typically the outdoor air.

## Determining the Chilled Water Supply Temperature

The first step in designing or evaluating a chilled water system is to determine the required chilled water supply temperature.

For any HVAC system to provide satisfactory control of both space temperature and space humidity, the supply air temperature must be low enough to simultaneously satisfy both the sensible and latent cooling loads imposed in that space. *Sensible cooling* is the term used to describe the process of decreasing the temperature of air without changing the moisture content of the air. However, if moisture is added to the room by the occupants, infiltrated outdoor air, internal processes, and so on, the supply air must be cooled below its dew point to remove this excess moisture by condensation. The amount of heat removed with the change in moisture content is called *latent cooling*. The sum of the two represents the *total cooling* load imposed by a building space on the chilled water cooling coil.

The required temperature of the supply air is dictated by two factors:

1. The desired space temperature and humidity setpoint
2. The *sensible heat ratio* (SHR) defined by dividing the sensible cooling load by the total cooling load

On a psychrometric chart, the desired space conditions represent one endpoint of a line connecting the cooling coil supply air conditions and the space conditions. The slope of this line is defined by the SHR. An SHR of 1.0 indicates that the line has no slope since there is no latent cooling. The typical SHR in comfort HVAC applications will range from about 0.85 in spaces with a large number of people to ~0.95 for the typical office.

The intersection between this "room" line and the saturation line on the psychrometric chart represents the required *apparatus dew point (ADP) temperature* for the cooling coil. However, since no cooling coil is 100% efficient, the air leaving the coil will not be at a saturated condition, but will have a discharge dry bulb temperature of about 1–2°F above the ADP temperature.

While coil efficiencies as high as 98% can be obtained, the economical approach is to select a coil for about 95% efficiency, which typically results in the supply air wet bulb temperature being about 1°F lower than the supply air dry bulb temperature.

Based on these typical coil conditions, the required supply air temperature can be determined by plotting the room conditions point and a line having a slope equal to the SHR passing through the room point, determining the ADP temperature intersection point, and then selecting a supply air condition on this line based on a 95% coil efficiency. Table 1.6 summarizes the results of this analysis for several different typical HVAC room design conditions and SHRs.

For a chilled water cooling coil, *approach* is defined as the temperature difference between the entering chilled water and the leaving (supply) air. While this approach can range as low as 3°F to as high as 10°F, a cost-effective value for HVAC applications is ~7°F. Therefore, to define the required chilled water supply temperature, it is only necessary to subtract 7°F from the supply air dry bulb temperature determined from Table 1.6.

**TABLE 1.6**
**Typical Supply Air Temperature Required to Maintain Desired Space Temperature and Humidity Conditions**

| Desired Space Conditions | | Supply Air DB/WB Temperature (°F) Required | |
|---|---|---|---|
| Temperature | Relative Humidity (%) | 0.90 + SHR | 0.80–0.89 SHR |
| 75 | 50 | 54/53 | 52/51 |
| 70 | 50 | 50/49 | 44/43 |
| 65 | 50 | 44/43 | 41/40 |

## Establishing the Temperature Range

Once the required chilled water supply temperature is determined, the desired temperature range must be established.

The required chilled water flow rate is dictated by the imposed cooling load and the selected temperature range. The larger the range, the lower the flow rate and, thus, the less energy consumed for transport of chilled water through the system. However, if the range is too large, chilled water coils and other heat exchangers in the system require increased heat transfer surface and, in some cases, the ability to satisfy latent cooling loads is reduced.

Historically, a 10°F range has been used for chilled water systems, resulting in a required flow rate of 2.4 gpm/ton of imposed cooling load. For smaller systems with relatively short piping runs, this range and flow rate is acceptable. However, as systems get larger and piping runs get longer, the use of higher ranges will reduce pumping energy requirements. Also, lower flow rates can also result in economies in piping installation costs since smaller-sized piping may be used.

At a 12°F range, the flow rate is reduced to 2.0 gpm/ton and, at a 14°F range, to 1.7 gpm/ton. For very large campus systems, a range as large as 16°F (1.5 gpm/ton) may be used (though low flow rates may introduce problems in selecting cooling coils).

# 2 Chiller Configurations

## THE SINGLE-CHILLER SYSTEM

The basic chilled water piping configuration for a single chiller is shown in Figure 2.1. Here, a single chiller provides chilled water to the cooling coils utilizing a single chilled water pump.

For small systems, this configuration has the advantage of lower initial cost, but does have some basic disadvantages:

1. With the single-compressor system, failure of any component (compressor, pump, or condenser) will result in no cooling being available. For most facilities, this is unacceptable and the use of multiple chillers allows at least some cooling (50% or more) be provided even if one chiller fails. In cases where cooling is critical to the facility (computer centers, hospital, laboratories, pharmaceutical or textile manufacturing, etc.), multiple chillers with at least one redundant chiller are often used. In this case, even if one chiller fails, 100% of the design cooling load can still be met.
2. As discussed in Chapter 1, once the cooling load imposed on a rotary compressor chiller falls to below about 30% of the chiller capacity, the efficiency of the chiller begins to decline. Thus, the use of multiple chillers allows a better overall capacity-to-load ratio and improved operating efficiency.

## MULTICHILLER SYSTEMS

For multiple chiller systems, there are two basic configurations that can be utilized, *series* or *parallel* flow.

In a series configuration with two chillers, as shown in Figure 2.2, each chiller is selected to produce half of the required cooling at the full system flow rate. Thus, half of the total design range is produced by each chiller. Load ratios other than 50/50 are possible, but this is by far the most common condition because of control problems with chillers at very small temperature differences.

Table 2.1 summarizes the temperatures at various load conditions for the configuration shown in Figure 2.2.

Series chiller systems are rarely utilized in present times because this configuration requires a constant chilled water flow rate at all times, resulting in high pumping costs. But, if a relatively large temperature difference is required or if there is a very steady base cooling load, the series configuration may offer some advantages.

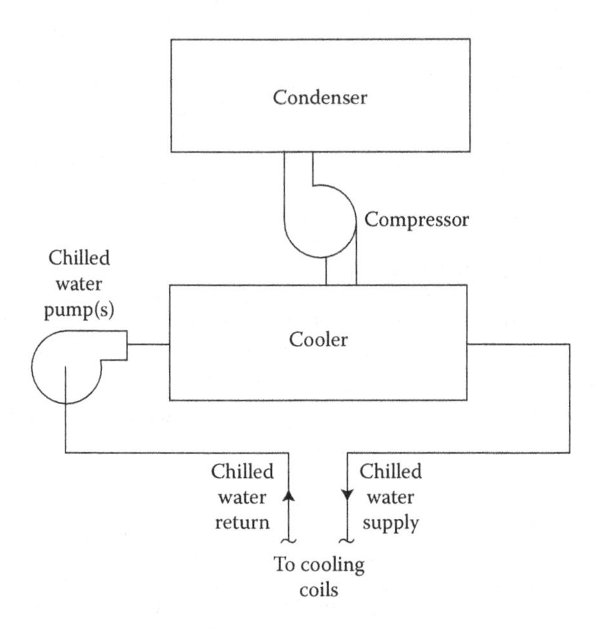


**FIGURE 2.1** Constant flow, single chiller configuration.

The parallel chiller configuration is far more common. In a two-chiller configuration, each chiller is typically selected to operate with the same design range, but with only a half of the total system flow requirement. This again results in a 50/50 load split, but other load ratios may be selected if dictated by operational requirements. And, there is no real limitation on the number of parallel chillers that can be utilized in one system.

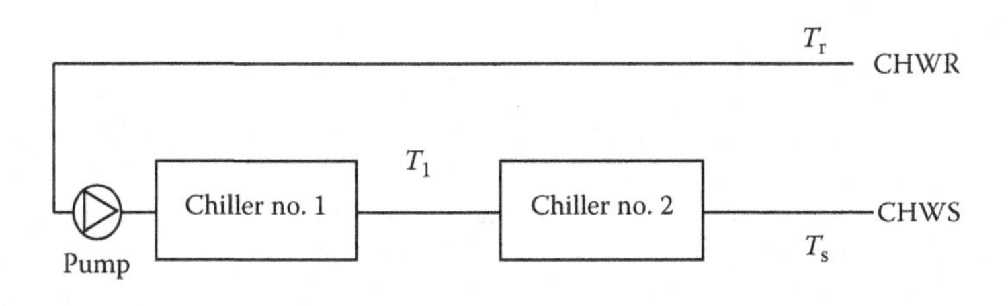


| Temp | Full Load | 75% Load | 50% Load | 25% Load |
|---|---|---|---|---|
| $T_s$ | 44 | 44 | 44 | 44 |
| $T_r$ | 56 | 53 | 50 | 47 |
| $T_1$ | 50 | 50 | 50 | 47 |

**FIGURE 2.2** Series chiller configuration.

**TABLE 2.1**
**Series Configuration Temperatures**

| Temperature | Full Load | 75% Load | 50% Load | 25% Load |
|---|---|---|---|---|
| $T_S$ | 44 | 44 | 44 | 44 |
| $T_R$ | 56 | 53 | 50 | 47 |
| $T_1$ | 50 | 50 | 50 | 47 |

## One-Pump Parallel Configuration

The one-pump parallel chiller configuration is shown in Figure 2.3 and the overall system performance and temperature conditions are summarized in Table 2.2.

With this configuration, there is an inherent problem. If both machines were operated for the full-load range (15–100% of peak capacity), by the time the total system load drops to 30% of full load, each individual chiller would be operating very inefficiently. Thus, most designers utilize controls to shut off one chiller when the total system load, as evidenced by the return chilled water temperature, falls below 40% of full load.

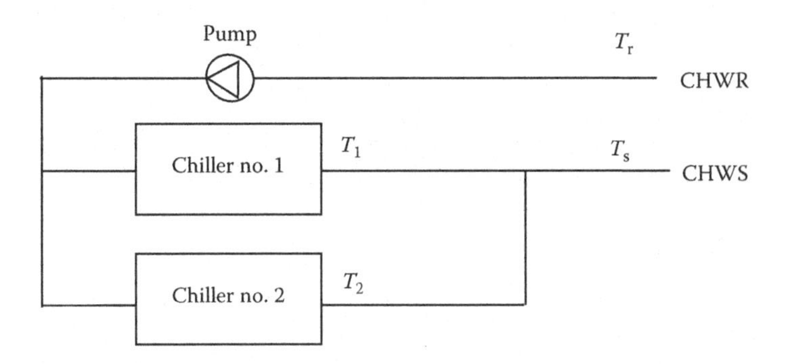


**FIGURE 2.3** One-pump parallel chiller configuration.

**TABLE 2.2**
**One-Pump Parallel Configuration Flows and Temperatures**

| | | Chiller No. 1 | | Chiller No. 2 | | |
|---|---|---|---|---|---|---|
| % Load | $T_r$ | Flow (%) | $T_1$ | Flow (%) | $T_2$ | $T_s$ |
| 100 | 56 | 50 | 44 | 50 | 44 | 44 |
| 75 | 53 | 50 | 44 | 50 | 44 | 44 |
| 50 | 50/56 | 50 | 44 | 50 | 56[a] | 50 |
| 25 | 50 | 50 | 44 | 50 | 50[a] | 47 |

[a] Chiller No. 2 off.

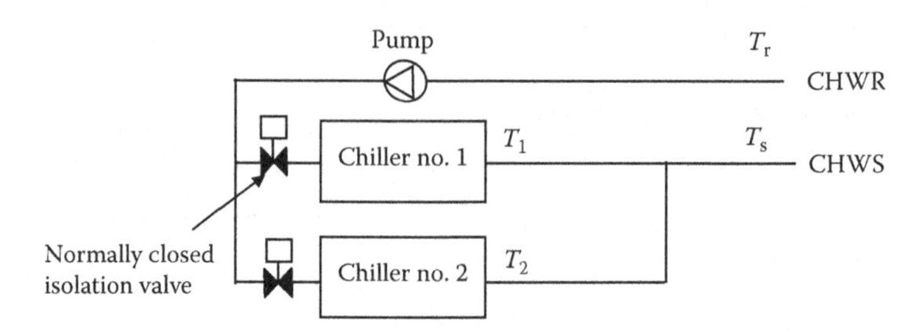


**FIGURE 2.4** One-pump parallel chiller configuration with isolation valves.

However, with this piping arrangement, if one chiller is not in operation, chilled water from the operating chiller will mix (blend) with the return water passing through the nonoperating chiller, effectively raising the system's chilled water supply temperature. In many cases, this may not be a problem. But, generally the interior spaces of large buildings still require more or less full cooling even when the perimeter spaces require no cooling at all and an elevated chilled water supply temperature may not satisfy these interior load conditions. In hot, humid climates, an elevated chilled water supply temperature may result in loss of humidity control.

To attempt to eliminate the blended supply water problem with the one-pump configuration, some designers have used chiller flow isolation valves, as shown in Figure 2.4. With this configuration, flow through the nonoperating chiller is closed off when the chiller is not in operation.

This arrangement results in increased flow through the operating chiller, but does reduce the blending problem, as illustrated in Table 2.3.

## Multiple-Pump Parallel Configuration

To ensure that the blended water condition does not occur, the multiple-pump parallel chiller configuration shown in Figure 2.5 is widely used.

**TABLE 2.3**
**One-Pump Parallel Configuration with Isolation Valves Flows and Temperatures**

| | | Chiller No. 1 | | Chiller No. 2 | | |
|---|---|---|---|---|---|---|
| % Load | $T_r$ | Flow (%) | $T_1$ | Flow (%) | $T_2$ | $T_s$ |
| 100 | 56 | 50 | 44 | 50 | 44 | 44 |
| 75 | 53 | 50 | 44 | 50 | 44 | 44 |
| 50 | 50/53 | 50/67 | 47 | 50/0 | N/A[a] | 47 |
| 25 | 47 | 67 | 44 | 0 | N/A[a] | 44 |

[a] Chiller No. 2 off. Thus there is no flow and $T_2$ does not affect $T_s$.

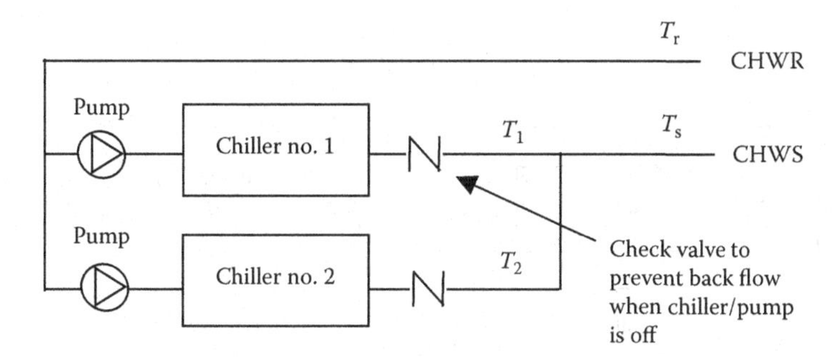


**FIGURE 2.5** Multiple-pump parallel chiller configuration.

With this configuration, each chiller has an individual chilled water pump. Thus, when one chiller is not operating, one pump is off, flow through the nonoperating chiller is zero, and no blending results.

Table 2.4 summarizes the performance and temperature conditions for this configuration at various load conditions.

## Primary–Secondary Parallel Configuration

Each of the configurations discussed above is essentially a constant flow system that utilizes three-way control valves at the cooling coils. Constant flow systems circulate the same amount of chilled water, no matter what the imposed cooling load, and, consequently, impose high pumping energy requirements.

To reduce these costs, the primary–secondary variable flow piping arrangement illustrated in Figure 2.6 is very commonly applied. Here, the *production loop* through the two chillers is hydraulically isolated from the *distribution loop* by a piping *bridge*. The bridge is a short section of piping shared by both loops and designed to have little or no pressure drop. Thus, the flow in one loop is not affected by flow in the other.

On the primary or production loop side, the system acts as multiple-pump parallel chiller installation, as described in the section "Multiple-Pump Parallel

**TABLE 2.4**
**Multiple-Pump Parallel Configuration Flows and Temperatures**

| | | Chiller No. 1 | | Chiller No. 2 | | |
|---|---|---|---|---|---|---|
| % Load | $T_r$ | Flow (%) | $T_1$ | Flow (%) | $T_2$ | $T_s$ |
| 100 | 56 | 50 | 44 | 50 | 44 | 44 |
| 75 | 53 | 50 | 44 | 50 | 44 | 44 |
| 50 | 50 | 50 | 44 | 50/0 | N/A[a] | 44 |
| 25 | 47 | 50 | 44 | 0 | N/A[a] | 44 |

[a] Chiller No. 2 off. Thus there is no flow and $T_2$ does not affect $T_s$.

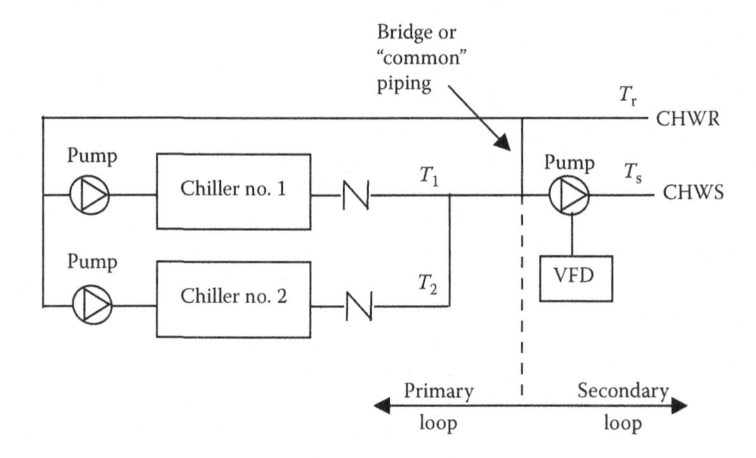


**FIGURE 2.6** Primary–secondary parallel chiller configuration.

Configuration." Flow in this loop varies in "steps" as the chillers are staged on or off and their respective pumps are started and stopped. However, in the secondary or distribution loop, the cooling coils utilize two-way control valves and the distribution pump(s) utilize a variable frequency drive(s) so that the chilled water flow rate is modulated from 0% to 100% of peak design flow as a function of the imposed cooling load. Thus, this loop has fully variable flow, but maintains a constant temperature range. At any load condition, the supply water temperature is the same as the water temperature leaving the chiller(s), as long as the production loop flow rate equals or exceeds the distribution loop flow rate.

Table 2.5 summarizes the performance and temperature conditions for this configuration at various load conditions.

**TABLE 2.5**
**Primary–Secondary Configuration Flows and Temperatures**

| | | | | | Chiller No. 1 | | Chiller No. 2 | | |
|---|---|---|---|---|---|---|---|---|---|
| % Load | $T_r$ | % Secondary Chilled Water Flow | % Primary Chilled Water Flow | $T_m$ | Flow (%) | $T_1$ | Flow (%) | $T_2$ | $T_s$ |
| 100 | 56 | 100 | 100 | 56 | 50 | 44 | 50 | 44 | 44 |
| 75 | 56 | 75 | 100 | 53 | 50 | 44 | 50 | 44 | 44 |
| 50 | 56 | 50 | 50 | 56 | 50 | 44 | 50/0 | N/A[a] | 44 |
| 25 | 56 | 25 | 50 | 53 | 50 | 44 | 0 | N/A[a] | 44 |

[a] Chiller No. 2 off. Thus there is no flow and $T_2$ does not affect $T_s$.

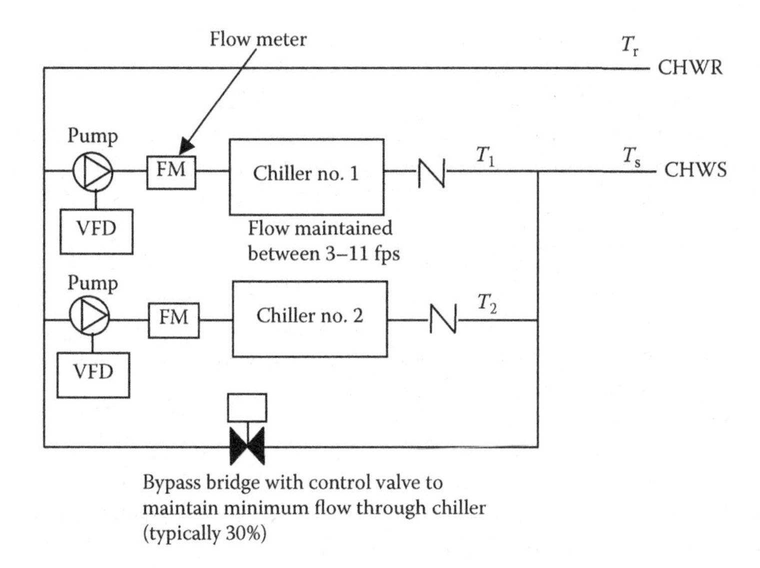


**FIGURE 2.7** Variable flow primary parallel chiller configuration.

## Variable Primary Flow Parallel Configuration

Since the early 1990s, rotary compressor water chillers have been manufactured with integral digital control systems. This represented a significant improvement in chiller control and allowed the use of the *variable primary* flow piping arrangement shown in Figure 2.7 to all the flow rate through each chiller to be varied and save even more in pumping energy (typically, ~15%).

The flow through the chillers is varied in response to the system load. A bypass valve opens to maintain minimum flow equivalent to about 30% of the flow through the largest chiller. This system has certain operational limitations since rapid load fluctuations and flow changes may result in fluctuation in the chilled water supply temperature, but this is a viable option where critical chilled water temperature control is not required.

Basic operating criteria for this system include the following:

1. Flow through each chiller must be maintained in the range of 3–11 fps. Consequently, good-quality flow meters, which must have periodic recalibration, are required for each chiller.
2. If flow rates through the chiller change too rapidly, the chiller controls cannot keep up. Therefore, load fluctuations must be limited to not more that 30% per minute.

One misconception about the variable primary flow configuration is that the chillers in this type of system will operate more efficiently. With variable flow through the evaporator, the logarithmic mean temperature difference

(LMTD) of the evaporator remains constant but the convective heat transfer coefficient decreases with flow, thus reducing the heat transfer effectiveness. With constant flow through the evaporator, the LMTD falls as the entering water temperature falls, but the convective heat transfer coefficient remains constant since flow remains constant. Thus, power consumption by a chiller will be essentially the same whether evaporator water flow is constant or variable.

Another misconception about these systems is their cost. The literature often states that variable primary flow systems are less costly compared to primary–secondary systems because one set of pumps is eliminated. However, the cost of high-quality flow meters (and the additional piping to ensure they are installed properly for accurate reading), the bypass flow control valve, and the more complex system controls will generally offset pump cost savings.

For large systems with high flow rates and long distribution piping runs, the additional savings in pumping energy produced by this configuration may be advantageous. However, it is a much more complex system and has the operating limitations previously discussed. So, for most systems, the simpler primary–secondary system configuration remains the recommended approach for the majority of multiple chiller systems. This recommendation was borne out by a December 2010 report in *HPAC Engineering* wherein the authors performed a study of primary–secondary systems versus variable primary flow systems, finding that the energy savings from primary flow systems, overall, was essentially nonexistent and that primary–secondary systems "have the advantage of more accurate, flexible, and stable control and operational conditions."

## SYSTEM PEAK COOLING LOAD AND LOAD PROFILE

One primary goal for chiller operation is to produce chilled water as economically as possible. To meet this goal, not only must individual chillers be efficient, but the entire multichiller system must also be designed to function efficiently.

The first step in evaluating the requirements for a chilled water system is to determine the peak cooling load requirement. However, the peak load will occur only for a very limited number of hours during the year. On an annual basis, the imposed load will vary based on the time of day due to occupancy patterns, solar heat gains, diurnal temperature swings, and so on and the time of year due to solar and outdoor temperature seasonal variations. Another factor that influences the load imposed on the chilled water plant is the amount of "core" building area that will require cooling year-round and whether an airside economizer cycle is installed to provide "free" cooling during the winter. *With all of these variables, every chilled water system will operate far more hours at a part load condition than it will at the peak load condition.*

Thus, in addition to determining the peak load to be imposed on the system, it is also necessary to determine the part load profile as a function of time, a function of outdoor temperature, or a function of both.

The process of estimating a system cooling load profile is best illustrated by using the following example:

1. Assume that a chilled water system serves a building with the peak load characteristics summarized in Table 2.6. This peak load is based on the building being fully utilized at 4:00 p.m. on an August afternoon.
2. The next step is to evaluate the peak day load profile by determining the loads imposed each hour of the design day as shown in Table 2.7. Note that there is an "occupancy variable" for lighting and other internal heat gains, a "solar variable" due to the day/night cycle, and a "temperature variable" due to the diurnal temperature profile over the 24-h period.
3. Next, examine the cooling load imposed during winter. While the occupancy variables will be the same as for the summer month, the solar and temperature variables will be markedly different. To determine the imposed chilled water load, the building may be considered to have a perimeter zone, in which heat losses through walls, roofs, and windows offset internal heat gains, and a core zone, in which only roof loads can be considered to offset internal heat gains. Obviously, the cold ventilation air can help offset any imposed space cooling loads. In Table 2.8, the building is assumed to have 50/50 core/perimeter area ratio and the loads are evaluated for different outdoor temperatures for both "occupied" and "unoccupied" conditions.
4. Finally, consider the impact of airside economizer cycles. If all facility air-handling units are equipped with economizers, there is no requirement for chilled water when the outdoor temperature is below the cooling

**TABLE 2.6**
**Example of Peak Heat Gain Components (4:00 p.m., August 31, 94°F DB/ 78°F WB Outdoor Conditions)**

| Component | Heat Gain (Btu/h) |
|---|---|
| Opaque walls and roof conduction | 165,000 |
| Window conduction | 210,000 |
| Window solar radiation | 490,000 |
| Internal | |
| People, sensible | 25,000 |
| People, latent | 20,000 |
| Lights | 900,000 |
| Equipment/appliances | 75,000 |
| Ventilation | |
| Sensible | 427,000 |
| Latent | 390,000 |
| Total | 2,702,000 |
| | 225 tons |

## TABLE 2.7
## Example of Estimated Design Day Load Profile

| | | Heat Gain (1000 Btu/h) | | | | | | | |
|---|---|---|---|---|---|---|---|---|---|
| Hour | Out Door Air Temperature (°F) | Walls/ Roof | Window Conduction | Window Solar | Internal | Ventilation | Total | Total Tons |
| 1 | 78 | 26 | 33 | 0 | 102 | 129 | 29 | 24 |
| 2 | 76 | 9 | 11 | 0 | 102 | 43 | 165 | 14 |
| 3. | 74 | –9 | –11 | 0 | 102 | –43 | 39 | 3 |
| 4 | 76 | 9 | 11 | 0 | 102 | 43 | 165 | 14 |
| 5 | 77 | 17 | 22 | 0 | 102 | 86 | 227 | 19 |
| 6 | 78 | 26 | 33 | 0 | 102 | 129 | 290 | 24 |
| 7 | 80 | 43 | 55 | 10 | 102 | 215 | 425 | 35 |
| 8 | 81 | 52 | 66 | 50 | 102 | 258 | 528 | 44 |
| 9 | 82 | 61 | 77 | 60 | 750 | 301 | 1249 | 104 |
| 10 | 84 | 78 | 99 | 80 | 900 | 387 | 1544 | 129 |
| 11 | 86 | 96 | 122 | 100 | 1020 | 473 | 1811 | 151 |
| 12 | 88 | 113 | 144 | 150 | 1020 | 559 | 1986 | 166 |
| 13 | 90 | 130 | 166 | 200 | 1020 | 645 | 2031 | 169 |
| 14 | 92 | 148 | 188 | 300 | 1020 | 731 | 2387 | 199 |
| 15 | 93 | 156 | 199 | 400 | 1020 | 774 | 2549 | 212 |
| 16 | 94 | 165 | 210 | 490 | 1020 | 817 | 2702 | 225 |
| 17 | 93 | 156 | 199 | 480 | 1020 | 774 | 2549 | 212 |
| 18 | 92 | 148 | 188 | 400 | 900 | 731 | 2367 | 197 |
| 19 | 90 | 130 | 166 | 350 | 900 | 645 | 2191 | 183 |
| 20 | 86 | 96 | 122 | 150 | 700 | 473 | 1541 | 128 |
| 21 | 84 | 78 | 99 | 90 | 500 | 387 | 1154 | 96 |
| 22 | 82 | 61 | 77 | 0 | 102 | 301 | 541 | 45 |
| 23 | 81 | 52 | 66 | 0 | 102 | 258 | 470 | 39 |
| 24 | 80 | 43 | 55 | 0 | 102 | 215 | 415 | 35 |

coil discharge temperature, typically 50–55°F. If airside economizers are not provided, chilled water must be provided by running the chillers and/or using a waterside economizer cycle (see the section "Waterside Economizer Cycle" in Chapter 14).

From this evaluation, both the peak load and the minimum load to be imposed on the chilled water system are determined.

1. In this example, the peak load is 225 tons.
2. If the example building were an office building, in all probability the air systems would be shut down when the cleaning crew is finished at about 8:00 p.m. (hour 20) and not started until about 6:00 a.m. (hour 6). However, for a lab building, the air systems will operate 24 h per

**TABLE 2.8**
**Example of Winter Cooling Requirements**

| Out Door Air Temp Bin (°F) | Occupancy | Perimeter Zones Heat Gain/Loss (1000 Btu/h) | | | | Core Zones Internal Heat Gain (1000 Btu/h) | Ventilation Heat Gain/Loss (1000 Btu/h) | Total Heat Gain/Loss (1000 Btu/h) | Total Tons |
|---|---|---|---|---|---|---|---|---|---|
| | | Walls/Roof | Window Conduction | Window Solar | Internal | | | | |
| 0 | Occupied | –618 | –788 | 490 | 510 | 510 | –1601 | –1497 | |
| | Unoccupied | –618 | –788 | 0 | 51 | 51 | –1601 | –2956 | |
| 10 | Occupied | –536 | –673 | 490 | 510 | 510 | –1388 | –1097 | |
| | Unoccupied | –536 | –673 | 0 | 51 | 51 | –1388 | –2495 | |
| 20 | Occupied | –454 | –578 | 490 | 510 | 510 | –1174 | –696 | |
| | Unoccupied | –454 | –578 | 0 | 51 | 51 | –1174 | –2104 | |
| 30 | Occupied | –371 | –473 | 490 | 510 | 510 | –961 | –295 | |
| | Unoccupied | –371 | –473 | 0 | 51 | 51 | –961 | –1703 | |
| 40 | Occupied | –289 | –368 | 490 | 510 | 510 | –747 | 106 | 9 |
| | Unoccupied | –289 | –368 | 0 | 51 | 51 | –747 | –1302 | |
| 50 | Occupied | –206 | –263 | 490 | 510 | 510 | –534 | 507 | 42 |
| | Unoccupied | –206 | –263 | 0 | 51 | 51 | –534 | –901 | |
| 60 | Occupied | –124 | –158 | 490 | 510 | 510 | –320 | 908 | 76 |
| | Unoccupied | –124 | –158 | 0 | 51 | 51 | –320 | –531 | |

day. Therefore, from Table 2.7, the minimum load imposed on the chilled water system would be as low as 3 tons if the building were cooled 24 h per day, but 24 tons if the building is shut down at night.

3. Based on Table 2.8, the building will require cooling by the chilled water system during the occupied period down to 40°F outdoor air temperature. If airside economizers are in use, the minimum winter-occupied cooling load is 76 tons, occurring at 60°F outdoor temperature.

## SELECTING WATER CHILLERS

### Basic Chiller Requirements

The peak cooling capacity and the corresponding input energy for water chillers are rated in accordance with the ARI Standard 550/590-98, *Water-Chilling Packages Using the Vapor Compression Cycle.* Chillers must be constructed and installed in accordance with ASHRAE Standard 15, *Safety Code for Mechanical Refrigeration.* Since the evaporator and condenser for R-134A chillers operate at pressures exceeding 15 psig, they are defined as "pressure vessels" and must be constructed in accordance with Section 8 of the American Society of Mechanical Engineers (ASME) *Boiler and Pressure Vessel Code.* And, all chillers must comply with the requirements of Underwriters Laboratory (UL) Standard 465.

The rated capacity of a water chiller, determined in accordance with ARI 550/590-98, is based on a standard set of operating conditions as follows:

| | |
|---|---|
| Chilled water supply temperature | 44°F |
| Chilled water temperature range/flow rate | 10°F/2.4 gpm/ton |
| Evaporator fouling factor | 0.0001 |
| Condenser water supply temperature/flow rate (water cooled) | 85°F/3.0 gpm/ton |
| Condenser fouling factor | 0.00025 |
| Entering air temperature (air cooled) | 95°F |

If any of these conditions vary for a specific project or location, the manufacturers will provide a specific cooling rating for the required conditions. Typically, if the chilled water temperature increases, chiller capacity increases. If the entering condenser water or air temperature decreases, chiller capacity increases. Generally, capacity does not vary with changes in range.

### Part Load Efficiency

As shown in the section "System Peak Cooling Load and Load Profile," chillers rarely operate for long at maximum capacity. To provide data to evaluate chiller

**TABLE 2.9**
**IPLV Rating Conditions (ARI Standard 550/590-98)**

| Rating Point | Percent of Full Load | Weighing Factor (%) | Water-Cooled Condenser Water Temperature (°F) | Air-Cooled Entering Air Temperature (°F) |
|---|---|---|---|---|
| A | 100 | 1 | 85 | 95 |
| B | 75 | 42 | 75 | 80 |
| C | 50 | 45 | 65 | 65 |
| D | 25 | 12 | 65 | 55 |

part load performance, ARI 550/590-98 requires chiller manufacturers to determine the part load operating performance characteristics as follows:

1. Integrated Part Load Value (IPLV)—This rating method requires the manufacturer to determine the input energy requirement for each chiller at the four operating conditions defined in Table 2.9. The "weighing factor" defined in Table 2.9 represents the percentage of the hours of the cooling season at which the chiller is expected to operate at the defined rating point.

   Since chiller energy performance will improve with reduced load and reduced condensing temperatures, the IPLV can be used to compare alternative chiller seasonal energy consumption using a common rating method. All other factors being equal, the chiller with the lower IPLV will have lower seasonal operating costs.

   The weighing factors used in the ARI standard are based on (a) weighted average weather data for 29 U.S. cities, and (b) weighted hours of use at different building operating scenarios.
2. Nonstandard Part Load Value (NPLV)—The IPLV rating is only directly applicable to the single chiller installation where the condenser water supply temperature is allowed to fall as the outdoor wet bulb temperature falls.

   Anytime multiple chillers are used, particularly if careful control allows the chillers to be used in or near their optimum efficiency range of 40–80% of rated capacity, the IPLV has little validity. Likewise, if the control of the condenser water temperature control method does not allow the supply temperature to go as low as 65°F, the IPLV is not valid. And, in fact, at least two chiller manufacturers recommend limiting the supply condenser water temperature to 65°F and 75°F, respectively.

   Therefore, to compare the seasonal energy performance of alternative chillers, it is necessary to define the specific weighing factors and condensing conditions to be applied to compute an NPLV. (See Chapter 8 for further discussion on this topic.)

As of December 30, 2008, state and local energy codes in the United States are, by Federal law, required to be equivalent to ANSI/ASHRAE/IESNA Standard

**TABLE 2.10**
**Minimum Water Chiller Efficiency Requirements from ANSI/ASHRAE/IESNA Standard 90.1-2004**

| Chiller/Compressor Type | Capacity (Tons) | Peak Load COP | IPLV (kW/ton) |
|---|---|---|---|
| Air-cooled, electric-drive | All | 2.80 | 1.256 |
| Screw or scroll compressor, water-cooled, electric-drive | <150 | 4.45 | 0.781 |
| | 151–299 | 4.90 | 0.710 |
| | ≥300 | 5.50 | 0.628 |
| Centrifugal compressor, water-cooled, electric-drive | <150 | 5.00 | 0.703 |
| | 151–299 | 5.55 | 0.634 |
| | ≥300 | 6.10 | 0.576 |
| Absorption, two-stage, direct- or indirect-fired, water-cooled | All | 1.00 | — |

90.1-2004, *Energy Standard for Buildings except Low-Rise Residential Buildings.* This standard requires that water chillers, when selected for ARI-rated conditions, have minimum peak load COPs and rated IPLVs as shown in Table 2.10. Typically, though, it is more cost effective to purchase chillers that are more efficient than these minimum values.

## Load versus Capacity

One of the major chiller manufacturers estimates that, for the typical chiller installation, 25% of the total owning and operating cost over the life of the chiller is related to the cost of designing, purchasing, and installing the chiller, while 75% of the total cost is consumed by energy expenses and maintenance costs. *Therefore, while each chiller must be selected to meet the peak cooling load, or its assigned part of the peak load, it is far more important to select each chiller to operate as efficiently as possible over the full range of part load conditions that can be expected.*

If the chiller "plant" is to consist of a single chiller, it is only necessary to select the chiller to meet the peak cooling load and then select the most efficient chiller (the chiller with the lowest IPLV or NPLV) within the scope of the available capital funds. But, when multiple chillers are to be selected, the selection of each chiller becomes more complex, since the basic requirement is to match capacity to load as closely as possible across the full-load profile.

As discussed in the section "The Single Chiller System," it is usually desirable to have multiple chillers available to satisfy the cooling load imposed on a chilled water system. There are several aspects that must be considered when selecting the chillers for a multichiller system:

*Compressor type.* For peak cooling loads below 100–125 tons, single rotary compressor chillers are expensive. Therefore, the typical choice is to select chillers using multiple scroll or screw compressors, or even multiple chiller modules described in Chapter 1, each with an independent refrigeration circuit to yield better reliability and redundancy.

Between 100 and 200 tons peak cooling load, two or more multiple scroll or screw compressor chillers can be used. Above about 200 tons, single rotary compressor systems begin to become cost effective.

*Condensing medium.* Water-cooled chillers are both more energy and water efficient than air-cooled chillers and are, consequently, the first choice. However, where an owner has a very limited budget or, even more importantly, lacks the ability to deal with the maintenance requirements of a cooling tower system, air-cooled condensing may use a chiller system of 200 tons or less total capacity.

*Symmetrical chiller capacity.* With this approach, all of the chillers are sized for equal capacity. The number of chillers and, thus, the size of the chiller "module" are based on the minimum anticipated load. Once the plant load is reduced to below the capacity of a single chiller, we want each chiller to operate in an efficient region (i.e., above 30% capacity) as long as possible.

For example, a building with a peak cooling load of 750 tons and a minimum cooling load of 100 tons could be served by three chillers, each rated at 250 tons. At 100 tons, a single chiller would operate at 40% capacity, which is still within the efficient region.

Had two 375-ton chillers been selected, the last chiller on line would operate at 27% of its capacity to meet the minimum cooling load requirement.

In this example, if the minimum cooling load had been 20 tons, even the 250-ton chiller would not have been able to fully respond. Since chiller minimum capacity is typically about 15% of the peak capacity, the chiller would have cycled off under its internal controls when the load fell below about 40 tons.

*Asymmetrical chiller capacity.* There is no engineering rule that says that all chillers in a multichiller system have to be of the same size. While there may be some maintenance advantages (common parts, etc.), different-sized chillers can be operated together.

From the previous example, the 750 peak load requirement could have been met with a 600-ton chiller and a 150-ton chiller. Both chillers would operate to produce 750 tons at peak load conditions, but the smaller chiller would meet the minimum load operating at 50% of its peak capacity, which would normally be a very efficient operating point.

Table 2.11 summarizes the load versus capacity relations for the symmetrical and asymmetrical examples.

The asymmetrical arrangement results in potentially lower operating cost and the ability to satisfy a lower minimum cooling load. If a single chiller fails with the symmetrical design, two-thirds of the peak capacity

**TABLE 2.11**
**Comparison of Symmetrical versus Asymmetrical Chiller Plant Performance**

| | | Symmetrical Plant Output (Tons) | | | Asymmetrical Plant Output (Tons) | |
|---|---|---|---|---|---|---|
| % Peak Load | Load (Tons) | Chiller 1 | Chiller 2 | Chiller 3 | Chiller 1 | Chiller 2 |
| 100 | 750 | 250 | 250 | 250 | 600 | 150 |
| 90 | 675 | 225 | 225 | 225 | 540 | 135 |
| 80 | 600 | 200 | 200 | 200 | 600 | Off |
| 70 | 525 | 175 | 175 | 175 | 525 | Off |
| 60 | 450 | 225 | 225 | Off | 450 | Off |
| 50 | 375 | 188 | 187 | Off | 375 | Off |
| 40 | 300 | 150 | 150 | Off | 300 | Off |
| 30 | 225 | 225 | Off | Off | 225 | Off |
| 20 | 150 | 150 | Off | Off | Off | 150 |
| 10 | 75 | 75 | Off | Off | Off | 75 |

is still available. However, if the 600-ton chiller in the asymmetrical arrangement fails, only 20% of the plant capacity is available.

Another common asymmetrical configuration is the 60/40 split with two chillers. In this case, one chiller is sized for 40% of the load while the larger chiller is sized for 60% of the load. In many applications, this split will allow better capacity-to-load matching and improved performance.

## Atmospheric Impacts

Gases that trap heat in the atmosphere are often called "greenhouse" gases. Some greenhouse gases, such as carbon dioxide, occur naturally and are emitted to the atmosphere through natural processes and human activities. Other greenhouse gases (e.g., fluorinated gases) are created and emitted solely through human activities. The principal greenhouse gases that enter the atmosphere because of human activities are:

*Carbon dioxide ($CO_2$).* Carbon dioxide enters the atmosphere through the burning of fossil fuels (oil, natural gas, and coal), solid waste, trees and wood products, and also as a result of other chemical reactions (e.g., manufacture of cement). Carbon dioxide is also removed from the atmosphere (or "sequestered") when it is absorbed by plants as part of the biological carbon cycle.

*Methane ($CH_4$).* Methane is emitted during the production and transport of coal, natural gas, and oil. Methane emissions also result from livestock and other agricultural practices and from the decay of organic waste in municipal solid waste landfills.

*Nitrous oxide* (*$N_2O$*). Nitrous oxide is emitted during agricultural and industrial activities, as well as during combustion of fossil fuels and solid waste.

*Fluorinated gases.* Hydrofluorocarbons, perfluorocarbons, and sulfur hexafluoride are synthetic, powerful greenhouse gases that are emitted from a variety of industrial processes. These gases are typically emitted in smaller quantities, but because they are potent greenhouse gases, they are sometimes referred to as "High Global Warming Potential gases" ("high GWP gases").

The first three gases are products of combustion and are released indirectly during the production of electricity used for electric-drive chillers and directly from engine-drive and absorption chillers. The last, of course, may include refrigerants used in chillers, as discussed in Chapter 1.

To make computations somewhat easier, all greenhouse gases can be defined in terms of *carbon dioxide equivalent*, and the goal for any chilled water system designer, contractor, and owner should be to consider the greenhouse gas release associated with each chiller system.

In 2006, the electrical power industry released about 2,500,000,000 metric tons of $CO_2$ into the atmosphere, along with 9,500,000 metric tons of $SO_2$ and 3,800,000 tons of $NO_x$. ASHRAE estimates that these greenhouse gas emissions amount to 1.76 lbs/kWh of $CO_2$ or its equivalent.

On a comparative basis, this high greenhouse gas emission level makes grid electricity a poor sustainable choice for any chiller prime mover and, at the very least, puts great pressure on designers and owners to select the most efficient electric-drive chiller configuration possible.

The electrical power industry is now under mandate in many states to increase the percentage of power generated from renewable energy sources (hydro, solar, wind, etc.) New generation capacity is anticipated to be new technology coal plants, which reduce their emissions by 50% or more, and/or nuclear, which has essentially zero emissions. This is already at work—the $CO_2$ emission level in 2006 was 4.5% lower than in 2005. With further support and mandates from a viable national energy policy, it is expected that greenhouse emissions per kWh of generated power will be reduced by 25% within 15 years.

Currently, while there are regional variations due to the mix of primary fuels for electric power generation (e.g., coal, natural gas, fuel oil, nuclear, hydro, etc.), typical greenhouse gas emissions for chiller systems, based on $CO_2$ equivalent, are summarized in Table 2.12.

## MIXED ENERGY SOURCE CHILLER SYSTEMS

For smaller chilled water systems (less than about 1000 tons), the lower average operating cost for electric-drive chillers generally eliminates alternative energy source considerations. However, larger plants, particularly where electrical

**TABLE 2.12**
**Greenhouse Gas Emissions for Various Chiller Systems**

| Chiller Systems | Approximate Emissions (lb of $CO_2$/ton-hour) |
|---|---|
| Electric-drive, helical or rotary screw compressor, air-cooled | 1.78 |
| Indirect-fired, two-stage, absorption, water-cooled | 1.50 |
| Direct-fired, two-stage, absorption, water-cooled | 1.44 |
| Electric-drive, rotary compressor, water-cooled | 1.16 |
| Engine-drive, rotary compressor, water-cooled | 0.86 |

demand and/or ratchet charges are high, have the potential of using absorption and/or engine-drive chillers as part of the multichiller "mix."

As discussed in Chapter 1, both absorption chillers and engine-drive chillers are less efficient than electric-drive chillers and, generally, have higher operating and maintenance costs. However, if the seasonal cooling load profile is such that higher-cost cooling methods can be used for a limited number of hours to reduce the peak electrical demand and associated demand charges, usually imposed on an annual basis, the net result is an operating cost savings, but with increased energy consumption. The first step in determining whether an energy source in addition to electricity should be considered is to understand electric utility rate schedules and the costs of alternative fuels.

All electric rates divide the allocation of cost into two components, demand and consumption. *Demand*, in terms of kW, represents the highest use of electricity over a specified time period, typically 15–30 min, during each monthly billing period. *Consumption*, in kWh, is the total use of electricity during the monthly billing period.

For the electric utility, the demand charge represents the cost incurred by the utility in having sufficient capacity available to meet each customer's maximum need, even though the maximum need may occur only for a limited number of hours each year. In essence, it represents the utility's attempt to equitably distribute the cost of capital recovery for their plants, distribution lines, substations, and so on to those requiring the availability of this capacity.

For most electric utilities, their peak system demand occurs during the summer because of air-conditioning use. Thus, demand charges are computed not only on the basis of the maximum use of electricity in each month but these charges may be imposed on an annual basis via ratchet clauses in the electric rate.

The typical *ratchet clause* will define the "billing demand" in each month as the greater of the actual demand incurred during that month or the maximum demand incurred in previous months. For most utilities, the ratchet period is 11 months. Some rates may use only an 80% ratchet, setting the billing demand in each month as the greater of the actual demand incurred during that month or 80% of the maximum demand incurred in the previous months. Some will separate summer peak demands and winter peak demands to define the minimum billing demand in each month.

To further address the impact of a customer's peak demand on the utility's system demand, on-peak and off-peak periods are typically specified by the utility on a daily basis. For example, one utility defines their on-peak summer period to be from 10 a.m. to 8 p.m., Monday through Friday, during the months of June, July, August, and September. During the on-peak period, demand charges may exceed $20.00 per kW, while off-peak demand charges might be only $3–$4.00.

Additionally, utilities impose further peak period cost penalties by establishing higher consumption charges during on-peak periods relative to off-peak periods. Power may be $0.14/kWh or more during the on-peak period, but only $0.06/kWh during the off-peak period.

Cooling systems, because of their high demand with correspondingly low energy consumption, have a particularly bad impact on the electric utility's overall load factor (the ratio of peak capacity to average capacity required). By reducing the peak load imposed on the electric utility or by shifting cooling energy consumption to off-peak periods, the user is effectively helping the utility improve its load factor. This benefit can be reflected in the rate schedules and contract terms that can be negotiated with the utility and can reduce the overall cost of chilled water production.

The most common approach to reducing electrical energy costs for cooling is to utilize an alternative energy source during on-peak periods. Alternative energy sources that can be considered for used in a mixed energy source chiller system include the following:

*Natural gas and/or fuel oil.* Natural gas can be used in direct-fired absorption chillers, smaller spark ignition engine-drive chillers, or very large gas turbine-drive chillers. No. 2 fuel oil can be used in direct-fired absorption chillers and diesel engine-drive chillers.

Natural gas rate schedules fall into two general categories: "firm" gas rates and "interruptible" gas rates. Firm gas rates are based on the utility having gas available at all times, regardless of the load on the gas distribution system. Interruptible rates are used when the facility allows the gas utility to cut off their gas supply during peak distribution load periods. Firm gas rates are typically much higher than interruptible rates. Since the gas utility has its peak distribution loads during the winter, gas used during the summer, even if purchased under an interruptible rate, is a very reliable energy source.

For most gas utilities, the summer distribution load is far less than their winter distribution load and it is of great economic advantage to the gas utility if it can add summer load without increasing the winter peak. Cooling is an ideal candidate for additional summer load and, therefore, some utilities will offer discounts for purchase of "summer gas."

Fuel oil usually is a consideration only if the fuel storage facilities are already in place to serve the boilers and/or emergency generators. The costs and environmental considerations associated with large fuel oil storage facilities are significant and generally eliminate this option for new installations.

Fuel oil is purchased in bulk lots (usually by the "tanker load" of ~8000 gallons) on the "spot" market. Therefore, the facility manager must continually shop for the best price.

*District steam or high-temperature hot water.* Larger-campus facilities such as universities, military bases, and hospitals may have steam or high-temperature hot-water (HTHW) heating distribution systems. Since these systems are lightly loaded during the summer, either steam or HTHW can be used in absorption chillers to provide chilled water during on-peak periods.

The cost of steam or HTHW must include the cost of the fuel, discounted by the firing inefficiencies and distribution losses in the system. Summer COPs for a central boiler plant may be as low as 0.5–0.7, so careful study is required to determine the real cost of steam or HTHW delivered to the chiller(s).

*Waste heat from industrial processes.* Heat, if sufficiently hot and available in the required quantities, can be used to produce steam or HTHW for absorption cooling.

As discussed in Chapter 1, no type of chiller is more efficient than the electric-drive chiller. Additionally, absorption and engine-drive chillers cost more than equivalent electric-drive chillers to purchase and install. So, the economic viability for use of an alternative fuel hinges on using an inefficient, high-cost method of cooling for relatively short periods to save enough on-peak and ratchet electrical charges associated with an equivalent amount of electric-drive cooling to pay for the increased capital requirements within an acceptable time frame.

Typically, this does not happen unless the cost of alternative energy sources is very low or additional cooling capacity is required anyway and only the incremental cost increase for alternative fuel systems must be recovered. In the first case, the energy cost must be validated and its reliability evaluated. In the second case, the capital cost difference between electric chillers and absorption or engine-drive chillers must be determined. In either case, careful analysis is required.

# *Part II*

## *Chiller Design and Application*

# 3 Chilled Water System Elements

## CHILLER PLACEMENT AND INSTALLATION

An air-cooled water chiller must be located outdoors to use atmospheric air as its heat sink. While some have designed outdoor chillers with a ducted air intake and/or discharge for indoor installation, these designs generally have failed dismally. Location and installation of an air-cooled chiller must address the following considerations:

1. Prevent *recirculation* of discharge air by locating the chiller with sufficient clearance to adjacent buildings, shrubbery, and other equipment to allow unobstructed air movement into and out of the condenser section. Discharge air is obviously warmer than ambient air and, if it allowed to recirculate and enter the condenser, chiller capacity is reduced.

   Screening, either by plantings or by screen walls, must be at least 10 ft away from the chiller, be no higher than the top of the chiller, and have at least 50% open area.
2. *Community noise* must be addressed. And, for rooftop applications, vibration must be considered. See the section "Noise and Vibration" in Chapter 6 for a detailed discussion on these topics.
3. *Service access* must be considered. Ground locations are preferable to rooftop locations simply because access is easier. Compressor replacement for a rooftop chiller can be complex, expensive, and aggravating due to the need to lift heavy compressors to the roof level.

Water-cooled chillers are located indoors in a mechanical equipment room or *chiller room*. Chiller rooms must be separate from boiler rooms or any other source of open flame and provide space for the chiller(s), chilled water pump(s), condenser water pump(s), and the condenser water treatment system (see Chapter 13). ASHRAE Standard 15, adopted by most state and local building code authorities, classifies a chiller room as a *refrigeration machinery room* if the type and amount of refrigerant in the largest chiller exceed certain specified levels. *If so classified, the room must be constructed to higher standard and a detailed code review for these spaces is critical.*

Refrigeration machinery room elements include the following:

1. The room shall have tight-fitting doors opening outward. Doors shall be self-closing if they open into the occupied building.
2. Each room shall be equipped with at least one refrigerant leak detector. The detectors shall be designed to sense the specific refrigerant utilized in the chiller(s) and must be wired to activate both an alarm (manual reset type) and operation of mechanical ventilation if the refrigerant concentration reaches the *threshold limit value-time weighted average* (TLV-TWA) that will not cause an adverse effect on humans.
3. Mechanical ventilation shall consist of one or more electric-drive fans capable of exhausting air from the refrigeration machinery room at the following rate:

$$Q = 100 \times G^{0.5}$$

where
$Q$ = exhaust airflow (cfm)
$G$ = mass of refrigerant in the largest chiller in the refrigeration machinery room (lb)
4. Where refrigerants other than A1 toxicity/flame rating is utilized, other refrigeration machinery room requirements apply as defined in ASHRAE Standard 15.

Chillers are big, heavy, and noisy, and all of these conditions must be taken into account when locating and sizing the chiller room:

1. Table 3.1 lists the operating weights for typical water-cooled rotary compressor chillers. Obviously, it is usually easier to deal with these concentrated weights in a ground floor or basement location than in the upper stories of a building.
2. Noise levels as high as 80–90 dBA can be produced by a chiller during part-load operation and the chiller room must be designed to contain the noise. And, while vibration by a rotary compressor chiller is very low, it must also be considered and addressed. Again, see the section "Noise and Vibration" in Chapter 6 for a more detailed discussion on these topics.
3. The chillers, pumps, water treatment equipment, and so on in the chiller room must be located and arranged to provide for adequate access for maintenance.

Figure 3.1 illustrates the piping and installation requirements for a typical rotary compressor chiller installed at or below grade level.

**TABLE 3.1**
**Approximate Chiller Operating Weight**

| Capacity (tons) | Approximate Operating Weight (lbs) | Condensing Medium |
|---|---|---|
| 25 | 3000 | Air-cooled |
| 50 | 5000 | |
| 75 | 7500 | |
| 100 | 10,000 | |
| 150 | 12,000–17,000 | Water-cooled |
| 200–300 | 15,000–20,000 | |
| 300–500 | 20,000–28,000 | |
| 500–700 | 22,000–30,000 | |
| 700–900 | 26,000–44,000 | |
| 900–1200 | 42,000–66,000 | |
| 1200–1500 | 57,000–71,000 | |
| 1500–2000 | 68,000–95,000 | |

## CHILLED WATER PIPING

### Piping Materials and Insulation Requirements

Chilled water distribution systems are assembled from commercially available piping materials, most commonly steel and copper.

*Steel pipe*. This is the most common above-ground piping type and is defined by its wall thickness, called *schedule*, and its finish. Up through 10″ pipe size, Schedule 40 piping is normally used for chilled water (and condenser) water applications. For piping 12″ and larger, most designers

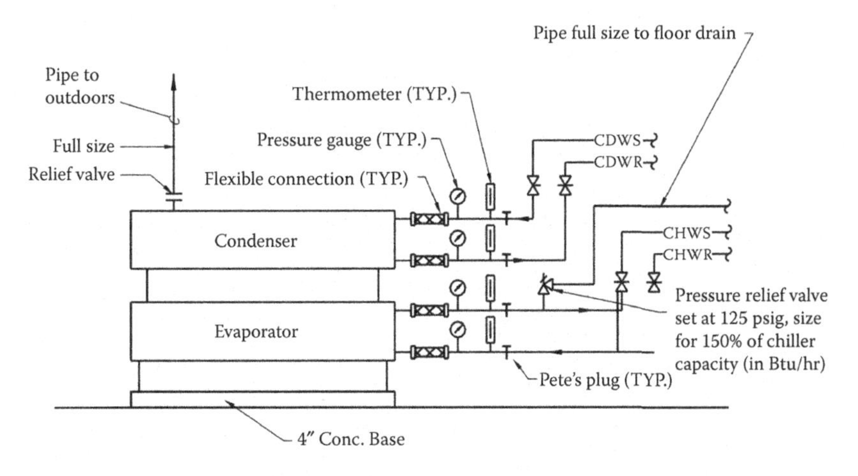


**FIGURE 3.1** Typical chiller piping.

standardize on piping with a 0.375″ wall thickness and ignore the pipe schedules. All chilled water piping is constructed of piping with a *mill finish* and is referred to as "black."

Piping 2″ and smaller is usually assembled by threaded joints, while piping 21/2″ and larger is usually assembled by welding.

Table 3.2 lists the physical properties of steel piping commonly used in water systems.

The fittings for steel pipe with screw connections 2″ and smaller are typically constructed of cast iron rated for up to 125 psig working pressure. Wrought steel "buttwelding" fittings are used for larger welded piping, while forged socket welding fittings may be used for smaller pipe.

*Copper tubing.* The cost of copper tubing is higher than that of a steel pipe, but the installation labor cost for smaller sizes, 2″ and smaller, is much less than that of steel. Therefore, most designers will allow copper tubing to be used in water systems for these smaller sizes at the contractor's option.

**TABLE 3.2**
**Properties of Commercial Steel Pipe**

| Pipe Size (in.) | Outside Diameter (in.) | Wall Thickness (in.) | Pipe Schedule |
|---|---|---|---|
| $\frac{1}{2}$ | 0.840 | 0.109 | 40 |
| $\frac{3}{4}$ | 1.050 | 0.113 | 40 |
| 1 | 1.315 | 0.133 | 40 |
| $1\frac{1}{4}$ | 1.660 | 0.140 | 40 |
| $1\frac{1}{2}$ | 1.900 | 0.145 | 40 |
| 2 | 2.375 | 0.154 | 40 |
| $2\frac{1}{2}$ | 2.875 | 0.203 | 40 |
| 3 | 3.500 | 0.216 | 40 |
| 4 | 4.500 | 0.237 | 40 |
| 6 | 6.625 | 0.280 | 40 |
| 8 | 8.625 | 0.322 | 40 |
| 10 | 10.750 | 0.365 | 40 |
| 12 | 12.750 | 0.375 | 30 |
| 14 | 14.000 | 0.375 | 30 |
| 16 | 16.000 | 0.375 | 30 |
| 18 | 18.000 | 0.375 | 30 |
| 20 | 20.000 | 0.375 | 20 |
| 24 | 24.000 | 0.375 | 20 |

To prevent galvanic corrosion that is initiated by dissimilar metals in direct contact, connection between steel and copper piping materials must be electrically separated by a *dielectric union*. This fitting uses rubber gasketing to separate the two materials.

The wall thickness of copper tubing is indicated by its *type*, defined as Types K, L, and M in decreasing order of wall thickness. Typically, Type K or L tubing, as summarized in Table 3.3, can be used for chilled water piping, but Type M copper tubing is not rated as for pressure service and can be used only for drain lines or other atmospheric pressure applications.

Fittings for copper tubing are wrought copper and the tubing and fittings are assembled by soldering the joints.

*Cast or ductile iron.* For underground piping, cast iron piping or ductile iron piping is routinely used. Iron piping is rated for application in terms of "pressure class" and for most chilled water systems Class 350 is used up through 12″ pipe size, Class 250 for 14–20″ pipe, and Class 150 for piping 24″ and larger.

Fittings are cast gray or ductile iron and the fittings and piping are assembled with gasketed mechanical pressure joints.

*PVC or CPVC.* As an alternative to iron piping for underground chilled water distribution, polyvinyl chloride (PVC) or chlorinated polyvinyl chloride (CPVC) piping can be used. PVC and CPVC piping wall thickness is defined in terms of schedules, much like steel pipe. For chilled water applications, Schedule 40 piping is normally used.

Fittings are also formed of PVC or CPVC and are joined by solvent soldering up through 2-1/2″ pipe size and by gasketed mechanical pressure joints for larger sizes.

**TABLE 3.3**
**Properties of Commercial Copper Tubing**

| Pipe Size (in.) | Outside Diameter (in.) | Type K Wall Thickness (in.) | Type L Wall Thickness (in.) |
|---|---|---|---|
| $\frac{1}{2}$ | 0.625 | 0.049 | 0.40 |
| $\frac{3}{4}$ | 0.875 | 0.065 | 0.045 |
| 1 | 1.125 | 0.065 | 0.050 |
| $1\frac{1}{4}$ | 1.375 | 0.065 | 0.055 |
| $1\frac{1}{2}$ | 1.625 | 0.072 | 0.060 |
| 2 | 2.125 | 0.083 | 0.070 |

For routine cleaning of evaporator and condenser tubes, piping connections at and near these heat exchangers must be a type that allows piping to be isolated, drained, and removed. To accomplish this, removable/replaceable pipe joining methods, such as flanged or grooved coupling connections, are required.

For all types of water pipe, flanged joints work well and is the recommended joining method to allow maintenance access. For steel piping, grooved piping and couplings may be used, but to prevent long-term problems and leakage, the following requirements must be met:

1. Grooved mechanical pipe couplings and fittings for use with rolled or swaged groove carbon steel pipe must be rated for water service up to 230°F. Joints shall be of rigid type.

   *Cut pipe grooves are not acceptable.* Gut grooves result in localized thin pipe walls and create a location for stress corrosion guaranteeing long-term failure.
2. Couplings may be malleable iron (ASTM A47) or ductile iron (ASTM A536), fabricated in two or more parts, securely held together by two or more bolts complying with ASTM A449 and A183.
3. Gaskets must be a rubber product recommended by the coupling manufacturer for the intended service.
4. Fittings may be malleable iron (ASTM A47), ductile iron (ASTM A536), or carbon steel (ASTM A53 or A106) that are *factory-fabricated* to accept grooved mechanical couplings.

Even after meeting these requirements, gaskets should be replaced each time a grooved joint is disassembled and couplings replaced every 5–7 years.

Chillers are required by ASHRAE Standard 15 to have pressure-relief protection to vent excess refrigerant pressure to the outdoors if the amount of refrigerant exceeds defined amounts (such as 110 lb for R-134a or other Group A1 refrigerant). Vent piping must be routed without valves to the outdoors, discharging at least 15 ft above grade or occupied area (such as a balcony or rooftop patio) and more than 20 ft from any window or ventilation air intake. Typically, vent piping must be steel with welded joints.

All above-ground chilled water piping, indoor condenser water supply piping, and all outdoor condenser water piping must be insulated to reduce unwanted heat gain and to prevent surface condensation. While as little as 1″ of insulation may serve to limit heat gains and meet the requirements of ANSI/ASHRAE/IESNA Standard 90.1-2004, more insulation may be required, particularly in southern climates, to prevent surface condensation in unconditioned spaces or outdoor locations. *Thus, at least 2″ thickness of insulation is recommended for indoor piping and 3″ thickness on outdoor piping.*

Designers and building owners in hot, humid climates have begun to ban the use of mineral fiber insulation, which includes fiberglass, on chilled water piping because of the almost universal occurrence of mold on installed insulation surfaces. Mineral fiber insulating materials, including fiberglass, are open-structured materials that rely largely on entrapped still air for much of their insulating properties. They have little long-term resistance to water and no resistance to vapor flow. *Thus, the potential for moisture absorption can be very high if the factory-applied vapor barrier is either damaged or inadequately sealed. Thus, the key to keeping fiberglass efficient is keeping it dry.*

The typical *all-service jacket* (ASJ) vapor retarder provided by most manufacturers of piping insulation has a rated permeance of 0.02 perm-inches, which is very low. However, some tests have shown that the ASJ may have an actual perm rating as high as 0.13 perm-inches, but even this rating is very acceptable. *However, all of these ratings are based on testing samples of ASJ in the laboratory and there are no data that defines ASJ performance as installed by the typical insulation contractor in the field.*

And, there are numerous opportunities for the contractor to fail in the installation of ASJ. First, the longitudinal sealing strip must be kept clean and be carefully aligned to provide a vapor-tight seal. The butt joints of the jacket must be carefully lapped and sealed with vapor barrier mastic and all joints at fittings, valves, equipment, wall penetrations, and so on must also be 100% sealed with vapor barrier mastic or tape. At hangers, rigid insulation inserts and sheet metal shields must be installed to protect the integrity of the vapor barrier. *The chances of all of this happening on the typical construction project are very poor.* Thus, every chilled water system insulated with fiberglass offers opportunities for vapor diffusion and the introduction of low-level moisture content.

Research has been undertaken into the effect of moisture on mineral fiber insulation by Achtziger and Cammerer of FIW in Germany. Their research concluded that 1% moisture content by volume could increase the thermal conductivity of the material by 36–107% with four of the five samples tested falling within the 95–107% increase range (*Forschungsvorhaben* Nr. 815-80.01.83-4 contained within CEN TC 88 WG 4 - N484). Tests by the National Insulation Manufacturers Association (NIMA) in 2000 showed that fiberglass insulation with ASJ can gain as much as 50% in weight from water absorption when installed in high-humidity locations.

Mold will grow on most surfaces if the relative humidity at the surface is above a critical value and the surface temperature is conducive to growth (from 40°F to about 140°F). The longer the relative humidity remains above the critical value, the more likely visible mold growth, and the higher the humidity or temperature, the shorter is the time needed for germination. The surface relative humidity is a complex function of material moisture content, material properties, and local temperature and humidity conditions. In addition, mold growth depends on the type of surface.

Fully recognizing the complexity of the issue, the International Energy Agency Annex 14 (1990), nevertheless, established a surface humidity criterion for design purposes: the monthly average surface relative humidity should remain below 80% RH. Others have proposed more stringent criteria, the most stringent requiring that surface relative humidity must remain below 70% RH at all times. Although there is still no agreement on which criterion is most appropriate, *mold can usually be avoided by limiting surface moisture conditions over 80% RH to short time periods and 70% RH for longer periods (even though some molds will begin to grow at a relative humidity as low as 60% RH).*

Based on the maximum 70% RH criterion from above and using the analysis methods outlined on pp. 23.20–23.21 of the 2001 *ASHRAE Fundamentals Handbook*, the minimum required fiberglass insulation thickness is 2″ for pipe smaller than 2″ and 1-1/2″ for larger pipe, *assuming dry insulation*. If 1% moisture content, by volume, has diffused into to the insulation due to poor ASJ installation or damage to the ASJ over time, the insulation thermal conductivity increases by 100% and the required minimum insulation thickness also increases by 100% to 3–4″ for all pipe sizes. *No designer currently requires 3–4″ thickness of fiberglass insulation on chilled water piping* and, thus, mold growth on chilled water piping in hot, humid climates is almost universal.

The performance of the vapor barrier is so important that one fiberglass pipe insulation manufacturer recommends that ASJ be covered with a second, welded PVC vapor-retarding outer jacket and that every fourth joint of fiberglass insulation be sealed with vapor barrier mastic when installed on chilled water piping in hot, humid climates.

Fiberglass insulation that is of insufficient thickness allows a boundary layer of high-humidity air to be created and mold immediately begins to grow on the surface. If the ASJ does allow moisture to invade, as time goes by the insulation moisture content rises to a level where the thermal conductivity increases enough to allow the surface temperature to drop to or below the ambient dewpoint and surface condensation forms. This liquid water further augments mold growth on the insulation surface and condensation that drips onto ceiling and wall materials below to create additional habitats for mold growth.

While secondary vapor barriers (usually welded PVC jacket or vapor-proof mastic coatings) can be added to ASJ/fiberglass insulation, the only real solution to this condition is to avoid the use of fiberglass insulation and instead use a non-hydroscopic insulation, insulation that will not absorb moisture, on chilled water piping in lieu of fiberglass. The insulation effectiveness, then, is not dependent on the quality of the installation of the ASJ vapor barrier to maintain a surface temperature that results in humidity below 70% RH.

*For indoor chilled water applications, cellular glass is the recommended material for insulating chilled water systems since this material has a perm rating of less than 0.005 perm-inches.*

For exterior applications, polyisocyanurate or expanded polystyrene can also be used. But, because both materials are combustible and release toxic gases when heated, their use indoors should be avoided (and can be a building code violation).

Recommended thickness of cellular glass insulation installed indoors is 2″ thick for pipe sizes up to and including 3″, 2-1/2″ thick for pipe sizes 4–10″, and 3″ thick for pipe sizes 12″ and larger. For outdoor applications, cellular glass insulation thickness should be increased by 1″ simply to reduce heat gain. The recommended thickness for polyisocyanurate or polystyrene insulation is 3″ for all pipe sizes installed outdoors.

For renovation of existing systems, improving the existing installation is the key and, typically, the following steps are required:

1. Shut down the chilled water system.
2. Remove all existing ASJ and inspect the underlying fiberglass insulation.
3. If it is wet to the touch, it can be allowed to dry, but the drying time can range from weeks to months. If the chilled water system can remain out of service over a winter, there is a good chance the insulation will dry and can be salvaged. If this long drying time is not feasible, the only option is to remove the existing insulation and reinsulate the system with cellular glass, after drying the pipe and cleaning it of scale and rust.
4. If the insulation feels dry to the touch, test it with a moisture meter and allow it to dry until the moisture content drops to below 1%. Then, add additional cellular glass pipe insulation, with ASJ, to achieve a final total insulation thickness as recommended above.
5. Typically, it will always be necessary to remove insulation at fittings, valves, and so on, and reinsulate these components with cellular glass.
6. At each hanger, crushed fiberglass insulation, wood blocks, and so on should be removed and a section of cellular glass insulation, as long as the insulation shield, installed to support the piping. Insulation shields (often called "pipe saddles") should be at least as long as 1.5 times the piping plus insulation outside diameter.

## WATER EXPANSION AND AIR REMOVAL

Water expands and contracts when heated and cooled in almost direct proportion with the temperature change. Since water is incompressible, a lack of expansion space in a closed piping system means that any volume increase will cause a pressure increase that may result in mechanical damage to system components. The most common way to accommodate water volume changes in a chilled water system is to use an *air cushion compression tank*, as shown in Figure 3.2.

To determine the size required for a compression tank, the following equation can be used:

$$V_t = \frac{(E_w - E_p - 0.02)V_s}{\left[(P_a / P_f) - (P_a / P_o)\right]}$$

where

$V_t$ = compression tank volume (gal)
$E_w - E_p$ = unit expansion of the system (typically 1% for chilled water systems)
$V_s$ = volume of system (gal)
$P_a$ = atmospheric pressure (14.7 psia)
$P_f$ = initial pressure in tank (usually atmospheric pressure)
$P_o$ = final pressure or relief valve setting (psia) (typically 100 psig or 114.7 psia).

| Tank connection line | | |
|---|---|---|
| System capacity (MBH) | Chilled water | Hot water |
| Up to 3000 | 3/4″ | 1″ |
| 3000–6000 | 1″ | 1-1/4″ |
| 6000–12,000 | 1-1/4″ | 1-1/2″ |
| 12,001 or Greater | 1-1/2″ | 2″ |

| CW makeup line | | |
|---|---|---|
| System flow (GPM) | Makeup size (IPS) | PRV flow (GPM) |
| ≤200 | 1/2 | 3 |
| 201–500 | 3/4 | 5 |
| 501–1200 | 1 | 10 |
| 1201–3000 | 1-1/4 | 20 |
| 3001–5000 | 1-1/2 | 30 |
| ≥5001 | 2 | 60 |

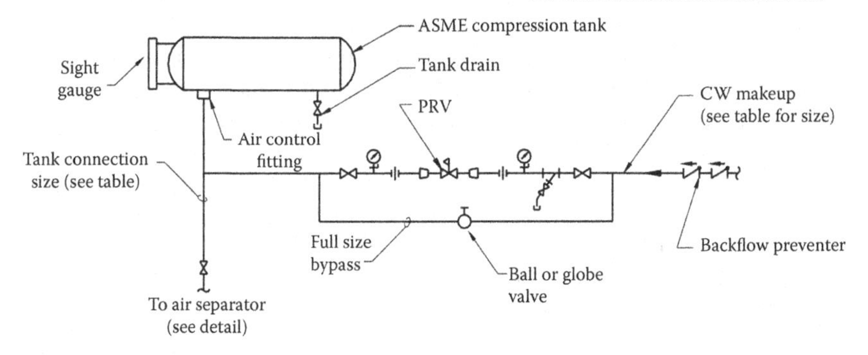


**FIGURE 3.2** Compression tank installation and piping.

The system volume, $V_s$, can be computed by adding the piping volume to the volume of water in the chiller(s) and coils. The water volume in the piping can be computed using the data from Table 3.4 and the lengths of each size of pipe in the chilled water system. Manufacturer(s) can provide chiller and cooling coil water volumes.

As water is heated or cooled, entrained air is released or absorbed in proportion to its temperature and pressure. To remove air from chilled water piping systems, an *inline tangential air separator*, as shown in Figure 3.3, is generally used. In the separator, the water velocity is reduced to less than 0.5 ft/s, which allows the entrained air to separate from the water. The tangential flow pattern in the separator creates a low-pressure vortex in the center where the entrained air can collect and rise through connecting piping to the compression tank above.

Air separators are sold with or without integral strainers. Since a strainer in a piping system is designed to filter and hold solids (scale, dirt, etc.), the separator strainer, coupled with the low water velocity, is the point in the chilled water piping system where "sludge" will tend to collect. However, most owners are unaware that their separator contains a strainer that must be routinely opened and cleaned. This maintenance item is difficult and results in significant water loss.

Since there are other strainers in the system (usually located at the pumps) that are designed for routine maintenance, a strainer in the separator is both superfluous and an unnecessary maintenance burden. *It is recommended that air separators without integral strainers be used.* However, the drain valve on the separator must be periodically opened and the accumulated sludge allowed to drain away.

**TABLE 3.4**
**Average Water Volume in Piping**

| Pipe Size (in.) | Outside Surface Area (sf/lf) | Water Volume (gallons/lf) |
|---|---|---|
| $\frac{1}{2}$ | 0.221 | 0.016 |
| $\frac{3}{4}$ | 0.275 | 0.045 |
| 1 | 0.344 | 0.078 |
| $1\frac{1}{4}$ | 0.435 | 0.106 |
| $1\frac{1}{2}$ | 0.497 | 0.174 |
| 2 | 0.622 | 0.249 |
| $2\frac{1}{2}$ | 0.753 | 0.384 |
| 3 | 0.916 | 0.514 |
| 4 | 1.18 | 0.661 |
| 6 | 1.73 | 1.501 |
| 8 | 2.26 | 2.599 |
| 10 | 2.81 | 4.096 |
| 12 | 3.37 | 5.875 |
| 14 | 3.93 | 7.163 |
| 16 | 4.44 | 9.489 |
| 18 | 4.71 | 12.04 |
| 20 | 5.24 | 15.01 |
| 24 | 6.28 | 22.11 |

## Water Treatment

Chilled water systems are "closed recirculating water" systems. For water treatment, closed systems have many advantages as follows:

1. There is no loss of water in the system (except when a leak occurs, strainers and air separators are drained, or chiller tube bundles are drained during cleaning) and, thus, little need for makeup water. Therefore, deposition or scaling is typically not a problem.
2. Once filled and entrained air is removed, the closed system creates an anaerobic environment that eliminates biological fouling as a problem.
3. Finally, closed systems reduce corrosion problems because the water is not continuously saturated with oxygen (an oxidizer) as in open systems. The low temperatures inherent with chilled water systems further reduce the potential for corrosion.

Corrosion in a closed system can occur due to oxygen pitting, galvanic action, and/or crevice attack. To prevent these conditions, the "shot feed" method of

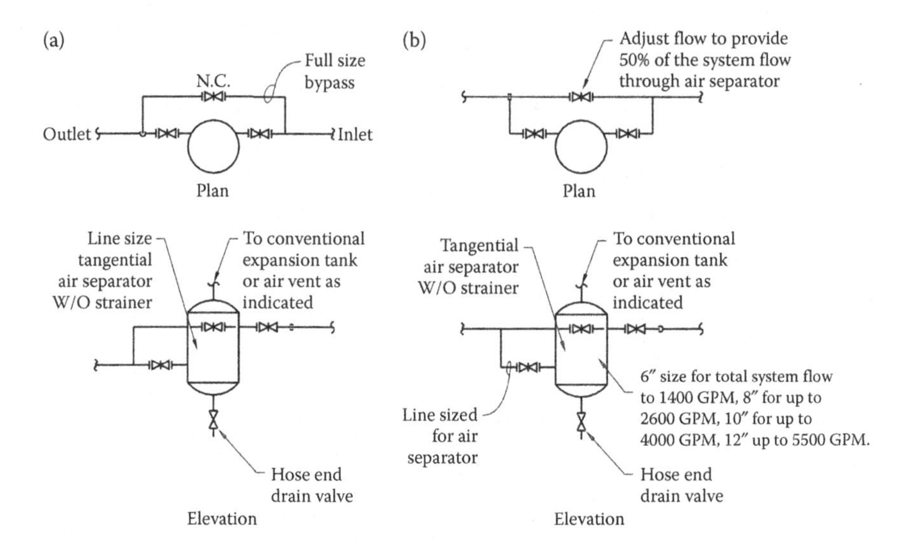


**FIGURE 3.3** Tangential air separator installation and piping. (a) System flow 300 GPM or less (4" and smaller air separator). (b) System flow greater than 300 GPM (6" and large air separator).

chemical treatment is used. With this method, a bypass chemical feeder, as shown in Figure 3.4, is used to add treatment chemicals to the system in a one-time "shot" just after the system is fil led with water. For the mixed metallurgy (steel and copper) systems typical for chilled water systems, a molybdate corrosion inhibitor is best. Recommended treatment limits are 200–300 ppm molybdate.

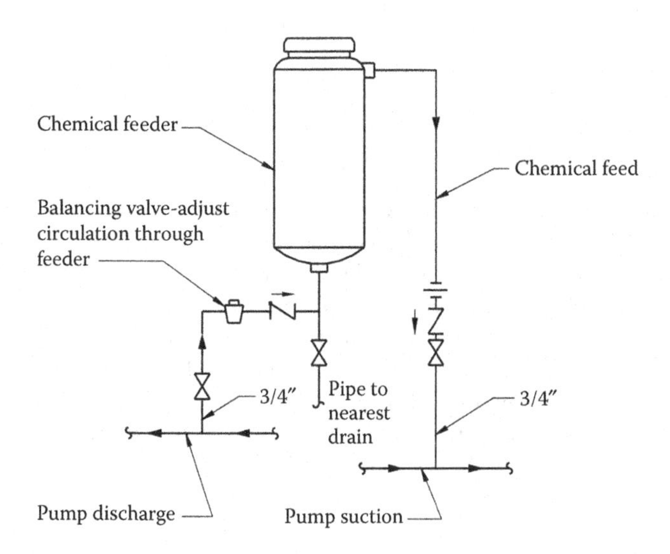


**FIGURE 3.4** Bypass chemical shot feeder.

Water treatment to eliminate any potential for corrosion is required only when the chilled water system is initially filled with water, or when any portion is drained and refilled for maintenance, repair, or modification.

## PUMP SELECTION AND PIPING

### Pump Basics

Chilled water pumps are almost universally *centrifugal* pumps. A pump converts the energy provided by a prime mover, most often an electric motor, to energy within the liquid being pumped. This energy within the liquid is present as velocity energy, pressure energy, static elevation energy, or a combination of two or more of these. The rotating element of a centrifugal pump, which is turned by the prime mover, is called the *impeller.* The liquid being pumped surrounds the impeller, being contained by the pump *casing*, and, as the impeller rotates, its motion imparts a rotating motion to the liquid, as shown in Figure 3.5. The entire pump assembly is illustrated in Figure 3.6.

Liquid enters the impeller at its center at low pressure and low velocity. As the liquid travels outward along the *vanes* of the impeller, it acquires velocity and an associated increased velocity pressure. But, when the liquid leaves the end of the vanes, as shown in Figure 3.5, it slows rapidly and the majority of the velocity pressure is converted to static pressure or *pump head.*

The centrifugal pumps used in water systems are typically as follows:

*Line-mounted pumps.* These pumps can be installed directly in the piping since the suction and discharge connections are arranged 180° apart. The motor and pump shafts, typically, are mounted vertically. The pump

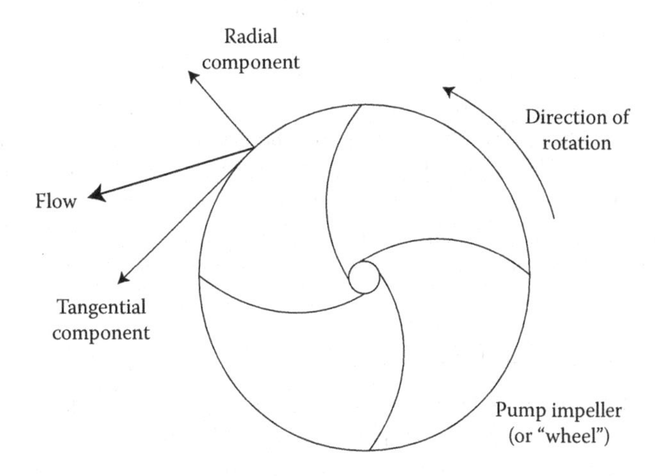


**FIGURE 3.5** Operation of a centrifugal pump impeller.

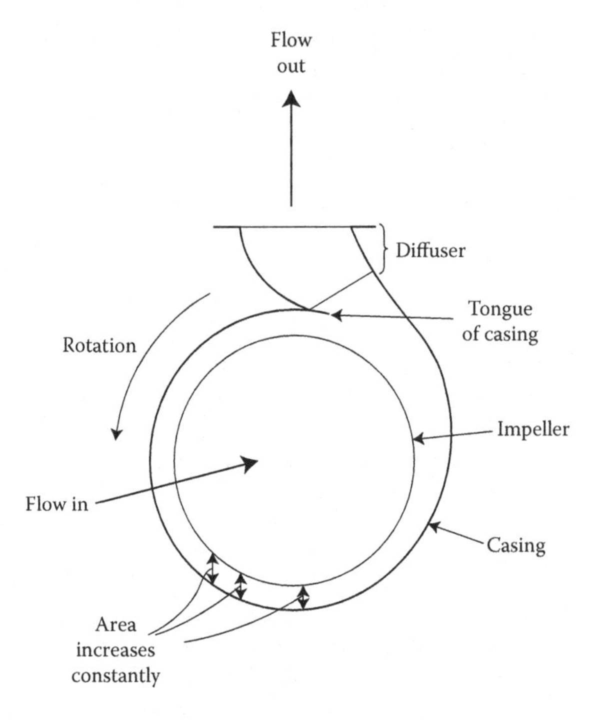


**FIGURE 3.6** Centrifugal pump configuration.

may be supported by the piping and/or by additional hangers or a foot stand.

*Base-mounted pumps.* Base-mounted pumps have the motor and pump shafts mounted horizontally, with both the pump and the motor bolted to a common frame or *base*. These pumps are available in two configurations, as illustrated in Figure 3.7.

Pumps are selected on the basis of the required flow and head, the pump efficiency, and the application or location requirement. Line-mounted pumps are normally limited to about 600 gpm maximum flow (about 75 hp), while end-suction pumps have a maximum flow of about 1000 gpm (about 60 hp). There is essentially no flow limit for horizontal split case, vertical split case, and vertical turbine pumps since these pumps can be fabricated as large as required.

## Pump Head and Horsepower

The required flow rate for the chilled water pump(s) is defined by the cooling load imposed and the chilled water temperature range. The next step in defining the pump requirement is to compute the required pump *head*, the pressure loss that the pump must overcome at the required flow rate.

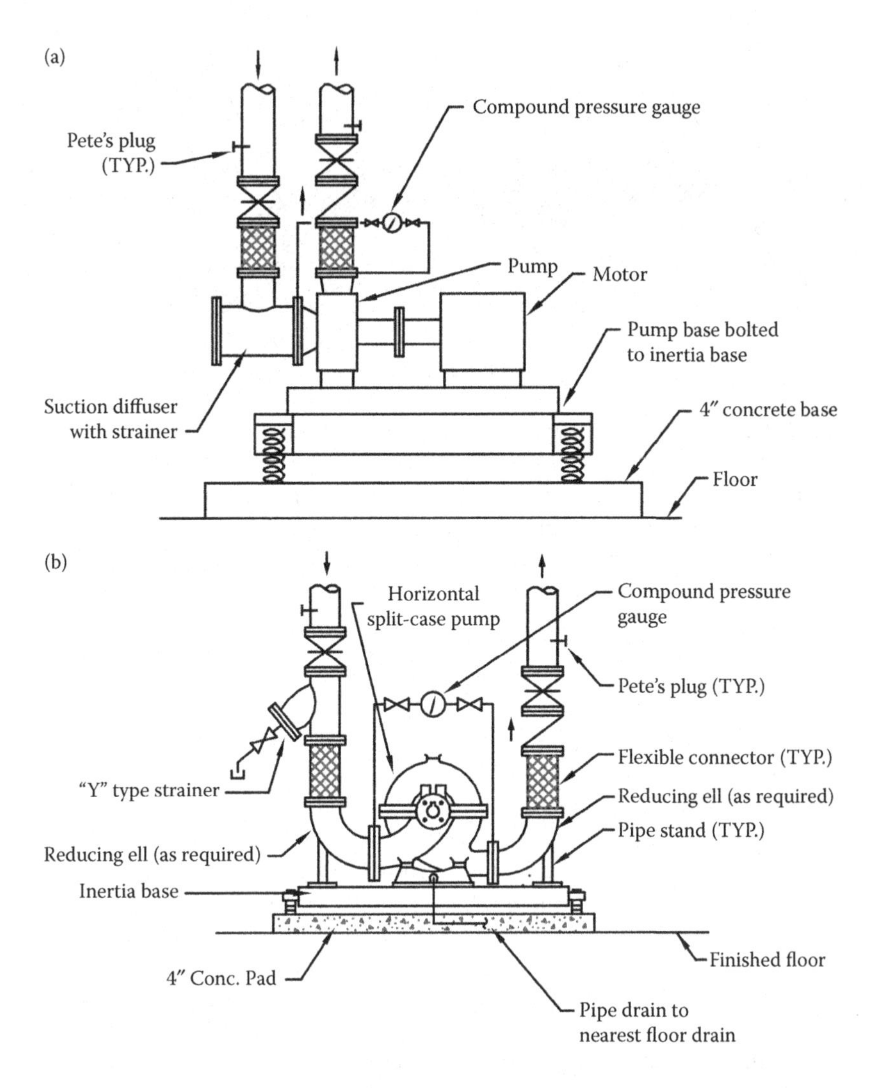


**FIGURE 3.7** (a) Recommended end-suction (single-suction) base-mounted pump installation. (b) Recommended horizontal split case (double-suction) base-mounted pump installation.

The pump head in a closed chilled water system is the sum of two pressure components in the piping system:

1. Head loss due to friction in the piping system. Water piping is typically sized utilizing Figure 3.8, based on a recommended design head loss factor of between 3 and 6 ft of water/100 ft of piping and a maximum flow velocity of 10 fps. Thus, a water flow of 1000 gpm would, from Figure 3.8, require an 8″ pipe size, resulting in a friction head loss of 1.5 ft of water/100 ft of piping. Additionally, a common method for

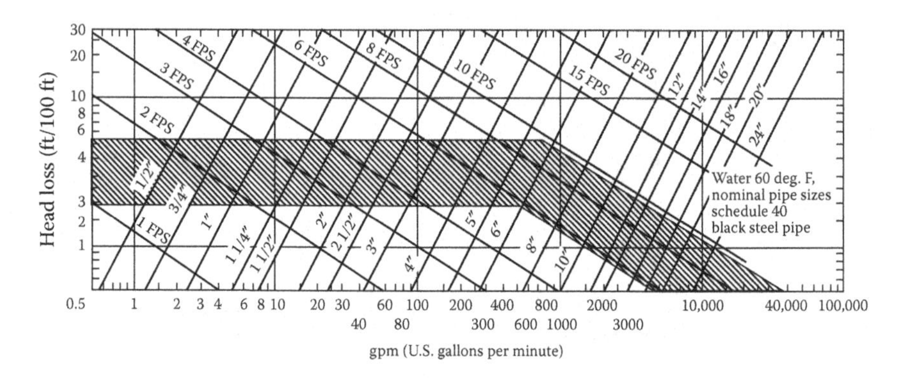


**FIGURE 3.8** Friction loss for water in Schedule 40 commercial steel pipe.

accounting for losses in piping fittings and specialties, including elbows, tees, valves, strainers, air separators, and so on is to use 50% of the straight pipe pressure losses.

Thus, the friction head loss can be conservatively estimated from the following:

$$H = 1.5 \times \left(\frac{L}{100}\right) \times F$$

where
$H$ = head loss due to friction (ft of water)
$L$ = piping length (ft)
$F$ = friction head loss factor from Figure 3.8 (ft of water/100 ft of pipe)

2. The pressure loss through the chiller evaporator, cooling coils, and control valves must also be included. For the chiller evaporator and cooling coils, the pressure loss at design flow can be provided by the manufacturer. Control valve losses are typically 5 psig or 13 ft (see Chapter 4).

The required brake horsepower for a water pump can be estimated using the following relationship:

$$\text{BHP} = \frac{\text{THD} \times F}{3960 \times \text{Eff}}$$

where
BHP = pump brake horsepower
THD = total pump head (ft of water column)
$F$ = water flow rate (gpm)
Eff = (typical) pump efficiency:

| | |
|---|---|
| vertical split case | 0.81 |
| horizontal split case | 0.82 |
| end suction | 0.70 |

**TABLE 3.5**
**Motor Synchronous Speed**

| Number of Poles | Synchronous Speed (RPM) |
|---|---|
| 2 | 3600 |
| 4 | 1800 |
| 6 | 1200 |
| 8 | 900 |

| | |
|---|---|
| inline | 0.65 |
| vertical turbine | 0.78 |

## Variable Flow Pumping

*Variable frequency drives* (VFDs) are applied to chilled pumps to create variable flow in primary–secondary and variable primary flow chilled water system configurations.

Pump motor speed (RPM) is defined by the following equation:

$$RPM = F \times \left( \frac{120}{\text{poles}} \right)$$

where

RPM = motor speed (revolutions/min)
$F$ = electrical frequency (0–60 Hz)
Poles = number of motor poles (2–8)

At full or *synchronous* speed at 60 Hz, the motor RPM is defined as by Table 3.5.

There are three major VFD designs commonly used: pulse width modulation (PWM), current source inverter (CSI), and variable voltage inverter (VVI). For HVAC applications, the vast majority of drives are the PWM type. These drives are reliable, efficient, affordable, and readily available from a number of manufacturers.

At 60 Hz, AC electrical service produces a sine wave of voltage versus time. To vary motor speed, the PWM VFD simulates a sine wave to supply the motor with various "time widths" of voltage that, when averaged together over the length of the cycle, looks like a sine wave to the motor. This is illustrated in Figure 3.9.

The typical PWM VFD consists of two major components: a *rectifier* converts the 60 Hz AC input to DC output to an *inverter* that converts the DC back to AC output at a controlled frequency. For the motor to supply constant torque, the output voltage is varied with the output frequency to maintain a constant relationship between the two (7.6 V/Hz for a 460-V motor and 3.8 V/Hz for a 230-V motor).

Efficiency of the PWM VFD is 92–96 + %. The VFD contains a solid-state reduced-voltage motor starter and starts the motor at low speed to "soft load" the

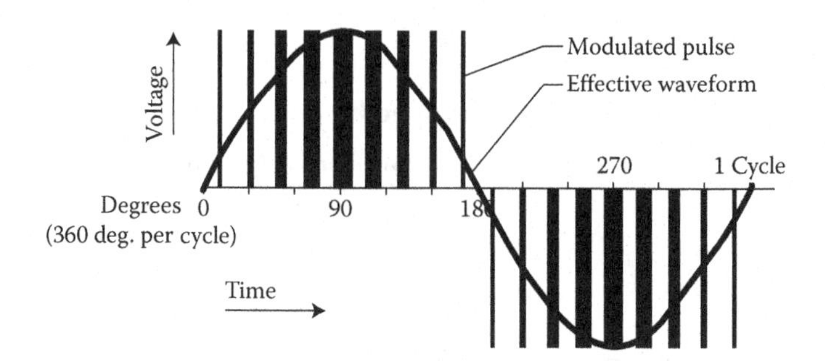


**FIGURE 3.9** Pulse width modulation as a frequency control method.

system. (Note, though, that the 4–8% inefficiency of a VFD manifests itself as heat released into the space that must be removed by ventilation or mechanical cooling to avoid drive overheating.)

*It is important that each VFD be provided with the following:*

*Main power disconnect ahead of drive.* If the motor disconnect switch is located between the drive and the motor, the drive can be destroyed if the disconnect is closed with the motor already spinning. The VFD will try to pull the motor down to a low, soft-start frequency, which can result in high current and an overload on the drive. This means that the VFD must be located close to the motor (to comply with National Electric Code requirements).

*Bypass switching.* In the event of VFD failure, an integral bypass switch, either manual or automatic, will allow the motor to be operated at full speed.

# 4 Chilled Water System Control and Performance

## START-UP CONTROL

There are two basic methods used for controlling the start-up of a chilled water system:

1. The most common method is "manual" initiation. With this method, the facility operating staff makes the decision to start the system on the basis of the time of year, outdoor temperature, and/or the number of "hot" complaints received from building occupants. This method is widely used in northern climates that have more distinctly separate heating and cooling seasons.
2. Outdoor temperature can be the trigger for starting a chilled water system, particularly if airside economizer systems are used in the facility. When airside economizer control is applied to building air-handling systems, outdoor air will satisfy imposed cooling requirements when the temperature is at or below the required air-handling units' discharge temperature setpoint, usually about 55°F. This temperature, then, may be set as the "change-over" temperature and the chilled water system is operated only when the outdoor temperature rises above this setpoint. A typical scheme would be to start the chillers when the outdoor temperature rises to 60°F and to stop them when the outdoor temperature falls to 55°F.

Starting a chilled water system means, first of all, starting the chilled water distribution pumps. After that, the chiller will start upon initiation by its factory controls if the chilled water supply temperature is above its setpoint condition.

For air-cooled chillers, the condenser fan(s) will start as required to maintain the correct condensing pressure. But, for water-cooled chillers, via hard-wired or software interlock, the factory chiller controls will start the condenser water pump and, then, if flow of both chilled water and condenser water is "proven" by flow switches (preferably, differential pressure type), the chiller will start.

Aside from the flow switches, every chiller is factory-equipped with basic operating safety controls to protect the refrigeration machine from damage due to any of the following conditions:

1. Low refrigerant temperature
2. Low chilled water temperature
3. High condensing pressure
4. Low oil pressure

In the event that any of these conditions exists, the chiller will shut down and require manual investigation and reset before operation can resume.

For rotary compressor chillers, there are typically two additional safety cutouts:

1. High motor winding temperature
2. Motor overload (high amps)

## CAPACITY CONTROL

Modern water chillers are normally provided with digital electronic controls designed and integrated by the chiller manufacturer. In addition to the operating safety elements outlined in the previous section, these controls are used to provide capacity control based on maintaining a supply chilled water temperature setpoint.

### Refrigerant Flow Control

For the vast majority of water chillers, capacity control means controlling the refrigerant flow rate through the evaporator. Depending on the type of compressor used, several methods are applied:

*Scroll compressors.* The most common method applied for scroll compressor capacity control is speed control utilizing an inverter or variable frequency drive to modulate the compressor rotational speed in response to the capacity required. Each compressor has a high (maximum)- and low (minimum)-speed requirement and these establish the control range for the compressor.

Less commonly, variable-displacement scroll compressor control is applied. The compressor is fabricated with "porting holes" in the fixed scroll member. The control mechanism disconnects/connects compression chambers to the suction side by opening/closing the porting holes. This creates a "refrigerant bypass" arrangement where full capacity is produced if all porting holes are closed and reduces capacity as more porting holes are opened.

Finally, pulse width modulation can be used as a control mode for modulating the axial pressure that maintains a seal between the scroll tips and bases. To modulate capacity while the motor speed remains constant, the seal pressure is modulated, allowing more or less refrigerant to bypass around the seals, loading or unloading the scrolls.

*Rotary screw compressors.* Since the rotary screw compressor is a positive-displacement compressor, suction throttling can reduce the refrigerant gas flow into the compressor. A modulating control method is desirable to produce essentially infinite capacity adjustment between the minimum and maximum flow rates and capacities.

A *slide valve* is a hot-gas bypass control valve with a sliding action arranged parallel to the rotor bores and located at the high-pressure discharge of the compressor. The valve is then modulated to return a variable portion of the discharge gas back to the compressor suction. This valve, in addition to controlling capacity, also adjusts the location of the compressor discharge port as the load changes. This "axial discharge port" then provides good part load performance without reducing full-load efficiency.

*Centrifugal compressors.* Historically, refrigerant gas flow into a centrifugal compressor has been controlled by adjustable inlet guide vanes, or *preswirl or prerotation vanes*, just as with a centrifugal fan. These vanes are arranged radially at the inlet to the compressor impeller and can be opened and closed by an external operator.

Since each vane rotates around an axial shaft, they affect the direction of the flow entering the impeller. When the inlet vanes are fully open, gas enters the impeller at 90° to the impeller. However, as the inlet vanes begin to close, flow enters the impeller at an increasing angle in the direction of the radial flow along the impeller blades. This "preswirl" condition reduces the ability of the impeller to impart kinetic energy to the refrigerant gas, thus reducing the flow rate.

Inlet vanes do not produce a pressure drop or "throttling" to reduce refrigerant flow through the centrifugal compressor.

A minimum volumetric rate flow through a centrifugal compressor is required for stable operation. If the volumetric flow rates fall below this minimum, the compressor will become unstable and *surge.* When this happens, the refrigerant flows alternatively backward and forward through the compressor, producing noise and poor operation. Extended operation under surge conditions will cause mechanical damage to the compressor. The surge envelope will vary from compressor to compressor but usually occurs when the volumetric flow rate is reduced by 40–60%.

To prevent surge from occurring, internal hot-gas bypass may be used to allow capacity to be reduced while maintaining sufficient gas flow through the compressor. (This condition, along with increasing windage losses and motor inefficiencies, accounts for the part-load performance characteristics shown in Figure 1.7.)

In recent years, capacity control of centrifugal chillers by speed control has been applied. Here, a large variable frequency drive is applied to the chiller motor and the motor speed modulated to control capacity. Generally, speed control improves efficiency over inlet vane control down to about 55% of rated capacity; whole inlet vane control is more efficient below 55% of rated capacity.

Compressor speed is directly related to capacity, but the pressure (lift) produced by the compressor is a function of the square of the speed. This may produce unsatisfactory operation and surge, requiring the use of hot gas bypass, as the chiller unloads under speed control.

*Adding speed control significantly increases the price of the chiller and this option must be carefully evaluated to determine if this concept is cost effective for a given cooling load profile.*

## Sequencing Multiple Chillers

Where multiple chillers are utilized, the chillers must be sequenced on and off to meet the imposed chilled water system load efficiently. To prevent rapid stopping and starting, called *short cycling*, a time delay of about 30 min is required between the stopping and starting of an individual chiller. Additionally, to equalize the run times for multiple chillers, a lead–lag or "rotation" sequence should be implemented to change the order of stopping and starting the chillers from one time to the next.

1. For constant flow chilled water systems, the imposed cooling load is indicated by the return chilled water temperature, assuming the supply chilled water temperature is maintained at setpoint.

   For example, consider a system with a 44°F supply chilled water temperature, a 12°F range, and three equally sized chillers arranged in parallel. When the return water temperature is 56°F (or higher), all three chillers are required. However, as the imposed cooling load decreases, the return water temperature will also decrease and, when the return temperature falls to 52°F, one chiller can be stopped. As the load continues to decrease and the return temperature falls to 48°F, a second chiller can be stopped.

   As the cooling load increases, the chillers can be started at these same temperature conditions.
2. With primary–secondary chilled water systems, the flow in the secondary loop varies in proportion to the imposed cooling load since the temperature range in this loop is maintained at a fixed setpoint. Flow in the primary loop varies with the number of chillers and primary chilled water pumps in use.

   As discussed in Chapter 2, the flow rate in the primary loop must always to be equal to or greater than the flow in the secondary loop to prevent blending of system return water flow with the supply water flow, resulting in an elevated chilled water supply temperature. Thus, control of the chillers stopping and starting can be based on the primary return loop temperature downstream of the common piping or bridge.

Again, consider a system with a 44°F supply chilled water temperature, a 12°F range, and three equally sized chillers arranged in a primary–secondary configuration. At full cooling load the flow rates in the primary and secondary loops are exactly equal and the temperature in the return section of the primary loop is equal to the return temperature from the secondary loop, 56°F.

However, as the cooling load falls, the secondary flow rate decreases. Since the primary loop flow rate now exceeds the secondary loop flow rate, chilled water from the supply side of primary loop mixes with return water from the secondary loop to produce a reduced temperature in the return side of the primary loop. When this temperature falls to 52°F, one chiller can be stopped since the reduced return water temperature indicates that the cooling load has declined by one-third.

When a chiller is stopped, the flow rate in the primary loop is reduced by one-third and now again equals the secondary flow rate. Blending is eliminated and the temperature in the return side of the primary loop returns to 56°F.

If the cooling load continues to fall, the temperature in the return side of the primary loop will again begin to fall as blending is reinitiated. When it falls to 50°F, the second chiller can be stopped.

As the cooling load increases, the flow in the secondary loop will begin to exceed the flow in the primary loop. Under these conditions, the temperature in the primary loop will rise to above 56°F, which indicates that a chiller must be started.

The general algorithm for control of chiller sequencing in a primary–secondary system made up of any number of equally sized chiller modules (a chiller and a primary chilled water pump) can be stated as follows:

$$T_1 = T_c + T_d$$

where

$T_1$ = the secondary loop chilled water temperature setpoint

$T_c$ = the system supply chilled water temperature setpoint

$T_d$ = the system design temperature range

If ($T_r$ + 0.5), where $T_r$ is the primary loop return water temperature, is greater than $T_1$ for at least 30 min, start a chiller. $T_2 = T_d/N$, where $T_2$ is the range assigned to each on $N$ equally sized chillers. If ($T_r$ – 0.5) is less than ($T_1 - T_2$) for at least 30 min, stop a chiller.

3. Sequencing chillers in a variable primary configuration is relatively complex. Each chiller must be equipped with a high-quality flow meter that is used to measure flow through each chiller evaporator.

   Again, consider a system with a 44°F supply chilled water temperature, a 12°F range, and three equally sized chillers. At full load, all chillers operate at full flow, chilled water is supplied to the coils at 44°F, and the return temperature is 56°F.

As the cooling load falls, the pump speed on each chiller reduces to deliver less flow and maintain the return water temperature setpoint of 56°F. Once the flow through all three chillers is reduced to 67% of design flow, one chiller and pump can be stopped. Once the flow through the remaining chillers is reduced to 50% of design flow, the second chiller can be stopped.

As cooling load increases, the chillers can be started at these same flow conditions.

## Optimizing Chilled Water Supply Temperature

As discussed in Chapter 1, electric-drive chillers use less energy as the supply chilled water temperature is indexed upward. Control of this indexing must, however, take into account both sensible and latent cooling loads as independent variables:

*Sensible cooling load.* Chilled water supply temperature setpoint can be indexed upward in the same proportion that the sensible load decreases.

*Latent cooling load.* However, often the sensible cooling load will reduce faster than the latent cooling load. If the chilled water temperature setpoint is indexed upward solely on the basis of decreased sensible cooling load, space humidity may increase to unacceptable levels since the chilled water temperature may be too high to allow dehumidification to occur.

To properly establish the optimum chilled water supply temperature setpoint, temperature and humidity conditions in each control zone must be evaluated and the chilled water setpoint indexed upward only while both conditions remain acceptable. This can only be accomplished via the application of a building-wide direct digital control system as follows:

1. Monitor the temperature and humidity in each building control zone.
2. Sort inputs from all zone temperature sensors and determine which sensors are within the required temperature or humidity setpoint range.
3. If all the sensors are within their required temperature or humidity setpoint range, increase the chilled water supply temperature setpoint by 0.5°F.

   Monitor and sort conditions over the next 30 min. If all the sensors remain within their required temperature and humidity range at the end of 30 min, increase the chilled water supply temperature setpoint by 0.5°F.

   Repeat this process until one or more sensors indicate that temperature or humidity is outside of its required condition(s).
4. If one or more temperature or humidity sensors are above the maximum required zone temperature or humidity level, reduce the chilled water temperature by 0.2°F and monitor the system for 15 min. If all the sensors return to within their required range, return to Step 3. If one or more

temperature or humidity sensors remain above the maximum required zone temperature or humidity level, reduce chilled water temperature by 0.2°F and monitor system for another 15 min. Continue this process until all temperature sensors return to within their required temperature or humidity range and then return to Step 3.

## VARIABLE FLOW PUMPING CONTROL

Variable flow distribution loop pumping is used in both primary–secondary and variable primary flow configurations of chilled water systems. The starting point of variable flow control is the cooling coil control valve. This must be a two-way valve that throttles flow through the coil as the cooling load decreases.

Water coils are arranged for counterflow configuration and the final temperature of the leaving chilled water may exceed the outlet temperature of the air leaving the coil. This is particularly important for chilled water systems that have small temperature-difference requirements. Another advantage of the counterflow configuration is that less surface area is required for a given rate of heat transfer.

The overall heat transfer by the coil, $Q$, is determined by the mass flow rates of the two media, the surface area, the temperature gradients, and the overall coefficient of heat transfer, $U$, as defined by the following equation:

$$Q = -M_h C_{p,h} dT_h = M_c C_{pc} dT_c = U \, dA \, (T_h - T_c)$$

where

$M$ = mass flow rate (lbm/min)
$C$ = specific heat of the fluid (Btu/lb °F)
$T$ = temperature of the fluids, hot ($T_h$ ) and cold ($T_c$ ), respectively (°F)
$dA$ = differential area of the coil under study (ft$^2$)

Integrating this equation over the entire area of the coil yields

$$Q = \frac{UA(\text{TD}_i - \text{TD}_o)}{\ln(\text{TD}_i / \text{TD}_o)}$$

where TD = $T_h - T_c$ at the inlet (i) and outlet (o) conditions, respectively. The temperature term $(\text{TD}_i - \text{TD}_o)/\ln(\text{TD}_i/\text{TD}_o)$ is called the *logarithmic mean temperature difference* (LMTD).

Since $U$ is defined by the design of the coil and the materials used in construction, which is fairly consistent between manufacturers, the basic performance of any coil is defined by the coil surface area and the temperatures required, or LMTD. If large temperature differences are available, as for most heating applications, the required coil area is small. For small temperature differences, as for chilled water systems, the coil area is large.

Water coil surface area is defined by the number of rows of tubes and the spacing, or density, of the fins (in terms of fins/in. of coil width). Thus, a heating coil may have 1 or 2 rows and 8–10 fins/in., but a cooling coil will have 6–8 rows and up to 16 fins/in.

The capacity of any water coil is controlled by modulating the flow of liquid through the coil. However, as flow is reduced, the capacity of the coil is not reduced linearly. To compensate for this nonlinearity, *equal percentage* characteristic control valves are utilized, resulting in a linear relationship between coil capacity and valve stroke or percent open, as shown in Figure 4.1.

These control valves must be selected on the basis of the maximum design flow through the coil and a "wide open" pressure drop equal to or greater than the pressure drop through the coil. Additionally, the valves must be rated for tight shutoff against the maximum system pressure that is developed, which, for variable flow systems, is the pump "cutoff" pressure, the pressure the pump will develop at zero flow.

As load is reduced at a cooling coil, the control valve throttles flow to the coil to match flow capacity to the coil load as follows:

1. Secondary chilled water pumps in a primary–secondary configuration can reduce their flow rate by either increasing head (forcing a flow reduction along the pump curve) or decreasing pump speed. The second method is more efficient and produces significant energy savings compared to the first. Thus, except for very small systems (<10 hp), VFDs are used to modulate pump speed in response to flow needs (see the section "Variable Flow Pumping" in Chapter 3).

   Accurate control of pump speed, and thus flow rate, is critical.

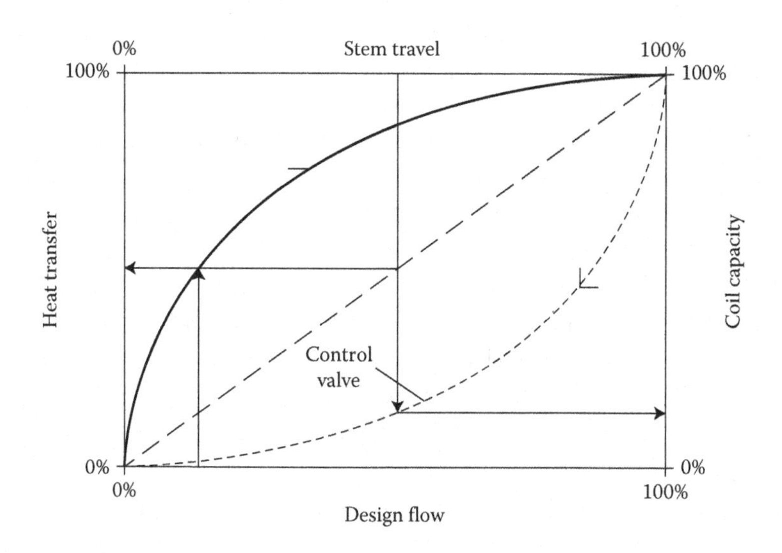


**FIGURE 4.1** Coil capacity control with equal percentage control valve.

When variable flow chilled water systems were introduced in the early 1980s, pressure control was typically used. A pressure sensor, located sufficiently downstream from the pump to be sensitive to control valve action, can respond to the pressure increase produced by throttling control valves by reducing the pump speed and, consequently, flow rate and pressure.

*But, it quickly became obvious that pressure control represented an attempt to respond to a "symptom" rather than to the "disease."* While pressure changes with flow changes, the change is not linear (pressure changes as the square of flow), multiple valves of different sizes may have different pressure/flow relationships, and the resulting pressure change in a larger system may have not real relationship to the cooling load. The basis of design for a primary–secondary system is to maintain a fixed chilled water supply temperature and a fixed temperature range in the secondary loop. If, by chiller capacity control and proper sequencing of chillers, the chilled water supply temperature is maintained at setpoint, the design range can be used to establish a return chilled water temperature setpoint to control pump speed. Any variation in the return chilled water temperature must result in an opposite linear change in the chilled water flow rate.

2. Variable flow control in a variable flow primary configuration is essentially the same as for the secondary pumps in a primary–secondary configuration.

As the pump speed is reduced, both flow rate and pressure are reduced. Ultimately, the available pressure at reduced flow rates may not be sufficient to ensure water flow to the more remote coils and, thus, these coils are "starved" and they fail to cool at all. To prevent this problem, one or more *differential pressure* sensors must be installed across the most remote flow circuit(s) and a minimum pressure setpoint determined for each circuit. Then, as the pump speed is reduced, if the pressure falls below the required setpoint in any circuit, further reduction of pump speed in inhibited. At this point, the system essentially becomes constant flow.

For variable primary flow systems, there is also a minimum flow requirement through the chiller, equal to 3 fps or about 25–30% of design flow. A *bypass control valve* must be modulated open when the flow rate through the last operating chiller is reduced to its minimum level, as indicated by the chiller flow meter.

# 5 Cooling Thermal Energy Storage

## ECONOMICS OF THERMAL ENERGY STORAGE

The goal for *thermal energy storage* (TES) in a chilled water system is to reduce the electrical demand during on-peak periods by shifting electric-drive chiller energy consumption to off-peak periods. This reduces electrical demand charges, often on a year-round basis. Aside from shifting of cooling energy consumption from the on-peak to the off-peak period, there *may* be some reduction in overall cooling energy consumption with TES.

Every TES system will have thermal losses:

1. Both water and ice storage systems have heat gains that increase cooling energy consumption.
2. Water systems have thermal mixing problems that require more cooling be stored than can actually be used.
3. Ice systems, due to the reduced temperatures, require greater compressor energy than an equivalent water storage system.

These losses may or may not be offset by improved chiller COP due to reduced condensing temperatures during off-peak periods when lower outdoor temperatures prevail.

Control of the thermal storage system is critical to ensure that enough cooling energy is stored during the off-peak period to offset the shortfall in chiller capacity during the on-peak period. Conversely, it is a waste of energy to store more cooling than can be used during the following on-peak period.

The analysis of thermal storage economics requires that an accurate "peak day" 24-h cooling load profile be developed. Basically, this consists of developing, for the cooling design day, the outdoor temperature and imposed cooling load at each hour. Then, the day must be divided into the on-peak and off-peak periods defined by the applicable electric rate schedule.

The total daily cooling requirement must still be met by the chiller, but, utilizing thermal storage, the chiller runs 24 h, not 12–14 h. The chiller, then, can be sized on the basis of the "split" between the energy required for *charging* during the off-peak period versus the energy consumed in *discharging* during the on-peak period, as shown in Figure 5.1.

Charging represents the off-peak period production of cooling energy by the chiller that is stored, while discharging is the use of that stored energy during the

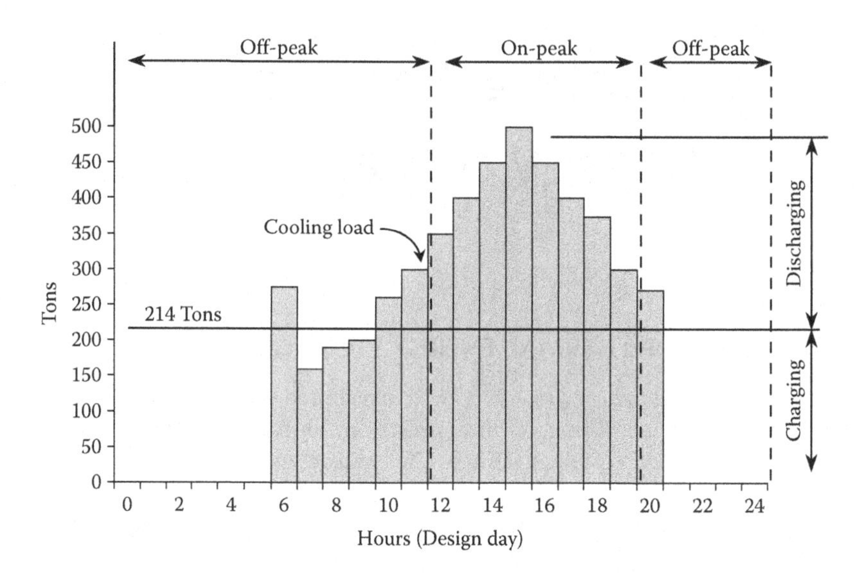


**FIGURE 5.1** Typical design day chilled water load profile.

on-peak period. The amount of cooling required during the on-peak discharge divided by the amount of cooling energy stored during the off-peak charging is defined as the *storage efficiency*, typically 0.95 or greater for ice systems and 0.90–0.95 for water systems.

Selecting chillers and the optimum amount of storage requires an evaluation of all of the cost factors under different chiller capacity/storage ratios and selecting the combination of chiller and storage that yields the best economic performance. To illustrate this process, consider the following example: A large building has a peak cooling load of 500 tons and a design day cooling load profile as shown in Table 5.1. The building cooling system is operated from 6:00 a.m. through 7:00 p.m., resulting in a total daily cooling requirement of 4880 ton-hours, 3495 of which occur during the on-peak period defined by the electrical rate as 12:00 noon through 8:00 p.m.

A conventional system without thermal storage, Option 1, would produce 4880 ton-hours of cooling by using a chiller without any storage. If the average performance of the chiller is 0.6 kW/ton based on 85°F entering condenser water, the energy consumption for cooling, including 0.2 kW/ton for pumps and the cooling tower, would be 3904 kWh, 2876 of which are consumed during the on-peak period. The on-peak demand for the 500-ton chiller and its auxiliaries is 400 kW.

The next possibility is to minimize the use of on-peak energy by doing all cooling off-peak, Option 2. This is sometimes called the "full storage" option. Thus, the chiller must produce 4880 ton-hours, plus storage inefficiencies, over a 16-h period (8:01 p.m. through 11:59 a.m.). Using 0.95 storage efficiency, the chiller size will be 321 tons and the storage capacity required is 5136 ton-hours. Lower

**TABLE 5.1**
**Example of Design Day Cooling Load Profile**

| Hour | O.A. Temperature (°F) | Cooling Load (tons) |
|---|---|---|
| 1 | 78 | System off |
| 2 | 76 | System off |
| 3 | 74 | System off |
| 4 | 76 | System off |
| 5 | 77 | System off |
| 6 | 78 | 275 |
| 7 | 80 | 160 |
| 8 | 81 | 190 |
| 9 | 82 | 200 |
| 10 | 84 | 160 |
| 11 | 86 | 300 |
| 12 | 88 | 350 |
| 13 | 90 | 400 |
| 14 | 92 | 450 |
| 15 | 93 | 500 |
| 16 | 94 | 450 |
| 17 | 93 | 400 |
| 18 | 92 | 375 |
| 19 | 90 | 300 |
| 20 | 86 | 270 |
| 21 | 84 | System off |
| 22 | 82 | System off |
| 23 | 81 | System off |
| 24 | 80 | System off |
| Total | | 4880 |
| Total on-peak | | 3495 |
| Total off-peak | | 1385 |

condenser water temperatures and resulting condensing temperatures during the off-peak period may reduce the off-peak energy requirement to 0.5 kW/ton. The resulting daily off-peak cooling energy use, including 0.2 kW/ton for pumps and the cooling tower, is 3595 kWh. The on-peak energy consumption is reduced to only about 0.1 kW/ton for pumping. Thus, an on-peak demand of only 50 kW is incurred, with on-peak electrical consumption of only 400 kWh.

Option 3 is to "level the load" by selecting a combination of chiller capacity and storage such that the chiller would operate over the entire 24-h design day at essentially full load. For each hour, the combination of chiller and storage must be sufficient to meet the imposed cooling load. If, during any hour, the chiller produces more cooling than required to meet the imposed load, the additional cooling is stored for use later in the daily cycle.

Under this scheme, the chiller must produce the total cooling requirement, plus storage inefficiencies, over a 24-h period. In this example, the total cooling requirement is 4880 ton-hours. A chiller sized to meet this capacity, plus storage inefficiencies, would be 214 tons, with a storage capacity of 2018 ton-hours. As shown in Table 5.2, the system performance results in on-peak demand and consumption of 171 kW and 1368 kWh, while off-peak demand and consumption are 171 kW and 2266 kWh.

**TABLE 5.2**
**Example of Design Day Cooling Energy Consumption Based on Load Leveling Option**

| Hour | O.A. Temperature (°F) | Cooling Load (tons) | Chiller Cooling (tons) | Storage Cooling[a] (tons) | Input Energy (kW/ton) | Total kW |
|---|---|---|---|---|---|---|
| 1 | 78 | 0 | 214 | −214 | 0.6 | 128 |
| 2 | 76 | 0 | 214 | −214 | 0.6 | 128 |
| 3 | 74 | 0 | 214 | −214 | 0.6 | 128 |
| 4 | 76 | 0 | 214 | −214 | 0.6 | 128 |
| 5 | 77 | 0 | 214 | −214 | 0.6 | 128 |
| 6 | 78 | 275 | 214 | +61 | 0.6 | 128 |
| 7 | 80 | 160 | 214 | −54 | 0.6 | 128 |
| 8 | 81 | 190 | 214 | −24 | 0.6 | 128 |
| 9 | 82 | 200 | 214 | −14 | 0.7 | 150 |
| 10 | 84 | 260 | 214 | +46 | 0.7 | 150 |
| 11 | 86 | 300 | 214 | +86 | 0.8 | 171 |
| 12 | 88 | 350 | 214 | +136 | 0.8 | 171 |
| 13 | 90 | 400 | 214 | +186 | 0.8 | 171 |
| 14 | 92 | 450 | 214 | +236 | 0.8 | 171 |
| 15 | 94 | 500 | 214 | +286 | 0.8 | 171 |
| 16 | 93 | 450 | 214 | +236 | 0.8 | 171 |
| 17 | 93 | 400 | 214 | +186 | 0.8 | 171 |
| 18 | 92 | 375 | 214 | +161 | 0.8 | 171 |
| 19 | 90 | 300 | 214 | +86 | 0.8 | 171 |
| 20 | 86 | 270 | 214 | +56 | 0.8 | 171 |
| 21 | 84 | 0 | 214 | −214 | 0.7 | 150 |
| 22 | 82 | 0 | 214 | −214 | 0.7 | 150 |
| 23 | 81 | 0 | 214 | −214 | 0.7 | 150 |
| 24 | 80 | 0 | 214 | −214 | 0.7 | 150 |
| Total | | 4880 | 5136 | −2018/+1701 | | 1368 on-peak<br>2266 off-peak |

*Note:* Shaded area represents on-peak period.
[a] indicates charging; + indicates discharging.

**TABLE 5.3**
**Comparison of Cooling Thermal Storage Options**

| Result | Option 1 | Option 2 | Option 3 |
|---|---|---|---|
| Chiller capacity (tons) | 500 | 321 | 214 |
| Storage capacity (ton-hours) | 0 | 5136 | 2018 |
| On-peak demand (kW) | 400 | 50 | 171 |
| On-peak consumption (kWh) | 2876 | 400 | 1368 |
| Off-peak demand (kW) | 280 | 280 | 171 |
| Off-peak consumption (kWh) | 1028 | 3595 | 2266 |
| Total consumption (kWh) | 3904 | 3995 | 3634 |

The results of these three options are summarized in Table 5.3. The consumption and demand charges for each option would have to be evaluated and computed on an annual basis.

Another option that may be considered in locations with extremely high on-peak demand charges is the "demand-limiting" option. With this option, storage is oversized so that the chiller can be operated at a reduced capacity during on-peak hours.

The next step in the analysis for each of these cases is to determine the capital costs for the chiller, cooling tower, pumps, piping, and so on. Then, using simple payback (or a more sophisticated analysis), the economic performance of each option would be evaluated to determine the most cost-effective approach.

In general, the full storage option results in the greatest capital requirement, but the greatest savings in electrical costs. The load leveling option reduces the capital requirements by minimizing both chiller and storage capacity. The demand-limiting option has demand savings greater than the load-leveling option, but lower than with the full storage option.

*There are economic risks associated with thermal storage.* If the utility rate schedule changes, there may be a negative impact on the cost performance of thermal storage and it is imperative that a long-term contract be negotiated with the utility. Another economic risk is system failure during the peak period, which can eliminate the demand savings for an entire year. This, however, is only a risk if other standby chillers are brought on line to provide cooling while the storage system is repaired. Again, the contract with the utility should allow a limited number of short-term demand increases with no significant rate penalty.

Utilities are experimenting with *real-time pricing* rate schedules, particularly in states that have implemented deregulation of their electric utilities. Under these rates, the cost of energy is based on the actual cost of energy production and/or power purchases by the utility on an hour-to-hour basis. Some authorities predict that these rates will damage the economic viability of thermal storage systems. Other experts point out that these rates will only impact on control methods and require variable chiller use versus storage use. Again, any method that increases the utility load factor reduces net costs to the utility and these savings can be passed on to users who contribute to achieving them.

## AVAILABLE TECHNOLOGIES

There are three general thermal storage system technologies: water, ice, and eutectic salts. These technologies can vary significantly in terms of size, efficiency, reliability, and cost, as described in the following sections and summarized in Table 5.4.

### Chilled Water Storage Systems

Chilled water storage systems use the sensible heat capacity of water, defined by its specific heat of 1 Btu/lb °F. These systems operate at temperature ranges

**TABLE 5.4**
**Comparison of Alternative Thermal Storage Technologies**

| Storage Medium | Volume (cf/ton-hour) | Storage Temperature (°F) | Discharge Temperature (°F) | Advantages | Disadvantages |
|---|---|---|---|---|---|
| Chilled water | 12–15 | 36–44 | 40–46 | Can use existing chillers<br>Stored water can be used for fire protection | Large storage volume<br>Higher heat gains<br>Difficult to maintain supply temperature setpoint |
| Ice | 2.4–3.3 | 32 | 34–36 | Low storage volume<br>High discharge rates<br>Easy to maintain supply temperature setpoint | Poor chiller COP due to low temperature requirements<br>Expansion problem<br>Poor thermal conductivity of ice |
| Phase change materials | 6 | 47 | 48–50 | Can use existing chillers | Relatively large storage volume<br>High chilled water supply temperature |

compatible with both reciprocating and rotary compressor water chillers and are generally the most cost-effective method for thermal storage in larger systems (2000 ton-hours or more).

The lowest temperature at which chilled water can be stored in the liquid state is 32°F. However, from a practical point of view, water temperatures lower than about 36°F are difficult to produce with rotary compressor chillers. As a result, the cooling storage capacity required is 12–15 $ft^3$ per ton-hour (about 100 gallons per ton-hour). But, chilled water storage costs per unit volume decrease with tank size and heat gains per unit volume also decrease as the tank size increases.

The most difficult problem associated with water storage is maintaining the required chilled water supply temperature, particularly near the end of the discharge cycle, since any mixing of return water with supply water results in an elevated supply water temperature. To address this condition, several approaches are available:

*Multiple tank storage.* This method keeps return water from mixing with supply water. If there are $N$ tanks and one tank is empty, the storage volume is $N$–1 times the individual tank volume.

During charging, the tanks are filled one at a time with chilled water. During discharge, one tank at a time supplies chilled water and only after it is empty does return water enter the tank.

The obvious drawbacks of this scheme are the space requirements for multiple tanks, the reduction in storage efficiency [$(N-1)/N$] by having one empty tank, and the high heat gains due to increased tank surface area.

*Storage tank with siphon baffles.* A single large tank may be divided into multiple sections by adding internal siphon baffles, as shown in Figure 5.2. Water is supplied from one end of the tank and returned to the other. Supply and return water mixing is reduced by the baffles and the difference in densities between the supply and return water.

*Experience with these tanks has not been good and their use is not recommended unless the exact tank configuration has been tested before installation.*

*Stratified water storage. Stratification* is a method of keeping warmer return water from mixing with colder supply water in a single tank. Water that is less dense tends to rise to the top of the tank, while denser water tends to fall to the bottom. Thus, supply water is removed from the bottom of the tank, while return water is injected at the top of the tank.

However, the overall density differences between supply and return water at typical chilled water temperature ranges is only about 0.03%, and so these tanks must be designed very carefully to prevent forced circulation, with the resulting mixing, as water flows into and out of the tanks.

Another condition must be considered: the density of water is greatest at 39.2°F. Therefore, cooling the supply water during the charging cycle to temperatures below 40°F is not recommended since there will

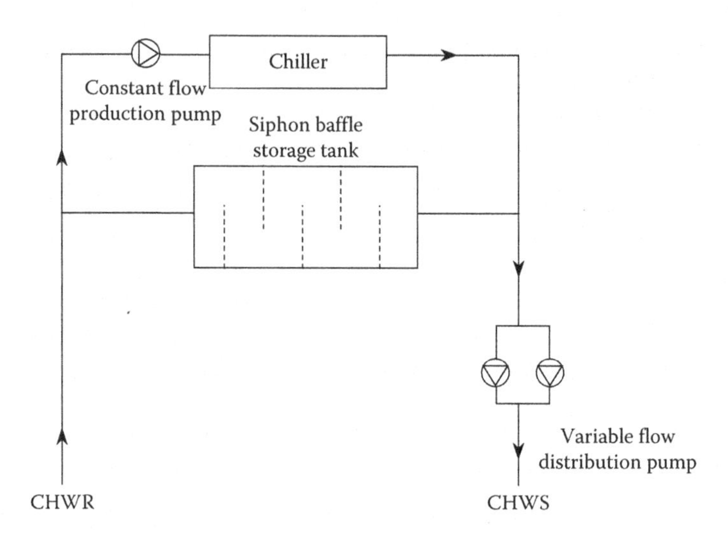


**FIGURE 5.2** Chilled water storage tank with siphon baffles.

be a tendency for the colder water to rise, effectively destroying the stratification.

## Ice Storage

Ice storage has two great advantages over chilled water storage. First, ice storage represents a *phase change* process with water absorbing or releasing 144 Btu/lb to melt or freeze at 32°F. Thus, ice storage systems require only about 20% of the storage volume of an equivalent water storage system. The second advantage is that ice melts at a constant temperature, making the delivery of a constant chilled water supply temperature very simple.

The major disadvantages of ice storage are the poor energy efficiency of the freezing process and the complexity of the ice storage systems themselves. These systems have relatively high cost and are most cost effective only in smaller storage capacities (<1000 ton-hours). Also, due to the low temperatures, reciprocating compressor chillers are typically required, although multiple rotary compressor chillers arranged in a "cascade" can be applied.

Additional drawbacks of ice storage are two problems associated with ice production:

1. Water expands as it freezes, producing a highly destructive force that must be addressed by the ice storage equipment. Ice also exerts a buoyancy force that must be addressed by some types of ice production systems.
2. Ice has poor thermal conductivity. As ice forms on an evaporator surface, even lower evaporator temperatures are required to freeze a greater thickness of ice.

There are a number of ice production and storage technologies currently available:

*Ice shedders.* A simple ice storage system consists of an evaporator located above a tank. This "ice shedder" system, as shown schematically in Figure 5.3, is designed to form ice on the evaporator surface, then, by mechanical means or by defrosting, to separate the ice from the evaporator surface so that it will fall into a tank. Chilled water can then circulate through the tank to serve the building cooling load.

Ice shedders aerate the water in the tank by pumping it continuously over the evaporator in an open configuration. Therefore, a heat exchanger is usually required to help isolate the chilled water system from the increased corrosion potential due to dissolved oxygen.

*External melt ice-on-coil.* A simple approach to ice storage is to insert evaporator coils into a water tank and freeze the water around the coils during the charging cycle. During discharge, the water surrounding the evaporator coil and its ice charge is circulated through the chilled water system to serve the building cooling load. This system is illustrated schematically in Figure 5.4.

However, because of the poor thermal conductivity of water, there are limits to the thickness of ice that can be formed on the coil surface. As this thickness increases, the evaporator temperature must be reduced to continue the freezing process and this, in turn, reduces the chiller COP.

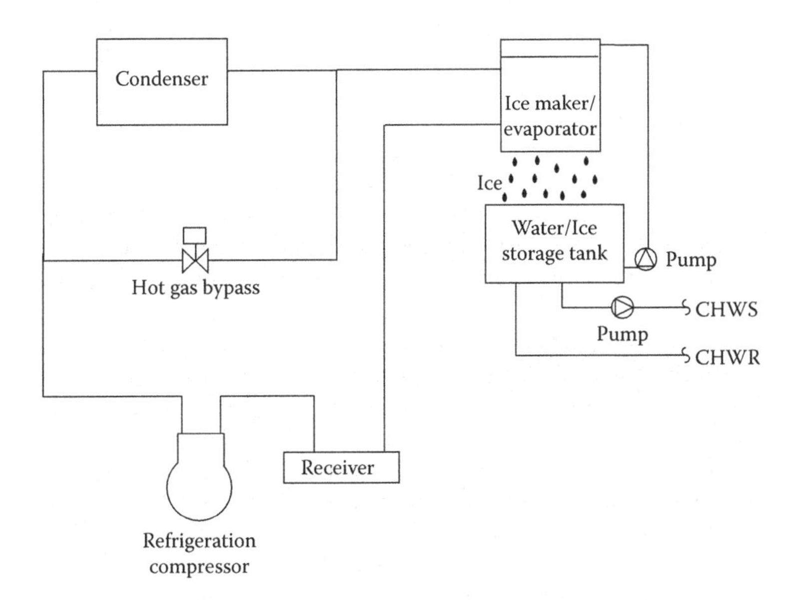


**FIGURE 5.3** Ice shedder thermal storage system schematic.

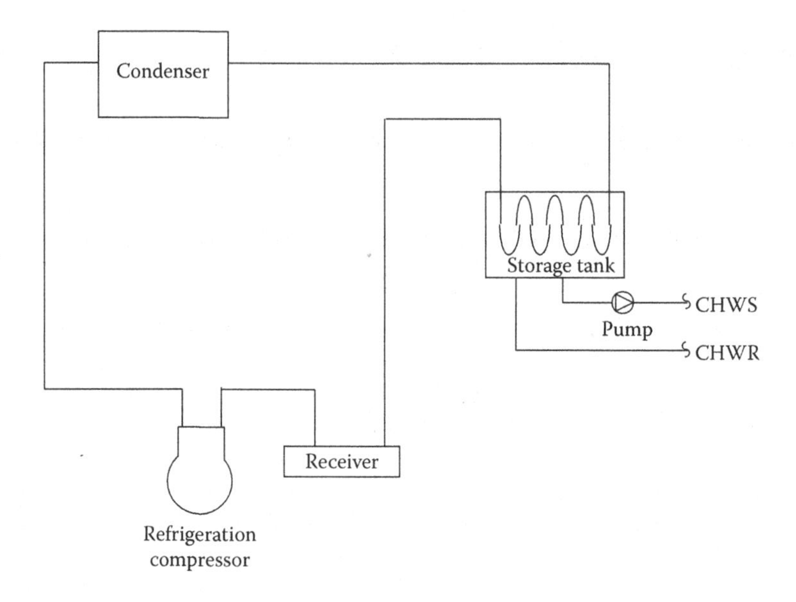


**FIGURE 5.4** External melt coil freezing thermal storage system schematic.

Another potential problem is *ice bridging*. To improve efficiency by reducing ice thickness, evaporator coils are closely spaced in the tank. However, if all of the ice is not melted during the discharge cycle, ice may grow from one evaporator to another, bridging the separation and cutting off flow around these evaporators, reducing discharge efficiency.

Overall, these systems have a lower overall efficiency than other types of ice storage systems.

*Internal melt ice-on-coil*. With this system, the freezing point of chilled water is lowered by adding propylene glycol or ethylene glycol. During the charging cycle, the chilled water (sometimes called "brine") is circulated through evaporator coils in water-filled tanks, and ice is formed on the coils' outer surfaces. During discharge, the ice is melted by circulating warmer return chilled water through the coils, cooling the water.

This system avoids the problems of the external melt system and eliminates the potential corrosion problems associated with ice shedders. There are numerous commercially available packaged storage systems based on the internal melt concept in the market today.

To minimize the heat gain by the large number of tanks typically used with this technology, the tanks must be well insulated. Often, it is cost effective to bury the tanks, leaving only their tops exposed, to reduce temperature gradients across the tank wall.

The basic operating modes of this technology are illustrated schematically in Figure 5.5.

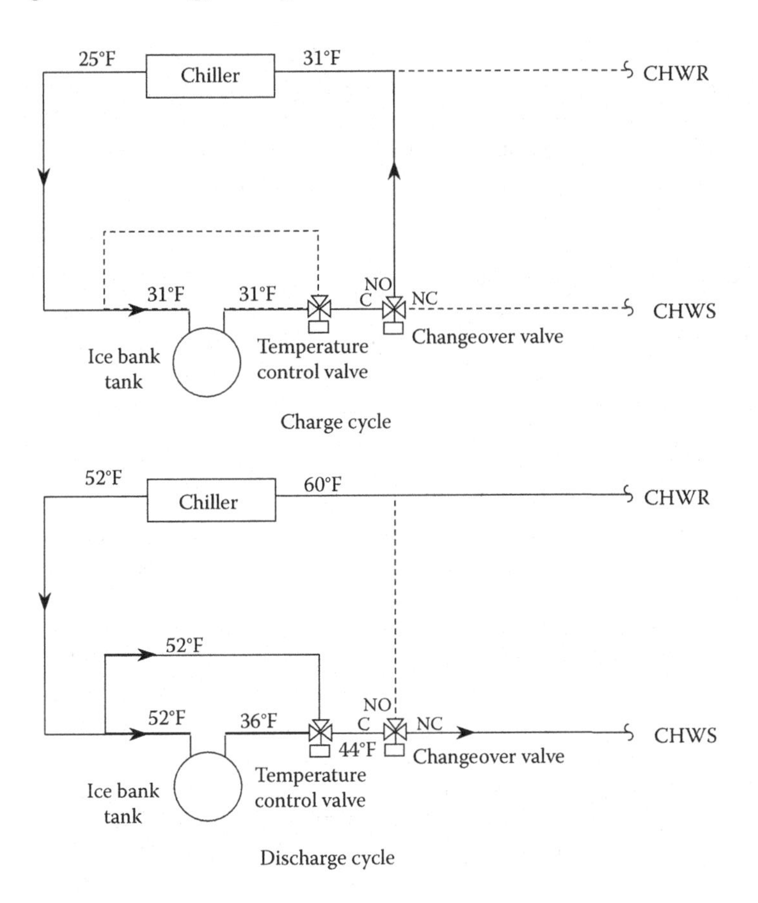


**FIGURE 5.5** Internal melt coil freezing thermal storage system schematic.

*Ice capsules.* Again, the freezing point of chilled water is lowered by adding a glycol compound. With this technology, a tank is filled with sealed containers of water submerged in the chilled water. During the charging cycle, the water-filled capsules are frozen by circulating cold chilled water through the tank. During discharge, as warmer chilled water is circulated through the tank, the ice melts, cooling the water.

The capsules are made of thin-walled polyethylene plastic and are designed to survive the expansion of freezing water. However, if water is kept still, as inside these capsules, it may cool to well below 32°F before it actually freezes. Therefore, to ensure complete freezing, the capsules must be cooled to about 25°F.

The basic operating modes of this technology are the same as for the internal melt storage system.

*Ice slurry.* With this technology, propylene glycol or calcium magnesium acetate (CMA) is added to water and the resulting brine is circulated over a special evaporator. As the brine flows down the tall evaporator surface that is at a temperature equal to the freezing point of the brine, ice begins to form. A mechanical scraper then removes the ice from the evaporator surface so that a mixture or "slurry" of ice and liquid is formed.

A heat exchanger is used to isolate the glycol solution from the building chilled water system.

The main disadvantage of ice storage is that ice freezes at a temperature much lower than is required for most HVAC applications, forcing compressors to operate at much lower evaporator temperatures and increasing energy consumption because of the lower COPs that result. The main advantage of ice storage is reduced storage volume because of phase change heat transfer.

## Phase Change Materials Storage Systems

Phase change material (PCM) eutectic solutions used in cooling TES systems are mixtures of two or more chemicals that, when mixed in a particular ratio, have a freezing/melting point in the range, ideally, of 40–45°F. Thus, PCMs are a latent heat storage material that stores energy as the material changes from liquid to solid ("freezes") and then releases energy as it changes back to liquid ("melts"). While ice is a PCM, the low temperature at which water freezes and thaws places an energy burden on ice storage systems that could be avoided with the use of a PCM with a phase change temperature more within the temperature range required by a chilled water system.

PCMs must have the following characteristics to be applied in a TES system:

*Thermal properties.* A suitable phase change temperature, high latent heat on a volumetric basis, and high thermal conductivity to ensure essentially uniform phase change

*Physical properties.* Phase stability during freezing/melting, high density to reduce storage volume, small volume changes during freezing/melting, and low vapor pressure

*Kinetic properties.* No "supercooling," which can significantly reduce with heat extraction from the storage volume

*Chemical properties.* Long-term chemical stability, compatibility with system tank, piping, and so on, low toxicity, and no fire hazards

While a vast number of PCMs have been developed, the materials most closely meeting the criteria specified above are *eutectics*, composition of two or more inorganic salts, each of which freezes and melts congruently, forming a mixture of component crystals during crystallization. Nucleating and stabilizing agents are incorporated and encapsulation is required due to the corrosive properties of hydrated inorganic salts. Currently, while research continues, the best PCM that has been developed has a freeze/melt temperature of 47°F and a latent heat of ~41 Btu/lb.

PCM technology eliminates the energy burden imposed by ice-freezing temperature requirements and reduces the storage volume required by 50% over chilled water storage systems. However, the disadvantage of this technology is having to deal with the high phase change temperature, which is typically too high for chilled water systems operating design loads unless latent loads are very low. Since the discharge temperature is not totally uniform, a real chilled water supply temperature of about 50°F can be expected. But, this temperature is well below normal chilled water return temperatures (particularly for systems with a larger design range), and so PCM TES can be used in a series configuration with chillers to precool return chilled water. In this configuration, energy savings over conventional cooling will always be at least 50% and as high as 100% during periods when dehumidification is not required.

## APPLICATION OF TES

The use of TES systems may be economically attractive only if the following conditions apply:

1. The peak cooling load is much higher than the average cooling load over the design day, that is, peak loads are of short duration.
2. The electric utility rate includes high demand charges, ratchet charges, or high differential between on-peak and off-peak rates.
3. Electric utilities offer a financial incentive for the installation of TES.

Owners or designers contemplating thermal storage as part of a chilled water system must do more evaluation than for conventional systems. The following steps are recommended to determine the feasibility of a TES system:

1. Consider the owner's ability to operate a chilled water system with thermal storage. These systems require ongoing monitoring and adjustment to work properly and efficiently. And, system failure can have significant negative financial impacts.
2. Determine the cooling load profile, hour by hour during the peak cooling period, in order to determine required chiller and storage capacities. Then, using a detailed annual cooling load profile for existing facilities or hour-by-hour computer modeling for new facilities, compute the annual operating cost for the system. *Compute electrical costs using the exact rate schedule(s) to be applied to this system.*
3. Select the TES technology to be used. Ensure that all operating modes are fully defined by the design and incorporated into the system controls.
4. Determine the capital requirements and anticipated economic performance of the thermal storage system. While simple payback methods can be applied, more sophisticated methods such rate of return or life cycle cost analysis should be applied.

5. Negotiate a long-term contract with the electric utility. This contract should establish the specific rates to be used and provide for exemptions from demand peaks caused by occasional system failure. The contract period must be longer than the payback period for the system.
6. While the TES system is being installed, throughout the commissioning period, and through the first year of operation, the operating staff must be fully involved, trained, and monitored to ensure successful system performance.

# 6 Special Chiller Considerations

## NOISE AND VIBRATION

Air-cooled chillers, and cooling towers associated with water-cooled chillers, are installed outdoors, presenting numerous problems relative to performance and *community noise* that can result from poor placement and/or incorrect screening. See sections "Tower Placement and Installation" in Chapter 11 and "Noise and Vibration" in Chapter 14 for discussions on these topics.

Water-cooled chillers are located indoors, which means that noise and vibration created by the chiller and its associated pumps may become a problem throughout the building. Therefore, careful attention to location and installation of water-cooled chillers is required.

Water chillers have broadband sound levels with characteristic frequency ranges depending on the type of compressor (scroll, centrifugal, or screw). And, because some centrifugal chillers produce more noise under part load operation, chiller noise must be analyzed for the entire load range.

*Sound* is the hearing sensation caused by a physical disturbance in the air. *Noise* is simply objectionable sound.

Sound waves in air arise from variations in pressure above and below the ambient pressure, and can be caused by repetitive pulsation in an air stream (due to rotating fan wheels or blades) and/or the cascading water. Some people can hear sound in the frequency range of 20–20,000 Hz, but in reality, few can hear below 30 Hz or above 8000 Hz.

Noise is defined in terms of the *sound pressure level*, measured in decibels (abbreviated as "dB"). Decibel values for sound represent a logarithmic scale, each 10 dB increase or decrease indicating a factor of 10 increase or decrease in the sound level, as shown in Figure 6.1.

The human ear is not equally sensitive to all sound frequencies. The sound pressure level from two sound sources may be identical, but if the sources have different frequencies, the "loudness" of one sound will be perceived higher than the other if it is concentrated within the frequency ranges within which the ear is more sensitive. Therefore, acoustics engineers have developed *frequency weighing* for sound-level meters, and the most common weighing standard is the "A-scale," which gives more "weight" to the middle-octave bands of the sound being measured. Table 6.1 tabulates the A-scale frequency response or weighting factors.

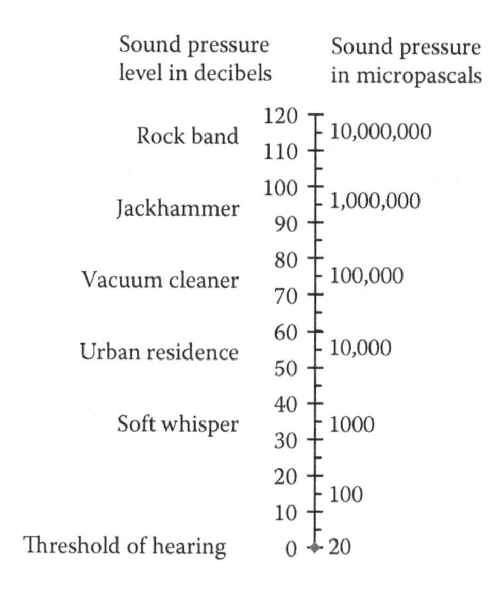


**FIGURE 6.1** Representative noise sound pressure levels.

Additionally, sound-level meters are usually provided with a method to time-average sound over a selected time period. Otherwise, it would be difficult to measure rapidly varying sounds at all.

To prevent the transfer of sound and vibration from chillers, pumps, and so on into occupied spaces, the following recommendations should be implemented:

1. *If feasible, locate chillers in a mechanical equipment building separate from occupied buildings.* If the chiller room is integrated into an occupied building, the walls and ceiling must be designed to prevent noise transfer.

**TABLE 6.1**
**A-Scale Sound Pressure Weighting Factors**

| Frequency (Hz) | Weighting Factor (dBA) |
|---|---|
| 31.5 | −39.4 |
| 63 | −26.2 |
| 125 | −16.1 |
| 250 | −8.6 |
| 500 | −3.2 |
| 1000 | 0.0 |
| 2000 | +1.2 |
| 4000 | +1.0 |
| 8000 | −1.1 |

Masonry construction and, perhaps, interior acoustical treatments will be required.

2. All piping and conduit penetrations must be sealed with flexible materials to stop noise and prevent the transfer of vibration to the building walls and floors.
3. Mount all equipment on vibration isolators. Pumps can be isolated with inertia bases and spring isolators as shown in Figure 3.7. If the chiller is installed in a ground floor or basement location, it can be isolated with cork-and-rubber pads designed to support the weight of the chiller without fully compressing. If located on an upper floor of a building, the chiller should be installed with spring vibration isolators as shown in Figure 6.2.
4. Make sure that all piping and electrical connections to rotating equipment are made with flexible connectors to prevent the transfer of vibration to electrical raceway and, then, the building structure.
5. For critical applications, chiller noise can be reduced by a sound-absorbing enclosure. Constructed of movable panels, these enclosures consist of lightweight panels with 1″–4″ of glass fiber sound insulation installed around and over the chiller. To allow air movement and access for servicing, the panels can be arranged with minimum 3′–0″ overlaps to eliminate straight-line noise paths.

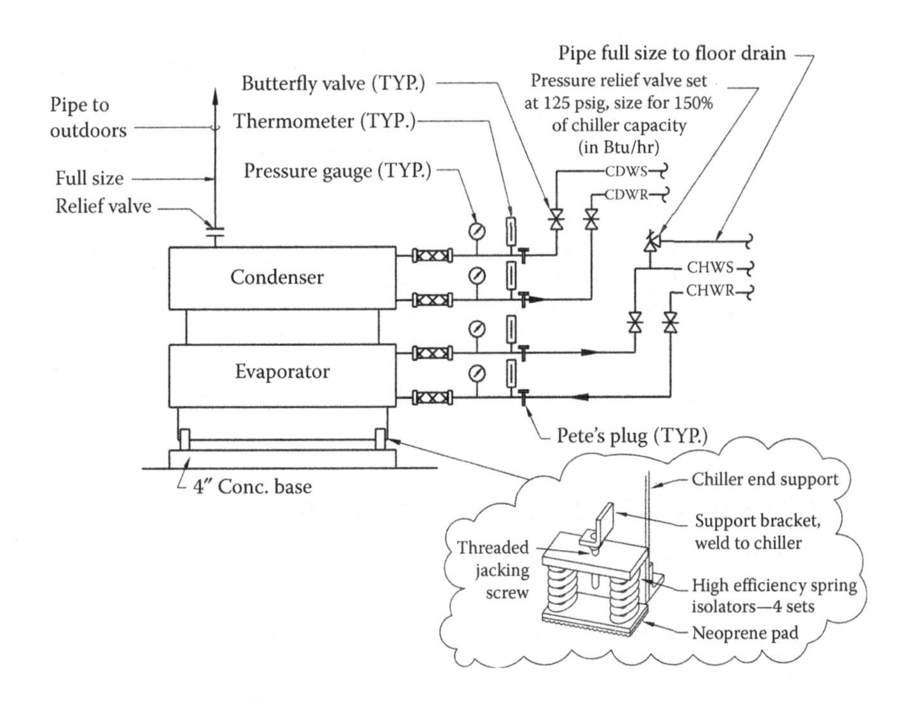


**FIGURE 6.2** Chiller installation for upper-level mechanical rooms.

Chillers are tested for noise level in accordance with ARI Standard 570. These data can be used to evaluate chiller selection and installation requirements to minimize noise impact.

## ELECTRICAL SERVICE

Electrical service for electric-drive vapor compression water chillers almost always consists of three-phase, four-wire service with the voltage-to-load ratios indicated in Table 6.2.

There are four methods of motor starting commonly used for electric-drive chillers. For small compressors (15 tons or less), *full voltage* or *across-the-line* starting is typically used. However, this type of starting imposes a starting or *inrush* current that is about six times greater than the normal full-load operating current of the motor.

High-current inrush and high starting torque can cause problems in electrical and mechanical systems; so, for larger motors (50 hp and greater), some type of *reduced voltage* starting is normally used. For scroll or small screw compressors, the *part-winding* or *solid-state* starters are used. For rotary compressors, *wye–delta* (closed transition) and *solid-state* starters are used. And, more commonly today, variable frequency drives with integral solid-state starters are utilized. Each of these methods has advantages and disadvantages in their application:

*Part-winding starters.* Part-winding starting is done by first applying power to a part of a motor's coil windings for starting and then applying power to the remaining coil windings for running. The motor must be a part-winding motor with two sets of identical windings that can be used in parallel. The windings then produce reduced starting current and reduced torque when energized in sequence.

The start winding draws about 65% of rated motor locked rotor current and develops about 45% of the normal motor torque. After about 1 s, the second winding is connected in parallel with the start winding and the motor runs with full current and full torque.

**TABLE 6.2**
**Typical Electrical Voltages for Water Chiller Service**

| Compressor Input (kW) | Typical Service Voltage (3-ph, 4-wire, 60 Hz) |
|---|---|
| 0–15 | 200–230 |
| 16–1000 | 460–600 |
| 1001+ | 4160 |

*Wye–delta starters.* Wye–delta starters can be used only on wye–delta motors that have sized leads that allow for the motor winding to be connected in either the wye configuration or the delta configuration.

Wye–delta starting begins by connecting the motor leads in a wye configuration for starting. With this configuration, the motor receives ~58% of line voltage and develops ~33% of normal current and torque. After a time delay (that is usually adjustable), the wye contactors in the starter open, momentarily deenergizing the motor, and the delta contactors close, reenergizing the motor in a delta configuration with full current and full torque.

This type of transition from wye to delta configuration is called "open" since there is a momentary period when no voltage is applied to the motor. A higher-cost version of this starter provides "closed" transition via another set of contractors and a resistor bank to maintain voltage to the motor windings during the transition period.

Wye–delta starters are called "electromechanical" starters since their action includes both electric aspects and mechanical contactors within the starter that must open and close during the starting sequence.

*Solid-state starters.* To provide reduced voltage starting without the potential operating and maintenance problems of electromechanical starters, solid starting can be used. This type of starter uses a silicon-controlled rectifier (SCR) that controls motor voltage, current, and torque as the motor accelerates during the starting sequence.

Starting current can be adjusted from about 25% to 75% of the full-load value.

The primary advantages of this type of starting are that it is almost infinitely adjustable (providing smooth, stepless motor acceleration), is small in size, is very rugged, and has no contacts to fail. Another advantage is that, due to their smaller size, the starter can typically be unit-mounted, while much larger wye–delta starters must be installed separately and require wiring runs between the starter and the chiller motor. The primary disadvantage of the solid-state starter is cost, which may be two or three times that of wye–delta starters.

*Variable frequency drives.* Whether applied to chillers or pumps, the integral solid-state starters and controls incorporated into VFDs allow for *soft starting* a motor, that is, starting the motor at low rpm and then ramping speed upward smoothly. VFDs avoid any inrush current and essentially start the load with an imposed amperage far lower than full-load amps.

Electric-drive compressors and their motors are classified as *hermetic* or *open* as follows:

*Hermetic motors.* These motors consist of a stator and rotor directly connected to a refrigeration compressor, both of which are sealed in a common enclosure or casing. In this configuration, refrigerant flowing from the evaporator passes through and around the motor as it enters the compressor to provide motor cooling.

*Open motors.* These motors are standard NEMA Design B motors that are mounted on the chiller and connect to the compressor via a shaft and coupling arrangement.

The open motor arrangement has the advantage of easier accessibility for service and, in the event of motor failure, the refrigeration circuit is not contaminated. However, careful alignment of the motor and compressor shafts is required (both initially and as an ongoing maintenance item) to prevent vibration and mechanical damage.

With hermetic motors, the heat generated due to motor inefficiency (5% or less of the rated motor kW) is rejected to the refrigerant and then to the condenser. However, motor heat from an open motor is rejected into the chiller room, requiring additional ventilation to limit heat buildup. Typically, the maximum allowable chiller room temperature is set at 104°F. With a 95°F ambient air temperature (design day), the required additional ventilation airflow, in cfm, to offset motor heat can be computed from the following, where "eff" is the motor efficiency (decimal value):

$$\text{cfm} = \frac{(\text{full-load kW})(1 - \text{eff})(3413)}{1.1 \times (104 - 95)}$$

The oil pump on an electric-drive chiller is a small load that uses full-voltage starting, but may be served separately from the main compressor motor. To avoid this problem, and the potential for errors during installation, it is recommended that the chiller be specified to have a *single-point electrical service connection.* Thus, only one wiring run from the power source to the chiller is required.

Absorption chillers have several small pump motors, each of which will almost always be 5 hp or less. Again, single-point electrical service connection should be specified so that the full-voltage starters and wiring between the motors and starters is part of the chiller package.

## CHILLER HEAT RECOVERY

The heat collected by a water-cooled chiller during cooling must be rejected through the condenser to a cooling tower (or evaporative cooler), as discussed in Chapter 1. However, if there is a simultaneous need for heating and cooling by the building, then this heat can be *recovered* and utilized rather than simply rejected to the outdoors.

During the 1960s, natural gas was not available in many parts of the country for commercial building heating and many *all-electric* HVAC systems were designed. However, due to the high cost of electricity for resistance heating, efforts were made to use rejected condenser heat for this purpose, essentially creating large-scale *internal source heat pump* systems. However, there were serious drawbacks of these systems:

1. To be applicable, the building had to have a high ratio of interior zones, which required cooling any time they are occupied, relative to perimeter zones, which require heating or cooling primarily as a function of outdoor conditions. Equally important, these interior zones had to have a high level of internal heat gain due to lights, people, and appliances.
2. The normal chiller leaving condenser water temperature of 95°F is very low for heating applications. The problem is compounded at reduced cooling load conditions as the leaving condenser water temperature decreases in direct proportion to the cooling load. To help this condition, heat recovery chillers were designed to operate with leaving condenser water temperatures as high as 110°F. This, however, significantly increased the chiller energy input and resulting operating cost.
3. Electrical costs must be high. The economics of heat recovery versus the use of an airside or waterside economizer cycle to provide "free" cooling during the winter must be carefully evaluated to determine if this type of heat recovery is cost effective.

Even in the 1960s, this type of heat recovery had marginal performance and economic benefits. And, as soon as natural gas became available, these

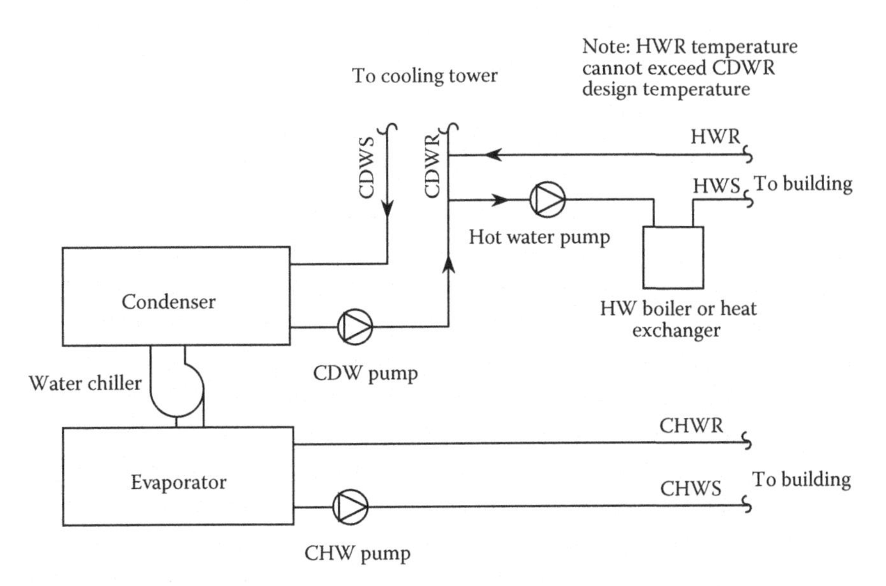


**FIGURE 6.3** Water-cooled chiller heat recovery schematic.

systems were abandoned. Since the 1960s, the energy cost picture has only gotten worse, while chiller heat recovery performance has not improved, essentially eliminating the internal source heat pump concept from consideration in new designs.

*However, if there is a need for low-level (85–95°F) heat during the cooling season, then recovery and use of condenser heat should be implemented as an energy-saving design element.* Typical applications include the following:

1. Domestic hot-water preheating can be applied as shown in Figure 6.3. Municipal supply water temperatures can range from 65°F to 85°F during the summer. For buildings with high domestic hot-water requirements (hospitals, research laboratories, dormitories, etc.), preheating the incoming water via a water-to-water heat exchanger will reduce hot-water heating costs.
2. Summer reheat loads can be met with condenser heat and the use of a water-to-water heat exchanger as shown in Figure 6.3. Since no space heating is required during the summer, the maximum reheat requirement is to bring the air to room temperature, typically 75°F. This temperature is within the range of the condenser water temperature available.
3. Process heating loads can be met with condenser heat if their thermal level is below about 80°F.

# Part III

## Chiller Operations and Maintenance

# 7 Chiller Operation and Maintenance

## CHILLER COMMISSIONING

Every chilled water system can be operated on the basis of meeting two goals:

*Goal 1: Satisfy imposed cooling loads*

- Maintain proper chilled water supply temperature to offset imposed sensible and latent cooling loads.
- Maintain proper supply water flow in variable flow systems to minimize pumping energy consumption.

*Goal 2: Minimize the cost of cooling*

- Maintain chilled water temperature as high as possible, but still low enough to satisfy both sensible and latent cooling loads.
- Maintain condenser water supply temperature as low as feasible. Most chillers operate satisfactorily with entering condenser water temperature as low as 70°F. And, at least one chiller manufacturer rates its machines for operation with an entering condenser water temperature of 55°F.
- Operate each chiller, to the maximum possible extent, in the 40–80% load range.

To ensure that this type of operation is attainable, the chilled water system must be properly installed and placed into service and its operation verified by test, a process called *commissioning.*

The first step in this process is to verify that the chiller installation is correct and in accordance with both the design and the manufacturer's recommendations. The design engineer should be retained to review the installation process and perform a final inspection to ensure compliance with the design requirements.

Before the chiller can be started for the first time, the chilled water pump(s) must be placed in operation and the chilled water flow "balanced" to within ±5% of design rates. The basic steps required to place the chilled water pump in service include the following:

1. Check pump installation, including mountings, vibration isolators, and connectors, and piping specialties (valves, strainer, pressure gauges, thermometers, etc.).
2. Check pump shaft and coupling alignment.

3. Lubricate pump shaft bearings as required by the manufacturer.
4. Lubricate motor shaft bearings as required by the manufacturer.
5. Turn shaft by hand to make sure the pump and motor turn freely.
6. "Bump" the motor on and check for proper rotation direction.

The air-cooled condenser or condenser water system must be placed into operation (see the section "Tower Commissioning" in Chapter 15). System controls must be tested and their proper operation must be confirmed. Chillers should be initially placed into service only by the manufacturer's service technicians, a process often called *factory start-up.*

The following checklist may be used for verifying both chiller installation and functional performance upon start-up:

*General Installation Checklist*

1. Test and balance completed (chilled water (CHW) and condenser water (CDW) flows with –5% to +10% of design).
2. Observe no visible water or oil leaks.
3. Observe no unusual noise or vibration.
4. Chilled water piping insulation in good condition.
5. Pressure gauges, thermometers installed and operable.
6. Confirm that O & M manuals are onsite.
7. Confirm that training was provided for operating staff.

*Additional Installation Checklist for Absorption Chillers*

1. Confirm that the steam control valve is installed and operating for an indirect-fired chiller.
2. Confirm that the chiller has been leak tested according to manufacturer's instruction.
3. Confirm that unit was evacuated properly before charging with refrigerant and solution.
4. Confirm that distilled water was used for refrigerant charge and proper amount of inhibitor added.
5. Confirm that the proper amount of lithium bromide was installed.
6. Confirm that the proper amount of octyl alcohol was installed.
7. Test electrical service to each pump for proper voltage/phase.
8. Test operating amps for each pump.

*Functional Performance Checklist*

1. Confirm that the factory start-up and tests were completed and the chiller is ready to be placed into service.
2. Determine entering CHW temperature and compare to design.
3. Determine leaving CHW temperature is within 1°F of setpoint.
4. Compute CHW range and compare to design.

Electric-Drive Chillers

1. Determine that the chiller full load amperes (FLA) is within 5% of design.
2. Test volts, phase-to-phase.
3. Compute voltage imbalance and determine that the maximum variation between phases is less than 2%.

Absorption Chillers

1. Determine that the high generator temperature is within the design range.
2. Test high/low generator solution level.
3. Determine that supply steam pressure is within 5% of design.
4. Test automatic overconcentration, dilution cycles, and dilution on shutdown.

Water-Cooled Chillers

1. Confirm that the cooling tower has been commissioned (see the section "Tower Commissioning" in Chapter 15).
2. Determine the entering CDW temperature and confirm that it is within 1°F of setpoint.
3. Determine leaving CDW temperature and compare to design.
4. Determine CDW temperature range and compare to design.

Air-Cooled Chillers

1. Determine ambient air temperature.
2. Determine chiller head pressure and compare to rated value.

Operations and Control

1. Confirm that the chiller appears to meet load.
2. Confirm that the chiller operates without alarm conditions or safety shutdowns.
3. Confirm that the chiller start sequence operates properly.
4. Confirm that multichiller staging sequence operates properly.

## CHILLER MAINTENANCE

When establishing a maintenance program for chillers, maintenance managers have three options:

1. Implement the program fully in-house.
2. Outsource the entire program.
3. Use a combination of in-house and outsourced functions.

One of the most important benefits of a program that uses in-house personnel is the *institutional knowledge* of those systems. Maintenance personnel who have been working with those chiller systems for years are most likely to know what many of the existing maintenance problems are.

Another benefit of using in-house personnel is long-term quality. Maintenance managers and in-house personnel are better able to focus their attention and efforts on both short- and long-term requirements, up to the expected life of the equipment. Outsourced programs tend to have a much shorter focus, that is, the period of the maintenance contract.

However, the cost of establishing a complete in-house program can be very high. Managers have to arrange for the training of maintenance personnel on the specifics of maintaining the chiller(s) installed in the facility, provide specialized training for infrequent testing and servicing, and purchase specialized test equipment for many of the inspection and maintenance activities that must be performed.

Combination of in-house and outsourced programs typically assign the routine and preventive maintenance tasks to in-house personnel, while contracting is utilized for specialized activities that require a higher level of expertise than is available in-house. This approach maintains in-house personnel being actively involved in the maintenance, while outside experts assist them by performing the more complex and infrequent tasks. This arrangement helps preserve the institutional knowledge and typically is the most cost-effective approach.

*No matter which maintenance scheme is utilized, a preventative maintenance program to ensure that the chiller operates reliably over its design life is required.*

The majority of chiller operating problems and maintenance needs are discovered by visual inspection and the monitoring of equipment operating parameters. Figure 7.1 is a form that can be used to guide chiller visual inspection and to collect operating data every 2 h during the day. These data will give a complete "snapshot" of the running conditions of the chiller, and variations in data between observations can be a prime indicator of operating problems.

Over the long term, these data can be used to predict maintenance requirements. These data, along with the data collected during periodic checks and routine maintenance procedures, can be plotted against time so that a trend or change in conditions can be identified. A decision on maintenance can then be made from the trend.

Recommended preventative maintenance procedures for air-cooled electric-drive water chillers are as follows:

*Daily monitoring/visual inspection*: The majority of chiller operating problems and maintenance needs are discovered by visual inspection and frequent monitoring of equipment operating parameters.

*Monthly, quarterly, and annual preventative maintenance*

1. Clean evaporator every 2–4 years (annually for chillers serving air washers or other "open" cooling loads).
2. Quarterly, calibrate pressure, temperature, and flow controls.

Date ________ Shift _______ Data Collected By ________________

Outside Air Conditions:

| Time | | | | | | | | |
|---|---|---|---|---|---|---|---|---|
| Dry Bulb Temperature | | | | | | | | |
| Wet Bulb Temperature | | | | | | | | |

Chiller No.

| Cooler: | Pressure Drop | | | | | | | | |
|---|---|---|---|---|---|---|---|---|---|
| | Refr. Temp | | | | | | | | |
| | CHWS Temp | | | | | | | | |
| | CHWR Temp | | | | | | | | |
| Condenser: | Pressure Drop | | | | | | | | |
| | CDWS Temp | | | | | | | | |
| | CDWR Temp | | | | | | | | |
| Compressor: | Amps | | | | | | | | |
| | Head Pressure | | | | | | | | |
| | Refr. Level | | | | | | | | |
| | Oil Temp | | | | | | | | |
| | Oil Level | | | | | | | | |
| Purge Unit: | Hours runtime | | | | | | | | |

Chiller No. _______

| Cooler: | Pressure Drop | | | | | | | | |
|---|---|---|---|---|---|---|---|---|---|
| | Refr. Temp | | | | | | | | |
| | CHWS Temp | | | | | | | | |
| | CHWR Temp | | | | | | | | |
| Condenser: | Pressure Drop | | | | | | | | |
| | CDWS Temp | | | | | | | | |
| | CDWR Temp | | | | | | | | |
| Compressor: | Amps | | | | | | | | |
| | Head Pressure | | | | | | | | |
| | Refr. Level | | | | | | | | |
| | Oil Temp | | | | | | | | |
| | Oil Level | | | | | | | | |
| Purge Unit: | Hours runtime | | | | | | | | |

**FIGURE 7.1** Chiller data logging form.

3. Annually, inspect starter wiring connections, contacts, and action. Tighten and adjust as required. Perform thermographic survey every 5 years.
4. Annually, test the operation of safety interlocks devices, such as flow switches, pump starter auxiliary contracts, phase-loss protection, and so on. Repair or replace as required.
5. Annually, perform dielectric motor testing to identify failures in motor winding insulation. For large chillers (100 tons or larger), additional

annual motor tests are required to test for imbalance of electrical resistance among windings, imbalance of total inductance with phase inductances, power factor, capacitance imbalance, and running amperage versus nameplate amperage.

6. Annually, check the tightness of the hot gas valve (as applicable). If the valve does not provide tight shutoff, replace it.
7. Annually, change the lubricant (oil) filter and the drier.
8. Laboratory analysis of the lubricant should be performed annually during the first 10 years of chiller life and every 6 months thereafter. [This oil analysis will define the moisture content (not to exceed 50 ppm), oil acidity (maximum 1 ppm) that may indicate oil oxidation and/or refrigerant degradation due to high temperatures, and metals or metal oxides that indicate chiller component wear and/or moisture in the oil.]
9. Valve and bearing inspection in accordance with manufacturer's recommendation.
10. Relief valves (both refrigerant and water) should be checked annually. Disconnect the vent piping at the valve outlet and visually inspect the valve body and mechanism for corrosion, dirt, or leakage. If there are problems, replace the valve; do not attempt to clean or repair it.
11. Every 6 months, inspect and clean air-cooled condensers as follows:
    a. Check unit casing and clear any leaf litter or organic matter in contact with the casing. Remove leaves, sticks, and so on, on or in the unit casing. Check for condition of paint, metal, and so on and repair as necessary.
    b. Check outdoor fans for proper rotation and that the fans do not run backward when off. Clean and, if needed, balance the fan blades.
    c. Lubrication: Typically, fan motors have sealed bearings and require no lubrication.
    d. Inspect condenser coils and clean, if necessary.
12. Perform chemical testing of chilled water at least quarterly. Treat as needed to ensure proper water chemistry.

*Extended (long-term) maintenance checks*

1. Every 5 years (or more frequently as the chiller ages), perform eddy current (electromagnetic) testing of heat exchanger tubes. [This testing will typically detect tube pits, cracks, and tube wear (thinning).] If only 1–5 tubes are found defective, plug tubes. If more than 5 are found defective, replace tubes. (Any tube that is replaced as a result of this testing should be examined and cross-sections cut so the cause of the defects can be evaluated.)
2. Other components must be serviced, inspected, and/or replaced at the intervals recommended by the chiller manufacturer.

Recommended maintenance procedures for water-cooled electric-drive rotary compressor water chillers are as follows:

*Daily monitoring/visual inspection*: The majority of chiller operating problems and maintenance needs are discovered by visual inspection and frequent monitoring of equipment operating parameters.

*Monthly, quarterly, and annual preventative maintenance*:

1. Clean the evaporator tubes every 2–4 years (annually for chillers serving air washers or other "open" cooling loads).
2. Clean the condenser tubes every year.
3. Quarterly, calibrate pressure, temperature, and flow controls.
4. Annually, inspect starter wiring connections, contacts, and action. Tighten and adjust as required. Perform thermographic survey every 5 years.
5. Annually, test the operation of safety interlocks devices, such as flow switches, pump starter auxiliary contracts, phase-loss protection, and so on. Repair or replace as required.
6. Annually, perform dielectric motor testing to identify failures in motor winding insulation. For large chillers (100 tons or larger), additional annual motor tests are required to test for the imbalance of electrical resistance among windings, imbalance of total inductance with phase inductances, power factor, capacitance imbalance, and running amperage versus nameplate amperage.
7. Annually, check the tightness of the hot-gas valve (as applicable). If the valve does not provide tight shutoff, replace it.
8. Annually, replace lubricant (oil) filter and drier.
9. Laboratory analysis of the lubricant should be performed annually during the first 10 years of chiller life and every 6 months thereafter. Analysis must address oil moisture content, acidity, and chiller wear as follows:
   a. Maximum moisture content of 50 ppm. Higher levels may indicate air leaks in low-pressure chillers or heat exchanger tube leaks.
   b. Maximum acidity of 1 ppm is normal for a new or reclaimed refrigerant. Higher levels may be caused by oxidation of oil during aging and/or degradation of refrigerant. If higher levels are found, contact the chiller manufacturer.
   c. Trace amounts of metals could be metal oxides formed from moisture in the oil or may indicate excess wear conditions. Due to variations in metals used by different chiller manufacturers, there are no standard limits. However, *if metal content increases from year to year, one or more of the following excess wear conditions may be responsible*:

      Aluminum: impellor or bearing wear
      Chromium: wear on rechromed shafts
      Copper: corrosion

Iron: corrosion and/or gear wear
Tin: bearing wear
Silicon: leakage of silica gel from dehydrator cartridge or dirt in system
Zinc: zinc left over from the manufacturing process or loss of galvanizing from some parts

10. Valve and bearing inspection in accordance with manufacturer's recommendation.
11. Relief valves (both refrigerant and water) should be checked annually. Disconnect the vent piping at the valve outlet and visually inspect the valve body and mechanism for corrosion, dirt, or leakage. If there are problems, *replace the valve; do not attempt to clean or repair it.*
12. Annually, inspect gearbox for wear and repair or replace it as needed.
13. Perform chemical testing of system water at least quarterly. Treat as needed to ensure proper water chemistry.

*Extended (long-term) maintenance checks*

1. Every 5 years (or more frequently as the chiller ages), perform eddy current (electromagnetic) testing of heat exchanger tubes. [This testing will typically detect tube pits, cracks, and tube wear (thinning).] If only a limited number of tubes are found defective, plug tubes. If more than few are found defective, replace tubes. (Any tube that is replaced as a result of this testing should be examined and cross-sections should be cut so that the cause of the defects can be evaluated.)
2. Every 3–5 years, perform vibration test and analysis to evaluate motor and rotor balance, bearing and gear alignment, and bearing and gear wear. Use accelerometer-type sensors for gears and bearings (high frequency) and piezoelectric velocity sensors for compressor motors, rotors, and bearings (low velocity).
3. Every 5 years, perform an acoustic emission test to identify potential stress cracks in pressure vessels, tubes, and tube sheets.
4. Other components must be serviced, inspected, and/or replaced at the intervals recommended by the chiller manufacturer.

Absorption two-stage direct-fired water chillers have significant maintenance requirements:

*Daily monitoring/visual inspection*: The majority of chiller operating problems and maintenance needs are discovered by visual inspection and frequent monitoring of equipment operating parameters.

*Mechanical components*

1. Preferably daily, but not exceeding weekly, the operation of the purge unit must be checked for both proper operation and *excess* operation,

indicating an air leak that can produce serious problems due to corrosion, contamination of the absorbent solution, and reduction in efficiency and capacity. Test for noncondensables (hydrogen).
2. Annually, refrigerant and solution pumps and the purge unit must be inspected and tested to ensure continued operation. Every 3 years, the pump seals and bearings will require inspection and, as indicated, replacement.

*Heat transfer components*

1. Annually, the condenser and absorber heat exchanger tubes must be cleaned. The evaporator tubes should be cleaned every 3–4 years.
2. Every 6 months (more frequently if runtime exceeds 4000 h per year), the lithium bromide solution must be analyzed for contamination, solids, pH, corrosion inhibitor level, and performance additive (typically octyl alcohol). Adjust the solution as necessary to meet the manufacturer's requirements. (Metals and metal oxides in the solution are typical indications that internal corrosion is occurring due to air leaks.)
3. Annually, perform leak test, using the pressure method.
4. Annually, perform eddy current testing of the high-stage generator tubes.
5. Every 3–5 years, perform eddy current testing of the absorber, condenser, low-stage generator, and evaporator tubes.
6. Every 3 years, service valves containing rubber diaphragms. These diaphragms should be replaced (which requires that the lithium bromide/water solution be removed from the chiller).

*Controls*: Annually, test the controls for proper operation at the beginning of the cooling season. Clean and tighten all connections, including field sensor connections. Vacuum control cabinets to remove dirt and dust. With microprocessor controls, have the factory service technicians test calibration and operation and ensure that the latest version of operating software is loaded.

*Burner*: Inspect, test, and maintain in accordance with the following:

*Monthly*

a. Inspect the venting system.
   Check the chimney or vent to make sure that it is clean and free of cracks, open joints, or other leaks.
   Check and clean combustion air intake.
b. Inspect burner and controls for proper operation.

*Annual shutdown*

a. Turn off the burner.
b. Open the main power disconnect switch to the burner. Close the fuel supply valve(s).

c. Clean all metal surfaces of the firebox, gas passages, and venting (breechings, vent pipe, etc.) with a wire brush and vacuum.
d. Check the condition of firetubes. Look for cracks, pitting, leaks, and so on. Replace firetubes every 5–8 years to avoid failure due to refrigerant-side corrosion.

*Chilled water*: Perform chemical testing of system water at least quarterly. Treat as needed to ensure proper water chemistry.

Each of these chiller maintenance programs requires routine cleaning of heat exchanger tubes. Modern chillers, to comply with ASHRAE Standard 90.1 efficiency requirements, are fabricated with evaporator and condenser tubes that have *enhanced* heat transfer surfaces on both their interior and exterior. Tubes are manufactured with "rifling" or spiral grooves on the inside to provide more surface area for heat transfer to increase water turbulence, but these enhancements make tube cleaning significantly more difficult.

Chiller heat exchangers must be cleaned to eliminate the buildup of scale (deposition) and biological films that form on their inside and effectively reduce heat transfer, reducing chiller capacity and efficiency. Evaporator tubes, being part of a closed system, require cleaning infrequently, usually every 3–4 years. However, condenser water systems, even ones that are well maintained and have good water treatment programs, will require cleaning annually.

There are several methods of cleaning chiller tubes:

*Rod-and-brush method.* The oldest and least pleasant method for chiller tube cleaning is the *rod-and-brush* method. A long-handled nylon or wire brush, slightly larger than the tube's inside diameter, is pushed/pulled manually through the tubes and the tubes are flushed with water between brushings. With enhanced surface tubes, this method is relatively ineffective.

*Chemical cleaning.* Less prevalent now than in the past years, this method uses acid solutions that are circulated through the tube bundles to break down or soften scale deposits in the tubes. Under the right conditions, tubes can be cleaned to bare metal using this method. In other cases, mineral scale can be softened enough to allow brush cleaning as a secondary operation. However, this process is costly, time consuming, and the chemicals represent a health and safety risk.

*Tube-cleaning guns.* This method uses compressed air and water or high-pressure water alone to propel a cleaning projectile through the tubes to remove deposits. The projectiles can be rubber bullets, brushes, or plastic or metal scrapers. This method is fast, little operator training or expertise is required, and tubes can be cleaned thoroughly with this method under the right conditions.

*Online systems.* Two basic online tube-cleaning systems are in use today. One uses plastic brushes that are "trapped" in each tube with plastic baskets that have been attached to each end of the tube using an epoxy. Periodically, the direction of the water flow through the tube bundle is reversed, causing the brush to travel to the other end of the tube. This back and forth brushing action is designed to keep the tubes clean. With the other type of system, a quantity of foam balls is circulated continuously through the system. In theory, a foam ball will travel through every tube in the bundle frequently enough to keep the tubes clean.

Provided the water treatment is adequate, these systems may work, but their record of cleaning and descaling enhanced tubes is very poor. Furthermore, these systems are very expensive, require regular brush/ball replacement, and are not suitable where mineral scaling commonly occurs or where the water treatment program is inadequate.

*Rotary tube cleaners.* Rotary tube cleaners utilize an electric or air motor to rotate a flexible shaft in a plastic casing that transports water to the cleaning tool. The cleaning tool might be one of a variety of brushes, a buffing tool, a hone, or a scraper tool. These systems are capable of cleaning almost all types of deposits, including hard scale. The operator simply feeds the flexible shaft through the tubes to brush and flush in one operation. The initial cost for this method is fairly low and the cost of consumables is reasonable. It also is simple to use; little training or expertise by maintenance staff is required.

Today's variable-speed rotary cleaners are designed for bidirectional operation, enabling the operator to spin the brush in the direction of tube internal spiral, and brush bristles are usually trimmed to two different diameters to improve cleaning in the spiral grooves. Water for flushing is delivered under high pressure for more effective cleaning.

*The use of rotary cleaners is, by far, the most effective and least costly approach to chiller tube cleaning available today.*

An impediment to adequate chiller tube cleaning is that the chiller may not be piped to facilitate easy removal of piping connections and/or water box heads as discussed in Chapter 3. Easily removable flanged or grooved-coupling connections are required, valves must be placed to allow the chiller water circuits to be isolated, and there must be a clear area at one end of each heat exchanger to provide access for cleaning equipment.

## CHILLER PERFORMANCE TROUBLESHOOTING

While both vapor compression cycle chillers and absorption chillers are reliable machines that require little "hands-on" attention by their operators, they are mechanical devices that are susceptible to failure and resulting operating problems. When operating problems do appear, they generally fall within the following three categories.

## Selection or Design Problems

Typically, there are two common chiller problems that can be traced to improper design and/or incorrect chiller selection:

1. Capacity is too low or too high for imposed load. The first problem, insufficient capacity, is fairly rare because most design engineers are conservative folks who would rather err on the high side. But, simple mistakes or changed conditions can result in a chiller that simply has too little cooling capacity.

   Unfortunately, there is no simple way of increasing capacity. However, the problem can be mitigated by keeping condensing temperatures as low as possible. Adding larger condenser coils for air-cooled applications or an additional cooling tower cell can be far less expensive than adding additional chiller capacity if the shortfall is relatively small.

   *High head pressure* in a chiller, which reduces capacity, is normally the produced by high condenser temperature conditions. This lost capacity, also, can be regained by improving condenser side performance.

   Chiller overcapacity is a far more common problem, resulting in inefficient performance and the potential for control point offset that may cause the chiller to shut down due to low refrigerant temperature or low chilled water temperature. Generally, this problem can be addressed by false-loading the chiller with hot-gas bypass. This will prevent the chiller from going off-line, but will not improve energy use.
2. Chilled water supply temperature is too high. As discussed in Chapter 1, for an HVAC system to simultaneously maintain the desired space temperature and humidity, the chilled water temperature must be low enough to allow supply air to be cooled to the required dewpoint temperature. If chilled water supply temperature is too high, high space temperature and/or humidity conditions will occur.

   The chiller's supply water temperature setpoint can be lowered from the rated value, but there will be a corresponding reduction in chiller capacity and increase in chiller energy consumption. Thus, there is a trade-off between water temperature and capacity that must be evaluated to determine how low the supply water temperature can be set without reducing capacity to below the imposed load.

   Again, both chilled water supply temperature and capacity can be improved by lowering the condenser side temperatures as much as possible.

## Installation Problems

Assuming that water flow rates are properly adjusted and controlled, control systems operate properly, and chillers are installed in accordance with the design requirements, there are a few operating problems. However, there is one common problem in larger, primary–secondary systems that results from installation problems with chilled water coils and control valves that show up at the chiller.

In most primary–secondary systems, the design chilled water range ($T_r - T_s$) for both primary and secondary loops is the same, that is, coils and chillers are typically selected for the same temperature rise. But, in many systems, a *low-range* problem (typically called "low delta-T problem") shows up, as follows:

Symptom: Having to operate more chillers than the load dictates in order to have enough chilled water flow to meet the needs of certain air-handling units

Result: System range ($T_r - T_s$) is lower than design. Thus, chiller plant capacity is actually "lost" due to the inability to obtain design range.

This lost chiller capacity is the problem and often requires that more chillers be operated at lower-load conditions to provide enough chilled water flow to the coils. This may result in chiller operating conditions of less than 30% of full load and chiller efficiency is also lost. Thus, it simply costs more to produce cooling, both in chiller operation and in pumping energy.

If the cooling coil heat transfer characteristics and the control valve flow control characteristics are not exactly matched, excess flow for the required part load capacity will result. If the coil is dirty, the heat transfer factor becomes poorer, requiring higher flow rates for a given part-load capacity. If the control valve leaks, excess flow results. All these conditions may be contributors to the low-range problem and all are difficult to find and fix in the field. However, there are a number of steps that can be taken to reduce the problem:

1. Clean all cooling coils on a regular basis to minimize fouling and airflow resistance. If the coil is old *and* dirty, consider replacing it with a new cooling coil selected for higher range, typically 25% higher than the system design range.

   Clean chilled water coils as follows:

   a. Use a mild dish detergent, such as Dawn® liquid, as the cleaning agent.
   b. *Keep the supply fan in operation* to provide differential air pressure across the coil to move the flushing water and cleaning solution through the coil.
   c. *Do not use a pressure washer—it is counterproductive since it just compacts the dirt into the coil passages.*
   d. Apply the detergent liquid (about a 10% solution in water) with a pump sprayer on the upstream face of the coil, allowing it to soak in. Then, flush repeatedly with clean, cold water.

   e. Repeat this process until suds and water appear on the downstream side of the coil, with the water off the coil finally running "clear."
   f. Inspect fins for aluminum oxidation and deterioration. Fins that are in poor condition indicate that coil replacement is required.
   g. If fins are bent or smashed, use a coil comb to straighten as much as possible.
2. Add "sensible cooling only" coils upstream of existing coils and pipe the chilled water flow in series with the existing coil. Thus, a higher water range can be obtained, with only a minor airside pressure drop penalty since these are added typically only 2–3 rows deep.
3. Control valves must be properly sized two-way *equal percentage* type. Make sure there are no three-way valves in the system. The valve wide-open pressure drop at full flow should be (a) 1.25–1.50 times the coil pressure, or (b) equal to the excess branch circuit pressure differential, whichever is greater. Select control valves for minimum 50:1 *rangeability*. Valve operators must be selected for tight shutoff against the pump cutoff pressure, the pressure delivered by the pump at zero flow.
4. All control valves must be normally closed and arranged to close if the air-handler is off, including fan coil unit control valves. Water bypassing through open control valves can be a root cause of a low-range problem.
5. Raise (reset upward) $T_s$ at reduced cooling loads as described in Chapter 4.
6. Calibrate and maintain airside temperature controls. If these read "high," excess chilled water flow will result.
7. Routinely monitor chilled water temperature range. *If you cannot measure it, you cannot manage it.*

## REFRIGERANT MANAGEMENT PROGRAM

Effective July 1, 1992, as required by Section 608 of Clean Air Act of 1990, the EPA established regulations (published in the *Code of Federal Regulations*, 40 CFR Part 82, Subpart F) that require the following:

1. Practices that maximize the recycling of ozone-depleting compounds (CFCs and HCFCs) during the operation, servicing, and disposal of refrigeration equipment
2. Certification of refrigerant recovery and recycling equipment, technicians involved with handling refrigerants, and reclaimers
3. Limiting of sale of refrigerants only to certified technicians
4. Repair of substantial leaks in refrigeration equipment with a refrigerant charge greater than 50 lb (equivalent to about 15-tons capacity or larger)
5. Safe disposal requirements to ensure removal of refrigerants from goods that enter the waste stream

*Vapor compression cycle water chillers using CFC/HCFC refrigerants fall within the jurisdiction of these regulations. Absorption water chillers and vapor compression cycle water chillers using HFC refrigerants are exempt from these requirements simply because these refrigerants have little or no ozone-depleting characteristics.*

For CFC/HCFC chillers, the certification of equipment and technicians, the limiting of leaks, and safe disposal of equipment are required parts of a refrigerant management program that must be implemented by the chiller owner.

*Equipment certification.* Equipment used to evacuate refrigerant from a chiller must be certified by an EPA-approved equipment testing organization (either ARI or UL). Refrigerant removed from a chiller can be returned to that chiller or to another chiller with the same ownership. However, if the refrigerant changes hands, the refrigerant must be "reclaimed" (i.e., cleaned to ARI Standard 700 purity level and chemically analyzed to verify that it meets this standard).

*Technician certification.* EPA requires certification of persons who perform maintenance, service, repair, or disposal that could be reasonably expected to release refrigerants into the atmosphere. The definition of "technician" includes anyone who would attach/detach hoses and gauges to measure pressure or would add or remove refrigerant from the system.

Four types of certification have been established by EPA, but for chiller applications, "Type II (high pressure)," "Type III (low pressure)," or "Universal (any type of equipment)" certifications would be required. Technicians are required to pass an EPA-approved test given by an EPA-approved certifying organization.

*Refrigerant venting and leaks.* Small amounts of refrigerant released during the purge of low-pressure systems is allowed by the regulations. Also, small leaks when hoses or connections are disconnected, if meeting low-loss standards, are also allowed. However, any chiller containing 50 lb or more of refrigerant must not leak more than 15% of the total charge in a year's time. If a leak is discovered (evidenced by the need to add refrigerant to maintain capacity and/or prevent evaporator low-temperature problems), owners have 30 days to repair the leak.

*Disposal.* A chiller that is removed from service must have all refrigerant removed prior to disposal. This refrigerant can be retained for use in other chillers under the same ownership or sold to a reclaimer for reclamation and resale. (Note that if refrigerants are recycled or reclaimed, they are not considered "hazardous" by the EPA.)

The regulations also establish requirements for recordkeeping. Technicians are required to provide owners with an invoice that indicates the amount of refrigerant added to any system that contains 50 lb or more. The owner, then, is responsible for maintaining these invoices and using them to track leakage rates from equipment. Additionally, owners must maintain servicing records documenting the date and type of service, as well as the quantity of refrigerant added.

Every organization that owns one or more chillers should review the following checklist to establish and maintain a refrigerant management program:

1. Obtain copies of current EPA regulations and become familiar with their requirements. Establish a program to collect, distribute, and communicate updates and amendments that are issued regularly by the EPA. (The EPA maintains a Web site at http://www.epa.gov that provides significant information and links to specific documents that apply to a refrigerant management program.)
2. Develop a "mission statement" specifically documenting your "intent to comply" with EPA regulations.
3. Develop a written job description for a "Facility Refrigerant Manager" who is responsible for refrigerant management and compliance with EPA regulations.
4. Designate and train your "Facility Refrigerant Manager." (This is the individual that the EPA will speak with when conducting a refrigerant compliance inspection.)
5. Develop written policies and procedures for refrigerant usage record-keeping, including a specific method for collecting and maintaining records and making them available to the EPA upon request.
6. Maintain copies of EPA certifications for all in-house and contracted technicians working at the facility.
7. Develop written policies and procedures for each of the following aspects of refrigerant handling and use:
   - Unintentional venting or leaking, including required reporting
   - Refrigerant purchase requirements with a designated technician or contractor responsible for all purchases
   - Labeling for refrigerant cylinders
   - Refrigerant inventory and storage, including in–out control and auditing
   - Leak testing process and service procedures for positive-pressure chillers
   - Disposal of refrigerant equipment, parts, and lubricants
   - Refrigerant safety
8. Develop a written "Emergency Response Plan" to address major venting incidents, maximum exposure levels, and evacuation procedures.
9. Include EPA regulation compliance language in the contracts with all service organizations to ensure that they assume all liability in the event of noncompliance.

10. Make sure that your refrigerant policies and procedures are "effectively communicated to all affected personnel" through posting in the workplace and by documented compliance training programs.
11. Annually, conduct a survey to ensure that your facility is in compliance with EPA requirements.

# 8 Buying a Chiller

## DEFINING CHILLER PERFORMANCE REQUIREMENTS

The first step in purchasing a new or replacement chiller is to define the performance requirements for the chiller and the type(s) of chillers that are applicable to the specific installation limitations.

As discussed in Chapter 2, the types of chiller are defined as follows:

Vapor compression cycle, scroll compressor, electric-drive, air-cooled or water-cooled
Vapor compression cycle, rotary compressor, electric-drive, air-cooled or water-cooled
Vapor compression cycle, rotary compressor, engine-drive, water-cooled
Absorption cycle, two-stage, indirect-fired, water-cooled
Absorption cycle, two-stage, direct-fired, water-cooled

To define capacity and performance requirements for any chiller, the following parameters must be specified:

1. Peak load parameters
    - Maximum cooling capacity (tons)
    - Chilled water supply temperature (°F)
    - Chilled water temperature range (°F)
    - Chilled water flow rate (gpm)
    - Maximum evaporator pressure drop (ft wg)
    - Evaporator fouling factor (usually 0.00010)
    - Condenser water supply temperature (°F)
    - Condenser water temperature range (°F)
    - Condenser water flow rate (gpm)
    - Maximum condenser pressure drop (ft wg)
    - Condenser fouling factor (usually 0.00025)
    - Electrical service: V/ph/Hz
    - Maximum sound power level (dBA)
2. Part-load parameters
    - Cooling load profile (tons), time- and/or temperature-based
    - Condenser water supply temperature (°F) as a function of outdoor temperature (°F)

For electric-drive chillers, the motor requirements and allowable input energy requirements must be specified:

Allowable motor type (hermetic and/or open)
Motor FLA
Maximum kW/ton at full load

For engine-drive chillers, the engine parameters must be specified:

Engine type and fuel(s)

For indirect-fired absorption chillers, the steam or hot-water parameters must be specified:

Steam operating pressure (psig)
Maximum steam flow rate (lb/h)

or

High temperature hot-water temperature (°F)
High temperature hot-water flow rate (gpm)

For direct-fired absorption chillers, the burner parameters must be specified:

Burner type
Burner horsepower (hp)
Fuel(s)
Maximum firing rate (Btu/h)
Minimum firing efficiency (%)

To compare the seasonal energy performance of alternative chillers, the necessary requirement for chiller vendors is to provide the input energy or fuel requirement at, at least, 25%, 50%, 75%, and 100% cooling loads for each alternative chiller, based on condenser water supply temperatures estimated from weather data and the method of condenser water temperature control to be applied (see Table 8.1 for this format). Then, using the annual cooling load profile, annual

**TABLE 8.1**
**Chiller Utilization Profile**

| Load % | Load (tons) | Condenser Water Temperature Entering Chiller (°F) | Chiller Input Energy Requirement (kW) |
|---|---|---|---|
| 100 | | | |
| 75 | | | |
| 50 | | | |
| 25 | | | |

cooling performance and energy consumption by the chiller can be modeled by either manual or computer-based methods.

As discussed in the section "Load versus Capacity" in Chapter 2, IPLV for a chiller is only directly applicable to the single-chiller installation where the condenser water supply temperature is allowed to fall as the outdoor wet bulb temperature falls. Anytime multiple chillers are used, particularly if careful control allows the chillers to be used in or near their optimum efficiency range of 40–80% of rated capacity, the IPLV has little validity. Likewise, if the chiller manufacturer does not allow the entering condenser water temperature to go as low as 65°F, the IPLV is not valid. In either case, more detailed modeling is required to accurately compare alternative chiller performance and energy consumption.

## ECONOMIC EVALUATION OF CHILLER SYSTEMS

The designer and/or owner should evaluate the total owning and operating cost of each alternative chiller that will satisfy the performance requirements.

A variety of chiller components (tube quantity, shell length, compressor size/quantity) can be selected by a chiller manufacturer in order to best satisfy a given load profile. Because chillers can be customized for optimum performance, *life cycle cost analysis* is the best way to compare alternative chiller selections. It provides for a consistent method of analyzing the economic aspects of each chiller and allows realistic comparison so that the most cost-effective, for example, lowest life cycle cost chiller can be selected. The process requires only that each potential chiller be evaluated using the same criteria, resulting in an "apples-to-apples" comparison, not "fruit salad."

The computations required to determine life cycle cost utilizing the *total owning and operating cost* methodology are simple. However, the methodology's accuracy depends wholly on the accuracy of the data utilized. Anyone can calculate "garbage" life cycle costs simply because they use data and/or assumptions that are "garbage." Two different individuals, faced with the same evaluation, may compute wildly different life cycle costs because they use significantly different data and/or assumptions in their computations.

The following subsections define the basic elements that make up the life cycle cost.

### FIRST COSTS

The initial capital costs associated with each potential chiller are all of the costs that would be incurred in the design and construction of that chiller installation. Equipment costs can be obtained directly from the prospective equipment vendors. Installation cost estimates can be obtained from local contractors or, lacking that, from cost data published by R. S. Means Co., Inc. (Construction Plaza, 63 Smiths Lane, Kingston, MA 02364-0800, 781/585-7880 or 800/334-3509).

The construction cost estimate must include the following, in addition to the cost of the chiller itself:

1. Vapor compression cycle chiller ancillary costs
   - Refrigeration room safety requirements (ASHRAE Standard 15)
   - Additional ventilation required for open motor/compressor configuration
   - Noise and vibration control
   - Fuel piping, exhaust venting, and noise abatement for engine-drive chillers
   - Additional mechanical room space required for engine-drive chillers
   - Power wiring for electric-drive chillers
2. Indirect-fired absorption cycle chiller ancillary costs
   - Additional mechanical room area
   - Additional floor slab or foundation structural requirements
   - Additional condenser water system elements, including larger piping, pumps, and cooling tower
3. Direct-fired absorption cycle ancillary costs
   - Same as for indirect-fired absorption chiller
   - Additional mechanical room height
   - Combustion air inlets
   - Combustion venting (breeching, stack, etc.)
   - Fuel gas piping
4. General conditions and overhead costs
   - Concrete housekeeping pad
   - Rigging
   - Demolition
   - Contractor overhead (insurance, bonds, taxes, and general office operations, special conditions), typically 15–20%
   - Contractor profit, typically 5–20%

Where air-cooled and water-cooled condensing are alternative options under consideration, the associated capital costs for condenser water piping, cooling tower(s), condenser water pumps, and so on must be included for water-cooled chillers. And, where air-cooled chillers are installed in coastal areas, condenser coil coatings must be considered (otherwise, salt air corrosion of aluminum coil fins can shorten the chiller life to 5 years or less).

Other costs that may be included in the capital requirement are design fees, which may increase or decrease as a function of the selected alternative; special consultants' fees; special testing; and so on. Also, chilled water and/or condenser water piping may change configuration and cost between alternative chillers.

Unless the cost of the chiller is being met from operating revenues, at least a portion of the capital expense may be met with borrowed funds. The use of this money has a cost in the form of the applied *interest rate*, and information about the amount of borrowed funds, the applied interest rate, and the period of the loan must be determined for the analysis. The total capital cost, including principal plus interest, can then be computed using the following:

$$C = A + \left\{ B \times n \times \left[ \frac{i}{1 - (1 - i)^{-n}} \right] \right\}$$

where

$C$ = total capital cost ($)
$A$ = first cost out-of-pocket ($)
$B$ = borrowed funds ($)
$i$ = interest rate (decimal value)
$n$ = number of years of loan

## Annual Recurring Costs

Once the chiller is placed into operation, two annual recurring costs must be met each year of its economic life: energy costs and maintenance costs.

The economic life for an alternative is the time frame within which it provides a positive benefit to the owner. Thus, when it costs more to operate and maintain a piece of equipment than it would to replace it, the economic life has ended. Economic life (or "service life") is the period over which the equipment is expected to last physically. The economic life for scroll compressor chillers is normally selected as 15 years, while rotary compressors and absorption chillers have an expected economic life of 23 years.

The computation of energy cost requires that two quantities be known: (1) the amount of electrical energy or fuel consumed by the chiller on an annual basis, and (2) the electrical rate schedule (for electric-drive chillers) or fuel unit cost (for engine-drive or absorption chillers) applied to that energy consumption. The second quantity is relatively easy to determine by contacting the utilities serving the site or, for some campus facilities, obtaining the cost for steam or power that may be furnished from a central source.

The computation of annual energy costs for electric-drive chillers requires that two quantities be known: (1) monthly electrical power utilization, both consumption (kWh) and demand (kW), by the chiller, and (2) rate schedule that applies to this utilization. The cooling load profile discussed in Chapter 2 is the starting point for this analysis. The part-load energy input characteristics of the chiller, provided by the vendor in accordance with Table 8.1, can then be applied to the load profile to "model" chiller energy consumption. Then, costs are computed by applying the applicable rate schedule.

Annual recurring maintenance cost is a very difficult element to estimate. Lacking other information, the annual routine maintenance cost associated with chillers can be estimated as 1–3% of the initial equipment cost.

### Nonrecurring Repair and Replacement Costs

Nonrecurring costs represent repair and/or replacement costs that occur at intervals longer than 1 year. For example, a compressor may require significant repair (or even replacement) after 10–15 years of service. These costs must be determined and the year of their occurrence estimated.

There is also the issue of unequal service life between air-cooled and water-cooled chillers if these alternative options are considered.

### Total Owning and Operating Cost Comparison

The total owning and operating cost for a chiller alternative, over the system economic life, can be computed using the following relationship:

$$\begin{aligned}\text{Life cycle cost} = {} & C + (\text{sum of repair / replacement costs}) \\ & + [(\text{economic life}) \times (\text{annual energy cost} \\ & + \text{annual maintenance cost})]\end{aligned}$$

where $C$ = total capital cost computed as discussed in the section "First Costs."

## PROCUREMENT SPECIFICATIONS

Recommended specifications for both air-cooled and water chillers are provided in Appendices B1, B2, and B3. Each specification includes numerous options and, therefore, must be edited carefully before use.

# *Section B*

## *Cooling Towers: Fundamentals, Application, and Operation*

# *Part I*

## *Cooling Tower Fundamentals*

# 9 Cooling Tower Fundamentals

## COOLING TOWERS IN HVAC SYSTEMS

As discussed in Chapter 1, with water-cooled chillers, cooling towers have the role of rejecting heat collected during the space cooling process to the ambient atmosphere. In addition, the inefficiencies of the water chiller represent heat that is added to the condenser water circuit and must also be rejected by the tower.

Once a design temperature range for the condenser water is established, the condenser water flow rate can be determined from the following relationship:

$$F_{\text{cdw}} = \frac{\text{total heat rejection}}{500 \times \text{range}}$$

where

$F_{\text{cdw}}$ = condenser water flow rate (gpm)
Total heat rejection = cooling system load + heat of compression (Btu/h)
Range = condenser water temperature rise (°F)

The total heat rejection by an electric-drive chiller can be determined from the compressor power input, as shown in Table 9.1.

As the condenser water flows through the tower, this heat is rejected to the ambient air and the condenser water is cooled, primarily, through evaporation of a small percentage of the total water flow. Evaporation is a process by which heat

**TABLE 9.1**
**Refrigeration System Total Heat of Rejection**

| Compressor Input Power (kW/ton) | Approximate Heat of Compression [% of Cooling Load, Including 2% for Condenser Water Pump(s)] | Total Heat Rejection (Btu/ton) |
|---|---|---|
| 0.8 | 25 | 15,000 |
| 0.7 | 22 | 14,640 |
| 0.6 | 19 | 14,280 |
| 0.5 | 16 | 13,920 |
| 0.4 | 14 | 13,680 |

is absorbed by air and the remaining condenser water is cooled to the desired leaving temperature.

All of the data presented above are based on the use of electric-drive vapor compression cycle systems. With *absorption cycle chillers*, the amount of heat that is added to the cooling load, and which must be rejected through the cooling tower, is typically 20–50% greater than for an electric-drive chiller. Therefore, the condenser water flow rate, for the same range, will be 20–50% higher (or, for the same flow, the range must be 20–50% greater).

## CONDENSER WATER SYSTEM ELEMENTS

The four elements of any condenser water system (discussed in Chapter 11) include the following:

*Condenser.* The condenser is usually an integral part of the water chiller (though separate refrigerant condensers are sometimes used). Typically, the condenser is a shell-and-tube heat exchanger with the refrigerant in the tubes, surrounded by condenser water in the shell. Heat transfer takes place across the tube wall.

*Condenser water pump.* The pump circulates condenser water from the cooling tower basin, through the piping system and the condenser, finally returning the water to the cooling tower wet deck.

*Condenser water piping.* It is the piping for condenser water transport between the cooling tower and the condenser.

*Cooling tower.* It is a device designed to reject heat from the condenser water to the ambient air.

Every cooling tower, no matter how configured or constructed, must have the elements shown in Figure 9.1 (and discussed in Chapter 10).

*Fill.* Heat transfer media in the cooling tower. Fill is designed to maximize the "contact" between the return condenser water and the ambient air. The better the contact, obviously, the better the evaporation and heat transfer.

*Hot-water distribution ("wet deck").* Pans or basins with metering outlets or spray nozzles designed to provide an even distribution of the return condenser water entering the fill.

*Cold water basin.* The basin, either as an integral part of the tower or as a separate sump, collects the water passing through the tower for supply to the system by the condenser water pump. The basin must also be sized to contain enough water to supply the condenser water system until the pump returns water to the tower.

*Fan(s).* All cooling towers used for HVAC applications are *mechanical draft* towers that use one or more fans to provide airflow through the tower.

*Inlet louvers and drift eliminators.* Inlet louvers act to force the air entering the tower into as straight and even flow pattern as possible, while the

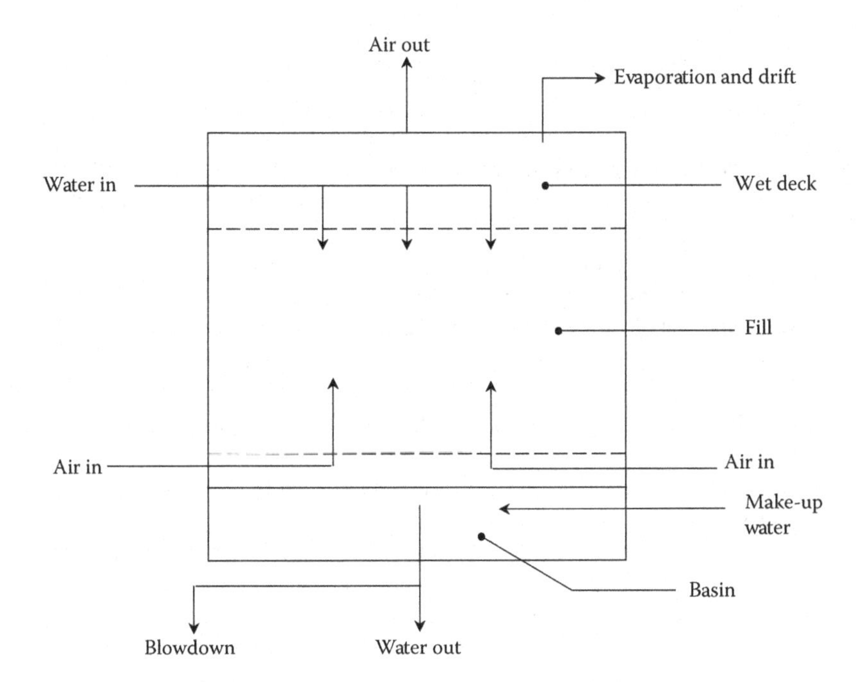


**FIGURE 9.1** Cooling tower elements.

drift eliminators are designed to trap and remove any entrained water droplets that may be in the tower's leaving air.

## NOMENCLATURE

As shown already, cooling towers and condenser water systems have their own vocabulary that must be mastered as follows:

*Approach*. Difference between the condenser water supply temperature and the entering wet bulb temperature (°F).

*Blowdown or bleed-off*. Condenser water purposely discharged from the system to control concentration of solids or other impurities in the water.

*British Thermal Unit (Btu)*. Amount of heat required to raise or lower the temperature of 1 lb of water by 1°F.

*Capacity*. Total amount of heat (Btu/h) that the cooling tower can reject at a given flow rate, approach, and ambient wet bulb temperature.

*Casing*. Exterior tower walls, excluding the louvers.

*Cell*. Smallest tower subdivision that can operate independently. Usually, a tower cell is designed with independent water and airflow and cells are controlled individually as a function of imposed overall cooling tower load.

*Counterflow*. Water flow path is configured at 180° to the airflow path, that is, the two flows are in opposite directions.

*Crossflow*. Water flow path is configured at 90° to the airflow path, that is, the two flows are perpendicular to each other.

*Double-flow*. A crossflow tower where two opposed fill banks and air intakes are served by common fan and air plenum (called a "twin-pack" in Great Britain).

*Drift*. Unevaporated water droplets that are lost from the cooling tower.

*Eliminators*. An assembly of baffles or other devices to remove entrained water droplets from the air leaving the cooling tower.

*Evaporation (loss)*. Condenser water that undergoes a phase change from liquid to vapor and exits the tower as part of an air–vapor mixture.

*Fill*. Heat transfer media or surface within the tower, designed to maximize the air–water contact surface area.

*Float valve*. A make-up water control valve activated by a mechanical float mechanism.

*Forced draft*. Airflow is "pushed" through the cooling tower by one or more fans at the air inlet(s), resulting in the tower being under positive pressure.

*Head*. Pressure required to be developed by the pump to overcome pressure losses due to friction through the condenser water piping and the condenser itself, the static lift required at the cooling, and any residual pressure required by the cooling tower nozzles (if any) (feet of water column).

*Induced draft*. Airflow is "pulled" through the cooling tower by one or more fans at the air outlet(s), resulting in the tower being under negative pressure.

*Inlet louvers*. Blades or other assemblies designed to (a) prevent water from splashing out of the tower, and (b) promote uniform airflow through the fill.

*Lift*. Static head represented by the height (in feet) between the water level in the cooling tower basin and the wet deck water entry point (feet of the water column).

*Make-up (water)*. Water added to offset water lost to evaporation, drift, and blowdown.

*Mechanical draft*. Airflow through the tower is produced by one or more fans.

*Natural draft*. Airflow through the tower is produced by air density differences produced by air temperature rise, that is, chimney effect.

*Nozzle*. Device to control water flow through a cooling tower. Normally, nozzles are designed to produce a spray pattern by either pressure or gravity flow.

*pH*. Measurement of the acidity or alkalinity of condenser water.

*Plume*. Water vapor and heated air mixture discharged from a cooling tower. (When condensation occurs, the water vapor becomes visible and this is sometimes considered objectionable.)

*Psychrometer*. A two-thermometer device for simultaneously measuring wet bulb and dry bulb temperatures of ambient air.

*Range.* Condenser water temperature increase (rise) in the condenser and decrease (drop) in the cooling tower (°F).

*Recirculation.* Air leaving the cooling tower that reenters the tower inlet, elevating the tower entering wet bulb temperature.

*Temperature (air).* Air temperature conditions that affect condenser water systems include:

Dry bulb temperature (°F) of the air upwind from the cooling tower, unaffected by recirculation or other cooling towers.

Web bulb temperature (°F) of the air upwind from the cooling tower, unaffected by recirculation or other cooling towers.

Wet bulb temperature (°F) of the air entering the tower, the product of the ambient air wet bulb temperature and any recirculation air wet bulb temperature. (Ideally, recirculation is zero and the entering wet bulb temperature is the ambient wet bulb temperature.)

*Wind load.* Structural load imposed by wind blowing on the tower casing (psf).

## COOLING TOWER HEAT TRANSFER

Heat is transferred from a water droplet to the surrounding air by both sensible and latent heat transfer processes. Figure 9.2 illustrates a typical water droplet and the heat transfer mechanisms.

This heat transfer process can generally be modeled using the Merkel equation:

$$\mathrm{T1} \rightarrow KaV/L = \frac{\int_{\mathrm{T2}} \mathrm{d}T}{(h_{\mathrm{w}} - h_{\mathrm{a}})}$$

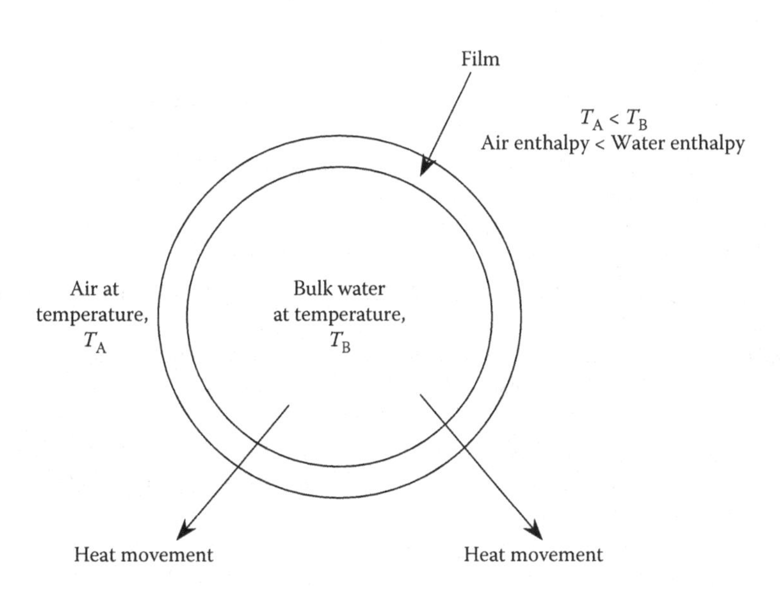


**FIGURE 9.2** Water droplet with surface film.

where

$KaV/L$ = cooling tower characteristic
$K$ = mass transfer coefficient (lb water/h ft$^2$)
$a$ = contact area/tower volume (1/ft)
$V$ = active cooling volume/plan area (ft)
$L$ = water mass flow rate (lb/h ft$^2$)
T1 = entering (hot) water temperature (°F)
T2 = leaving (cold) water temperature (°F)
$T$ = bulk water temperature (°F)
$h_w$ = enthalpy of air–water vapor mixture at bulk water temperature (Btu/lb of dry air)
$h_a$ = enthalpy of air–water vapor mixture at wet bulb temperature (Btu/lb of dry air)

The amount of heat removed from the water must be equal to the heat absorbed by the surrounding air as shown by the following equation:

$$L/G = \frac{(h_2 - h_1)}{(\mathrm{T1} - \mathrm{T2})}$$

where

$L/G$ = water-to-air mass flow ratio (lb of water/lb of air)
$h_1$ = enthalpy of air–water vapor mixture at inlet wet bulb temperature (Btu/lb of dry air)
$h_2$ = enthalpy of air–water vapor mixture at exhaust wet bulb temperature (Btu/lb of dry air)

While the tower characteristic $KaV/L$ can be calculated using numerical methods, it can also be represented graphically as shown in Figure 9.3. In this figure, the graphical variables are defined as follows:

C′ = entering air enthalpy at entering air wet bulb temperature $T_{wb}$ (Btu/lb of dry air)
BC = initial enthalpy driving force
CD = air operating line with slope $L/G$
DEF = projecting the leaving air point onto the water operating line and then onto the temperature axis yields the outlet wet bulb temperature (°F)
ABCD = the area within this region is the graphical solution to the cooling tower characteristic

For any given graphical solution, if the tower characteristic remains constant, an increase in imposed heat rejection load will have the following effects:

1. The length of the line CD will increase and shift to the right.
2. Both T1 and T2 will increase.
3. The range and the approach areas will increase.

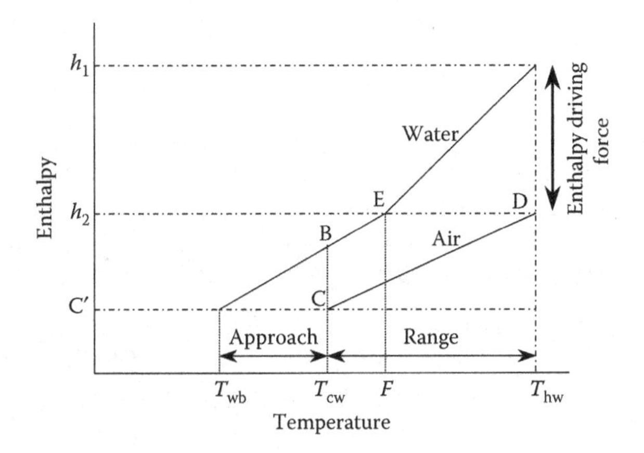


**FIGURE 9.3** Graphical representation of the cooling tower characteristic.

The Cooling Tower Institute (CTI) long ago published a compendium of graphical solutions to the equations given above for use by cooling tower designers and manufacturers. In recent years, computer software to aid tower design has become common and numerical methods have supplemented the CTI charts.

In HVAC applications, there are essentially no "custom" cooling towers (except in a few very special cases). Manufacturers have designed a range of cooling towers whose characteristics are well defined. For any HVAC application, it is only necessary to use the specified flow rate, range, and ambient wet bulb temperature to compute a required tower characteristic and then test that required characteristic against those available in the product line.

Manufacturers provide catalog data and selection software for the purpose of selecting a cooling tower for a specific application. *But, since most HVAC designers and the purchasers of towers are not expert in the use of a particular manufacturer's catalog or selection software, it is always prudent to have the tower manufacturer perform any required tower selection. Then, if an error is made, the manufacturer will have to make good on the tower, not the designer or buyer.*

The cooling tower's evaporative cooling process can be shown on a psychrometric chart as illustrated in Figure 9.4. The change in condition of a pound of dry air as it moves through the tower and contacts a pound of water (e.g., $L/G = 1$) is shown by the solid line. Ambient air, at 78°F DB and 50% RH, enters the tower at Point 1 and begins to absorb moisture in an effort to gain equilibrium with the water. This process continues until the air exits the tower at Point 2. During this process, the following conditions changed:

1. The air enthalpy increased from 30.1 to 45.1 Btu/lb of dry air, an increase of 15 Btu/lb of dry air. This heat came from the 1 lb of water, reducing its temperature by 15°F.
2. The specific humidity of the 1 lb of air increased from 0.0103 lb of moisture to 0.0233 lb of moisture. This increase of 0.013 lb of moisture

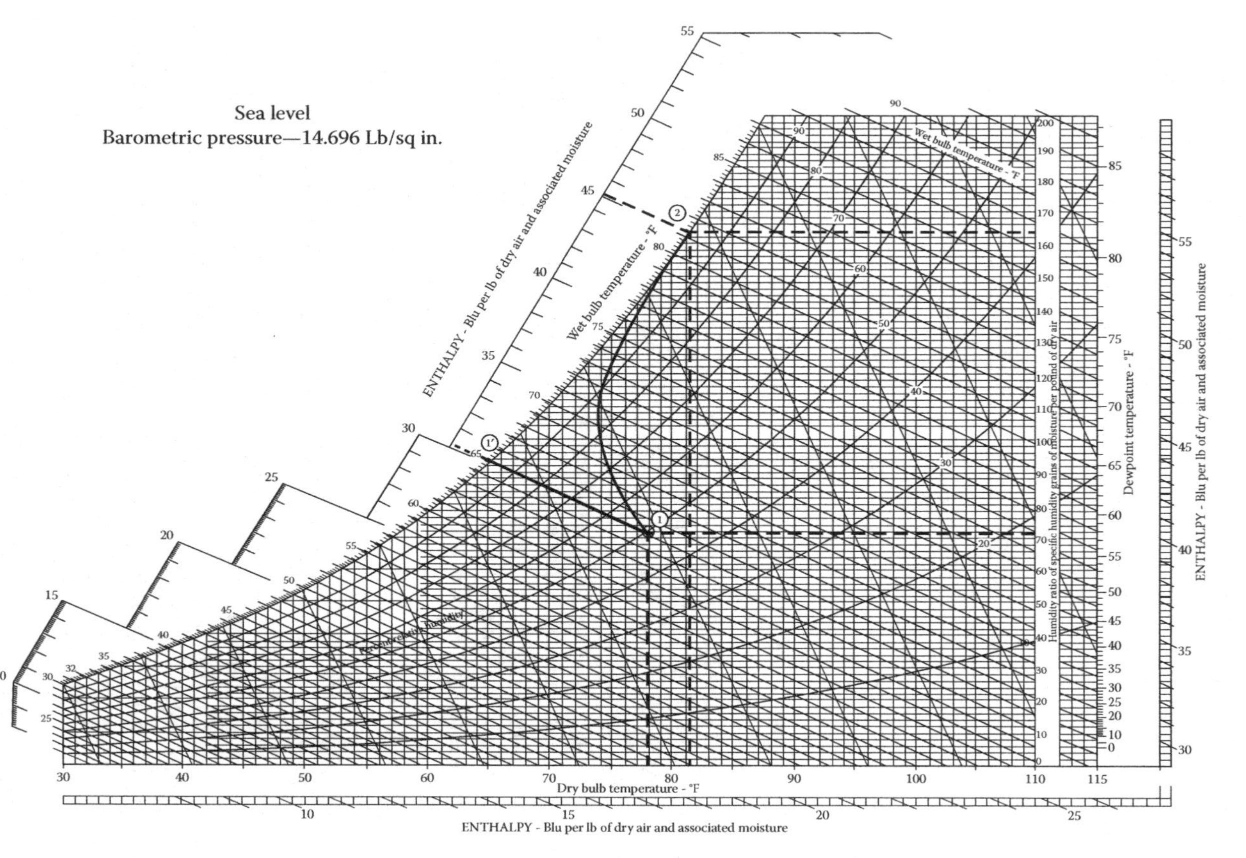


**FIGURE 9.4** Air–water temperature curve. (Courtesy of the Marley Cooling Tower Company, Overland Park, KS.)

represents the amount of condenser water that was evaporated. Using the latent heat of vaporization at 85°F, the heat transfer due to evaporation is 0.013 lb × 1045 Btu/lb or 13.6 Btu, 91% of the total 15 Btu of heat transferred to the air.
3. The condenser water temperature was reduced by 15°F, while the air temperature was increased by only 3.3°F (from 78°F to 81.3°F).

## COOLING TOWER PERFORMANCE FACTORS

The selection of a cooling tower for a specific set of required performance parameters (condenser water flow rate, selected range, and ambient wet bulb temperature) results in establishing a required cooling tower characteristic. But what is the impact if the required performance changes?

As shown in Figure 9.3, there are three factors that define the requirements for a specific cooling tower characteristic:

1. Entering (ambient) air wet bulb temperature
2. Condenser water flow rate
3. Approach

Once a tower is selected, that is, the cooling tower characteristic is established, changes in any of these three factors may necessitate a change in the cooling tower. Typically, the performance parameters that are subject to change include the following:

1. Increase of entering wet bulb temperature. Too often, the wrong ambient wet bulb temperature is selected for tower sizing.
2. Increase of rejected heat load. This may dictate increasing the condenser water flow rate and/or increasing the range.

A change in wet bulb temperature and/or a change in range will result in a change in approach. Therefore, there are three performance impact relationships as shown in Figures 9.5 through 9.7.

Figure 9.5 illustrates the impact of changing the tower approach by changing the design wet bulb temperature and/or changing the condenser water range. Obviously, a relatively small increase in required approach will require a much higher cooling tower characteristic.

Figure 9.6 illustrates the impact of changing the tower load by changing the water flow rate, while Figure 9.7 illustrates the impact of changing tower load by changing condenser water range (without impacting approach). These two variables have a smaller impact on the cooling tower characteristic.

## BASIC COOLING TOWER CONFIGURATION

Chapter 11 describes the various configurations of HVAC cooling towers and the particular advantages and disadvantages associated with each. However, basic

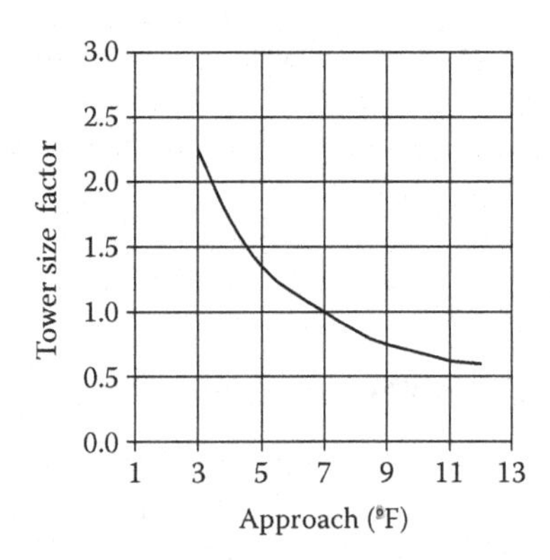


**FIGURE 9.5** Variation in tower size factor with approach.

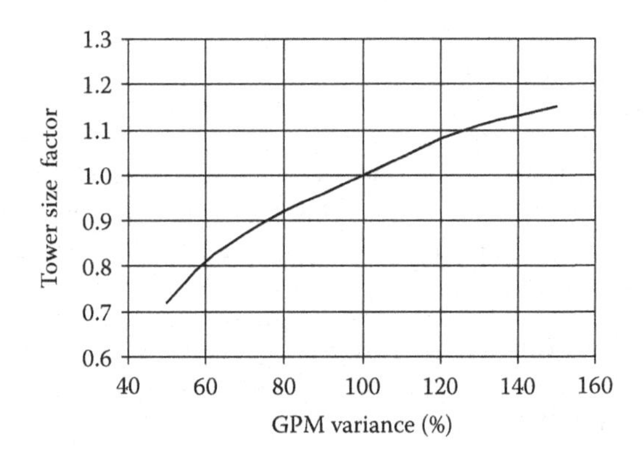


**FIGURE 9.6** Variation in tower size factor with condenser water flow rate.

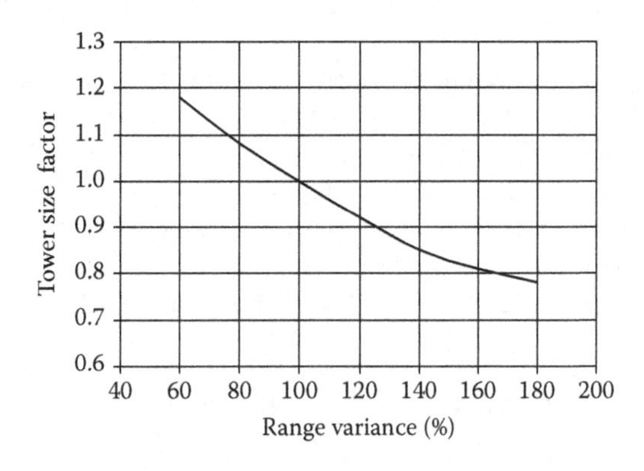


**FIGURE 9.7** Variation in tower size factor with range.

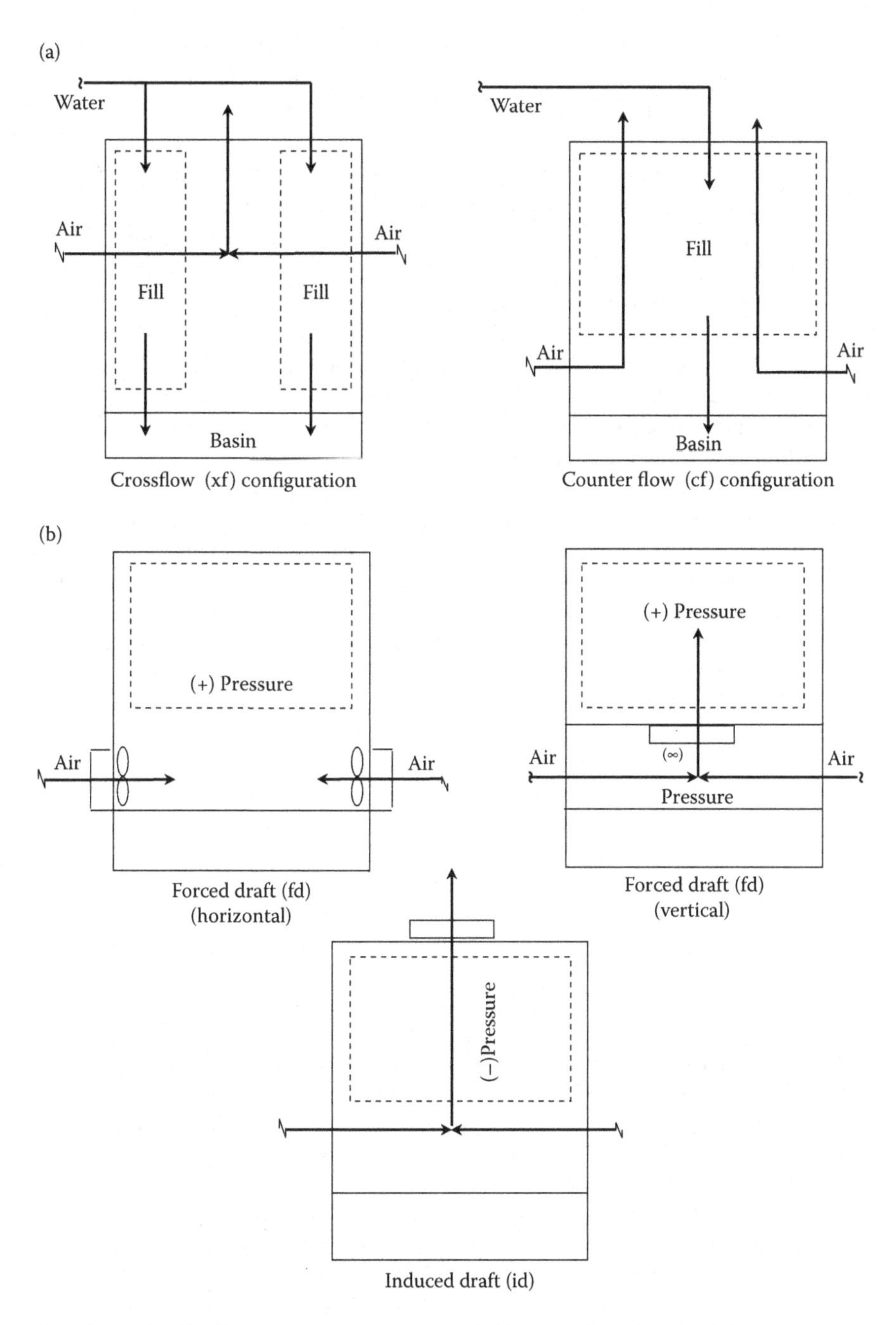


**FIGURE 9.8** Cooling tower configurations. (a) Air/water flow. (b) Fan(s) location.

tower configurations are very simple, as shown in Figure 9.8, and are dictated by (a) the direction of the air versus water flow through the tower fill, and (b) the location of the tower fan(s):

*Air/water flow*. In *counterflow* towers, the water and air flow in opposite directions, the water flows vertically downward and the air flows vertically upward. In *crossflow* towers, the two flow streams are arranged at 90° to each other, that is, the water flows vertically downward through the fill, while the air flows horizontally through it.

*Fan location*. The tower is defined as *forced draft* when its fan(s) is arranged to blow air through the tower. In this configuration, the fan(s) are located on the entering-air side of the tower and the fill is under positive pressure. In *induced draft* towers, the fan(s) is located on the leaving-air side of the tower and the fill is under negative pressure.

For many years, one manufacturer (Baltimore AirCoil, Inc.) offered an "induction draft" or *Venturi* cooling tower, as shown in Figure 9.9, for the HVAC market. In lieu of a fan, the return condenser water was sprayed horizontally through high-velocity nozzles into a Venturi chamber, much like the throat of an automobile carburetor. The high-velocity water flow through the Venturi created a low pressure region that (1) induced airflow through the tower, and (2) tended to vaporize the water. Since airflow and water flow were in the same direction, this tower

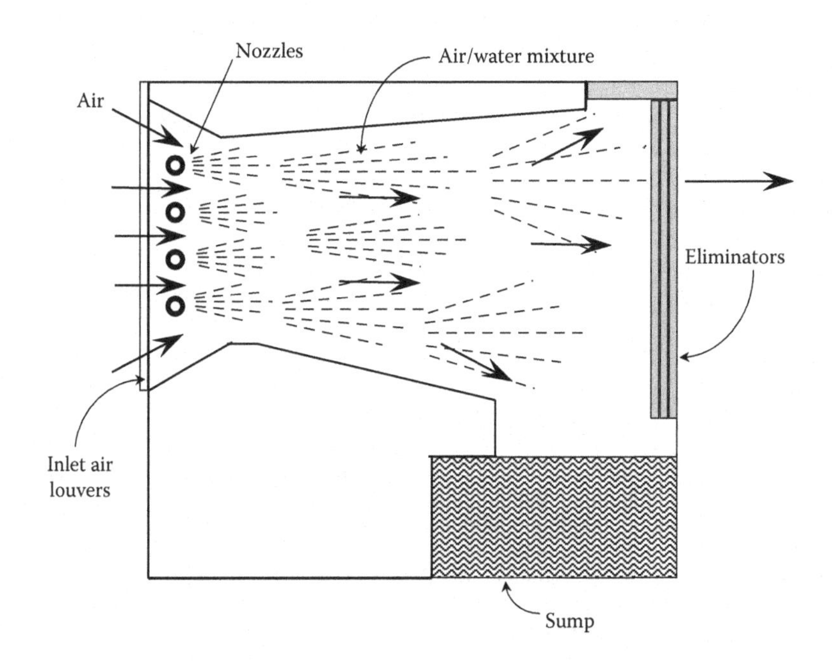


**FIGURE 9.9** Induction draft or Venturi cooling tower configuration.

would be classed as a "parallel" flow type. Since there is no fill, this was a "spray fill" tower. The high water pressure required by the tower significantly increased the pump horsepower requirement, more or less offsetting the fan energy savings. This problem was mitigated by using a two-speed pump motor and operating the tower at low speed when the outside ambient wet bulb temperature was low. This tower suffered from both performance and maintenance problems and is no longer available except by special order.

# 10 Cooling Tower Components

As described in Chapter 9, every HVAC cooling tower has six functional components: (1) fill, (2) wet deck, (3) basin, (4) fan(s), (5) inlet louvers, and (6) drift eliminators. To this list we must also add the two structural elements: the structural frame and the casing.

## FILL

The function of the tower fill is to provide a large "contact area" between the water flow and the airflow to promote evaporation and heat transfer.

### SPRAY FILL

Very early, it was recognized that by breaking the water flow into droplets the contact area between the water flow and airflow increased due to the increased water surface area exposed to the air. Thus, the use of water sprays represents the earliest concept used to improve the efficiency of evaporation and heat transfer.

In the mid-1800s, spray ponds were used to provide condensing for the ice-making refrigeration systems then in use. These ponds, however, had relatively low efficiency and high drift losses, which motivated a search for a better method. Then, in 1898, George Stocker is credited with building the first "cooling tower." This "atmospheric" tower consisted simply of wooden, louvered walls built around a spray pond and the enclosure significantly improved performance and reduced drift losses.

Over time, improvements such as draft chimneys and, then in the 1920s, the use of mechanical draft fans significantly improved airflow rates, decreasing the *L*/*G* ratios and improving performance.

Spray fill really is not "fill" at all. The water is sprayed through nozzles to create fine (small) droplet size. The spray, then, is contained by the tower casing (walls), creating a water-filled plenum through which the airflow is directed. Spray density ranges from 3 to 7 gpm/cf of tower volume.

The overall efficiency of spray fill is relatively low, requiring large towers with high airflow rates compared to other types of fill. Thus, spray fill is not used in HVAC cooling towers except for Venturi towers, which are no longer available except by special order.

## Splash Fill

The next step in improving cooling tower efficiency was made by introducing splash fill in the 1940s. Since most cooling towers of that era were constructed of wood, splash fill was fabricated from wooden slats that were then stacked in the tower air plenum to create a "cascading" path for the water as it fell by gravity through the tower. Fill slats were either flat or triangular shaped, which tended to reduce the airflow pressure drop, as shown in Figure 10.1.

Wooden splash fill remained in use until the 1980s, but modern splash fill is typically constructed of vacuum-formed PVC slats or V-shaped bars that are

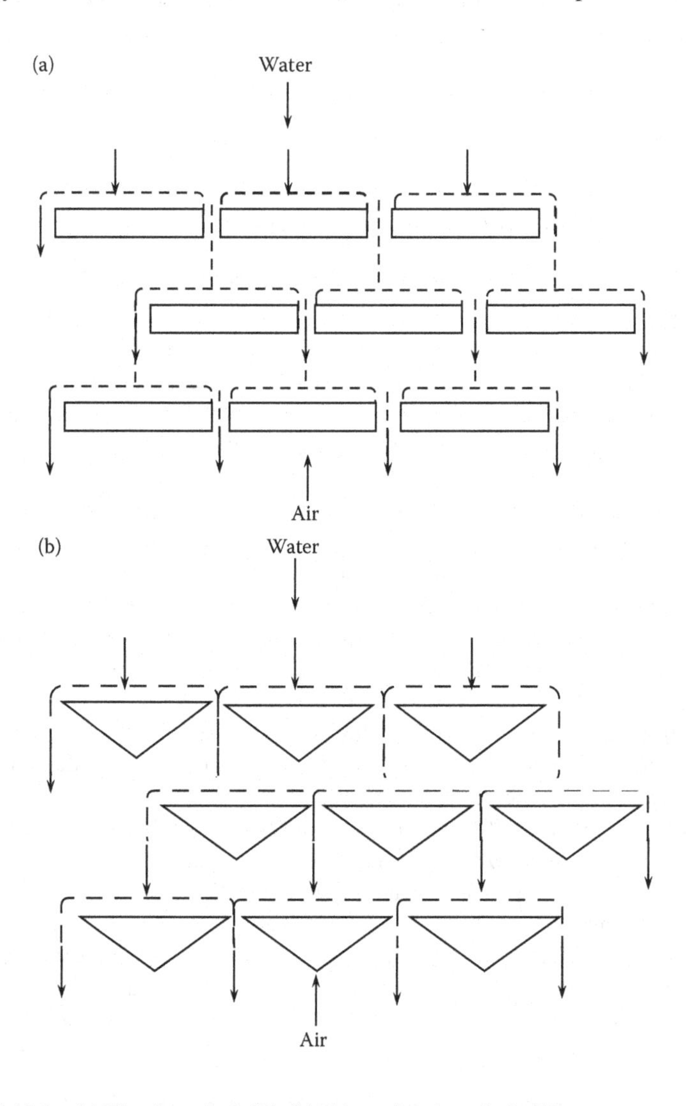


**FIGURE 10.1** (a) Flat slat splash fill. (b) Triangular slat splash fill.

designed to provide good droplet formation and eliminate the rot problem common to wood in wet environments.

Ceramic fill is an alternative splash fill material sometimes used in HVAC applications. Formed from fired, ceramic shapes, this fill is stacked to create a splash fill form. Ceramic fill is virtually impervious to rot or corrosion, but is very expensive compared to other types of fill.

## Film Fill

Film fill was introduced in the 1960s and created a new concept in cooling tower design. Instead of breaking the water flow into fine droplets, film fill provides large surface areas over which the water flows. With the large areas and slower "falling" speeds than in splash fill, water forms a thin sheet or "film" as it flows across the fill surfaces, creating large air–water contact area and efficient evaporative heat transfer.

Film fill is typically constructed of thin sheets of vacuum-formed PVC, stacked in vertical layers and formed with corrugations to ensure even water distribution over the entire fill surface. This configuration has low resistance to airflow; the air pressure drop through the tower is low. Typical film fill is shown in Figure 10.2.

Film fill is both less costly and more efficient than splash fill. The higher efficiency results in a significantly smaller tower (higher tower characteristic) for an equivalent capacity. Thus, most HVAC towers applied today utilize film fill.

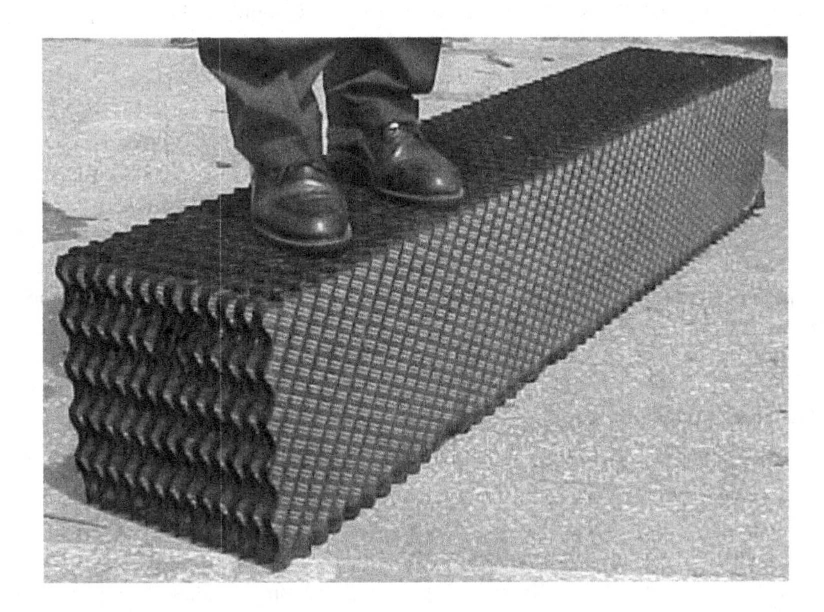

**FIGURE 10.2** Typical cooling tower film fill. (Courtesy of Evapco, Inc., Westminster, MD.)

The disadvantage of film fill is that the large surface area and correspondingly small water passages are sometimes susceptible to (1) clogging by debris buildup (from leaves, dirty water, etc.), and (2) microbiological fouling, since mechanical cleaning cannot be done unless the fill is removed from the tower and disassembled (an expensive and a time-consuming job).

## STRUCTURAL FRAME

A structural framing system is required to support the wet deck, fan(s), fill, intake louvers, and drift eliminators. Casing is required to enclose the fill and to create the tower air and water flow path.

### WOODEN STRUCTURE

Early cooling towers were constructed with wooden structural frames, and redwood, because of its natural resistance to rot under wet conditions, was the wood of choice. However, by the 1960s, the increased cost and reduced availability of redwood forced the industry to shift to West Coast Douglas fir.

Fir, however, has almost no natural resistance to wet rot and must be chemically treated for use in cooling towers. The use of wood preservatives has significant undesirable environmental side effects, and so, today, only under very special circumstances are wooden towers considered for HVAC applications, though they remain in common use for large industrial applications. If fir is to be used, it should be specified to meet the requirements of CTI Standard 114 and all fasteners should meet the requirements of CTI Standard 119.

One of the drawbacks of a wooden structure cooling tower is that most property insurance companies will require that it be protected by a fire sprinkler system installed in accordance with the National Fire Protection Association (NFPA) Standard 214 (see Chapter 14). While it may seem odd that a cooling tower needs fire protection, wood is highly combustible and fires in cooling towers that are out of service have been a relatively common occurrence. For this reason, and the fact that wooden towers in smaller sizes are no longer cost competitive, wooden structure cooling towers are rarely used for HVAC applications today.

### STEEL STRUCTURE

Since the 1950s, steel has become the structural framing system of choice for HVAC cooling towers. To reduce the effects of corrosion in a wet tower environment, two types of steel are used:

*Galvanized steel.* Mild carbon steel is coated with zinc, lead, and aluminum in a process called *galvanizing* to provide a corrosion-resistant barrier between the underlying steel and the air/water on the surface.

In the galvanizing process, the steel to be coated is first cleaned in a caustic bath followed by pickling in a dilute hydrochloric or sulfuric acid bath. The steel is then subjected to a fluxing operation with a zinc ammonium chloride solution.

Early galvanizing used the "wet kettle" method, in which a floating layer of zinc ammonium chloride flux was placed on the surface of molten zinc in a large open kettle. A section of sheet steel was then immersed in the molten zinc, passing through the layer of flux and remaining in the molten bath until the temperature of the steel was the same as the molten zinc. The zinc bath was then swept free of the remaining flux and the sheet removed. The sheet was then immersed in a water quench bath to cool it for handling. The quench water was usually treated with sodium dichromate to provide further protection.

Since the 1980s, galvanizing has typically been done by the "dry kettle" method. The steel sheet, now in continuous rolls, is first immersed in a kettle of zinc ammonium chloride flux and then removed and allowed to dry. The fluxed steel sheet is then dipped into the molten zinc bath. It is removed after heating and sent through a water quench bath.

While the dry kettle method has been found to provide a higher-quality surface finish, using a much cleaner process, there remains a significant argument in the industry relative to the quality of the resulting galvanizing (see the section "White Rust" in Chapter 13). Therefore, most cooling tower manufacturers and HVAC designers recommend that a protective coating of epoxy or polymer finish be added to all galvanized structural members.

One important note: sometimes literature will mention "hot dip galvanizing after fabrication" as a method of ensuring good corrosion protection. However, hot dip galvanizing after fabrication has not been available from U.S. tower manufacturers since the late 1970s.

The thickness of the galvanizing film is rated in terms of "ounces of zinc per square foot of metal surface." Typical cooling tower galvanizing is 2.35 oz/sf (called G235), but galvanizing as thick as 7.0 oz/sf (G700) is available from some manufacturers.

Galvanized steel structural frames are typically assembled with galvanized fasteners, although some tower manufacturers will provide an option for stainless steel. Galvanized steel is never welded since welding (1) destroys the galvanizing, and (2) establishes a point for differential corrosion to start (see Chapter 13).

*Stainless steel.* Stainless steels generally have lower structural strength than mild carbon steel but are far more resistant to corrosion in wet environments. Stainless steel consists of steel alloys (carbon steel plus molybdenum, nickel, and/or nitrogen) and chromium. The corrosion resistance credited to stainless steels results from an extremely thin layer of chromium oxide that forms on the steel surface exposed to air. Over time, if the layer is damaged, the underlying chromium (minimum 10.5% of the alloy) is reexposed and the oxide layer reforms.

Almost any galvanized steel cooling tower can be fabricated using stainless steel and most tower manufacturers offer this option.

One advantage of stainless-steel members is they can be welded and the corrosion integrity maintained. Where structural fasteners are required, they are also stainless steel. (Note, however, that stainless nuts and bolts tend to "creep" with stress and age, and thus are difficult to remove and often cannot be reused.)

### Other Structural Systems

In the never-ending search to improve HVAC cooling tower corrosion resistance and extend tower life, other materials are used for structural systems:

*Concrete.* Concrete has been used as combined structure and casing for large hyperbolic utility cooling towers for many years. However, the material is not widely used in HVAC applications due to its relatively high cost. One manufacturer does offer a line of factory-fabricated, field-erected concrete towers, but they have not had large sales.

Concrete, however, can become a desirable material for HVAC cooling towers when it is necessary to "blend" the tower into a larger architectural scheme. Using concrete, the tower can be constructed to almost any configuration to hide its utilitarian function.

*Fiberglass.* Fiberglass structural tower frames, joined with stainless-steel fasteners, are offered by a few tower manufacturers for smaller HVAC cooling towers. These towers utilize structural framing members constructed of 1/4″–1/2″ thick fiberglass.

*Stressed skin fiberglass/stainless steel.* In the 1980s, the "bottle" cooling tower, developed around the Pacific Rim, came to the United States. These towers were designed for field assembly of two sets of curved fiberglass panels, much like "pie slices," that were bolted together atop a simple, tubular stainless-steel space frame. One set of panels formed the basin of the tower, while the other set formed the enclosure/air plenum. Once assembled, the panels acted together to form a uniform stressed skin structural system to support the fill, the wet deck, and the fan. Unfortunately, these towers suffered from severe performance and reliability problems and have generally been replaced.

## CASING

A cooling tower's casing performs two roles. First, it forms an enclosure around the fill to create a contained air path or *plenum*, forcing the airflow through the fill. Second, it simply helps to keep the water inside the tower.

Wood-framed cooling towers originally used cedar or redwood planks as tower casing. However, to reduce expense, cement–asbestos panels came into wide use in the 1950s, only to be replaced by fiberglass or UV-inhibited plastic panels in the 1980s as the environmental problems with asbestos were identified.

Galvanized steel-framed cooling towers are typically encased by galvanized steel, fiberglass, or UV-inhibited plastic panels. If galvanized steel is used for the casing, it too should be coated with an epoxy or polymer final finish for corrosion protection. Stainless-steel- and fiberglass-framed cooling towers are normally encased by stainless-steel, fiberglass, or UV-inhibited plastic panels.

## WET DECKS/WATER DISTRIBUTION

The wet deck is located at the top of the tower and its job is to distribute the incoming warm condenser water as evenly as possible over the fill to ensure uniform heat transfer.

In crossflow towers, the wet deck is not in the air stream and normally consists of shallow basins or reservoirs with evenly distributed bottom "holes" consisting of plastic, metering orifices. Warm condenser water is dumped into the basin from the return piping so as to maintain a fairly uniform depth (4″–6″). Thus, the gravity water flow rate through each bottom outlet will be uniform and the flow rate over each section of fill will also be the same.

Since the wet deck on a crossflow tower is open to the atmosphere, removable basin covers must be installed to keep leaves and other wind-blown debris out of them.

The "wet deck" in counterflow towers is in the air stream. The wet deck consists of spray nozzles that are used both to evenly distribute the water flow over the fill and to provide initial water atomization (i.e., act as spray fill).

## BASINS

The basin is located at the bottom of the tower and its job is to collect the cold condenser water for supply to the condenser water pump. Most commonly, basins are provided as an integral part of the cooling tower and are typically 12″–18″ deep. Operating water depths are usually less than 12″.

In cold climates, where tower freezing is a problem, or when it is necessary to have a basin with more water volume capacity than normal, a remote basin (sometimes called a "sump") may be constructed, usually of concrete. Separate sumps are sometimes used to allow the use of vertical turbine pumps (see Chapter 11).

There are three critical aspects relative to the basin:

1. It must have sufficient capacity to contain the water that will flow to it when the condenser water pump shuts down.
2. The volume of water stored in the basin must be large enough to fill all empty piping upon start-up. Otherwise, the basin may be pumped dry.
3. There must be a sufficient height of water above the condenser water supply connection to prevent *vortexing* and the entrainment of air into the water flow. Typically, to avoid vortexing, a side-outlet sump is provided

for piping connection to the basin, ensuring that the piping is fully "flooded" and with a 90° direction change in flow to create turbulence that breaks up vortex flow.

See Chapter 11 for further discussion on these aspects.

## INTAKE LOUVERS AND DRIFT ELIMINATORS

Intake louvers are provided on all crossflow cooling towers to help control the airflow over the fill. The louvers are spaced and slanted to direct air evenly into the fill pack.

Drift eliminators are installed on the leaving side of the fill designed to "trap" entrained water droplets and prevent them from leaving the cooling tower. Fabricated of PVC or steel, these devices force the air to make a three or four high-velocity directional changes (about 90° each). Water droplets, then, will impinge on the eliminators and drain back into the fill.

While older towers may have a drift rate of 0.02%, modern (since about 1990), induced draft, crossflow cooling towers will have drift rates of 0.0001–0.005% of the condenser water flow rate. Forced draft towers, since about 2005, are also capable of these same low drift rates.

## FANS, MOTORS, AND DRIVES

One or more fans, connected to one or more motors via a drive assembly, provide the motive power for airflow through a mechanical draft HVAC cooling tower.

### FANS

Two types of fans are used, centrifugal fans and axial propeller fans. Both types may be applied forced draft towers, but only propeller fans are used with induced draft towers. Both types produce airflow through the tower by increasing the static pressure and kinetic energy of the air.

With the centrifugal fan, air enters through the central portion of an impeller wheel and exits through a scroll and outlet at right angle to the inlet paths. Axial flow fans, by contrast, have a straight-through airflow path. Figure 10.3 illustrates these two fan configurations.

Fan performance characteristics are plotted for a particular rotational speed in *revolutions per minute* (RPM), with airflow rate as the independent variable and power, static pressure, and efficiency as dependent variables. Figure 10.4 is a typical centrifugal fan curve. Since the fan curve is plotted for one fixed fan speed, airflow changes are affected only by changes in the system resistance.

For fans with *backward-inclined blades*, the mechanical efficiency curves peak at a higher efficiency and at higher fractions of maximum airflow rates than do fans with *forward-curved blades*. Also, the horsepower requirement continues to increase through full-load power as the flow rate is increased in a forward-curved

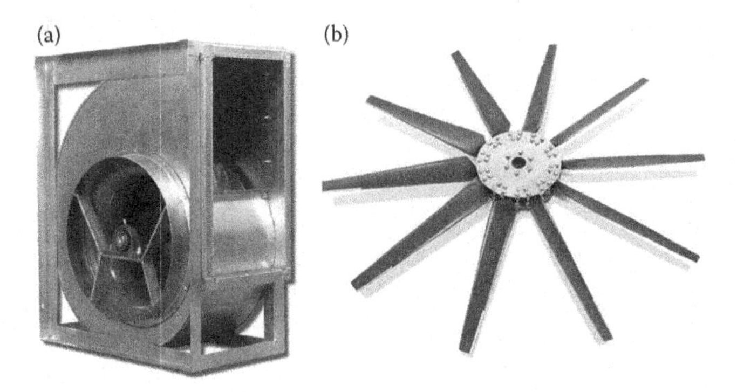


**FIGURE 10.3** (a) Typical centrifugal fan. (b) Typical axial propeller fan.

fan, whereas in the backward-inclined fan the horsepower requirement peaks at between 60% and 80% of maximum airflow capacity. Therefore, a forward-curved fan can overload its motor if allowed to operate without restriction, while the backward-curved fan's power requirement is self-limiting.

Unfortunately, most centrifugal fans used in cooling towers are the forward-curved type since (1) the fans are relatively small, and (2) forward-curved fans are far less expensive to manufacture.

In Figure 10.4, the shaded area represents the normal selection range for a centrifugal fan, ideally at the highest point of the fan efficiency to minimize fan losses and horsepower requirement for the given airflow conditions.

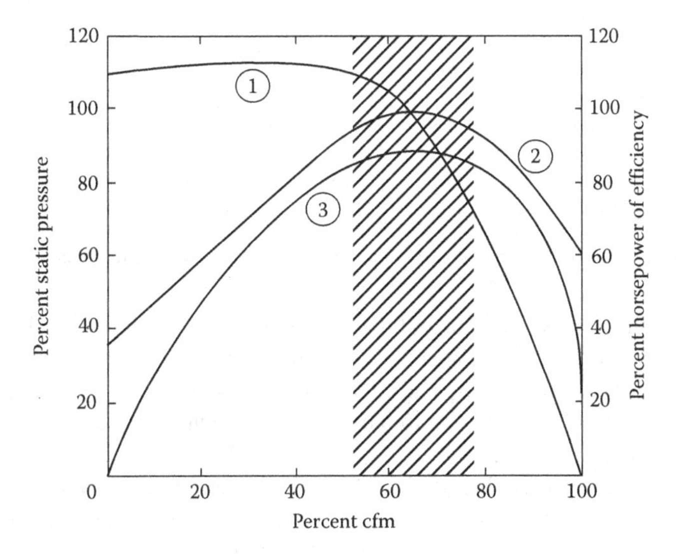


**FIGURE 10.4** Typical performance characteristics of fans: (1) CFM/static pressure curve; (2) brake horsepower curve; and (3) mechanical efficiency curve.

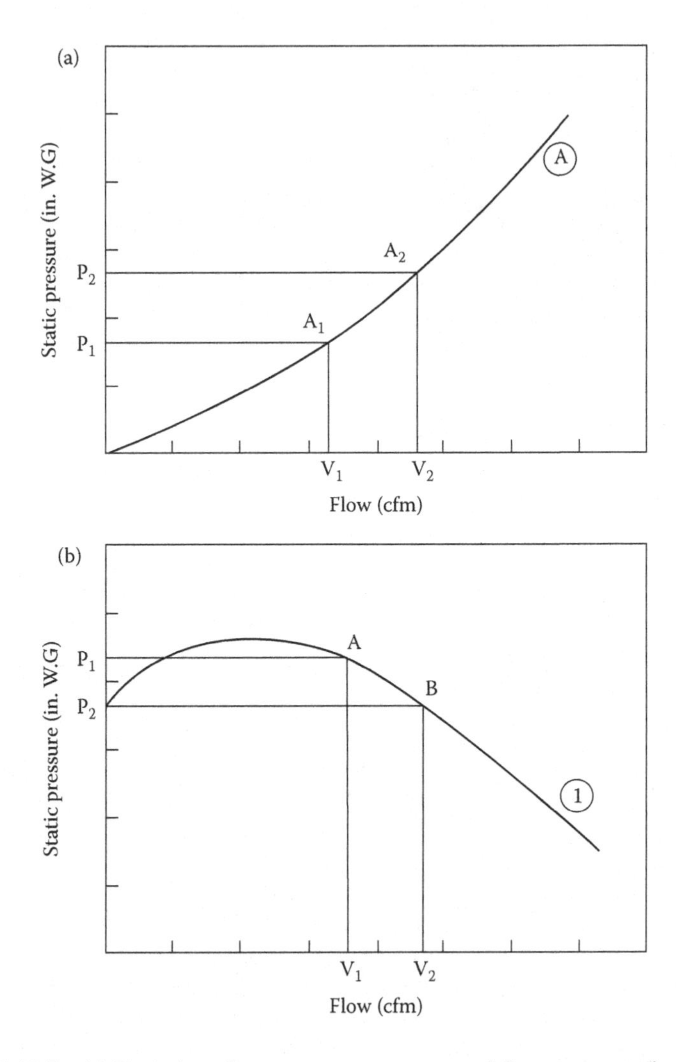


**FIGURE 10.5** (a) Typical cooling tower system curve: airflow versus static pressure. (b) Typical cooling tower fan performance curve: airflow versus static pressure.

Relationships between a fan and the system it serves are illustrated in Figures 10.5 through 10.7. First, consider the static pressure required to force a certain volumetric flow of air through a cooling tower. As the volumetric flow rate increases, the static pressure required also increases, but increases approximately as the *square* of the increased flow.

Figure 10.6 illustrates changes in the cooling tower resistance and fan characteristics. Anytime the resistance changes (due to dirty intake louvers, clogged fill, etc.), a new system characteristic curve is generated. Similarly, if the fan speed is changed, a new fan curve is generated. When a particular fan operating at a particular speed is installed in a particular cooling tower, the only stable operating

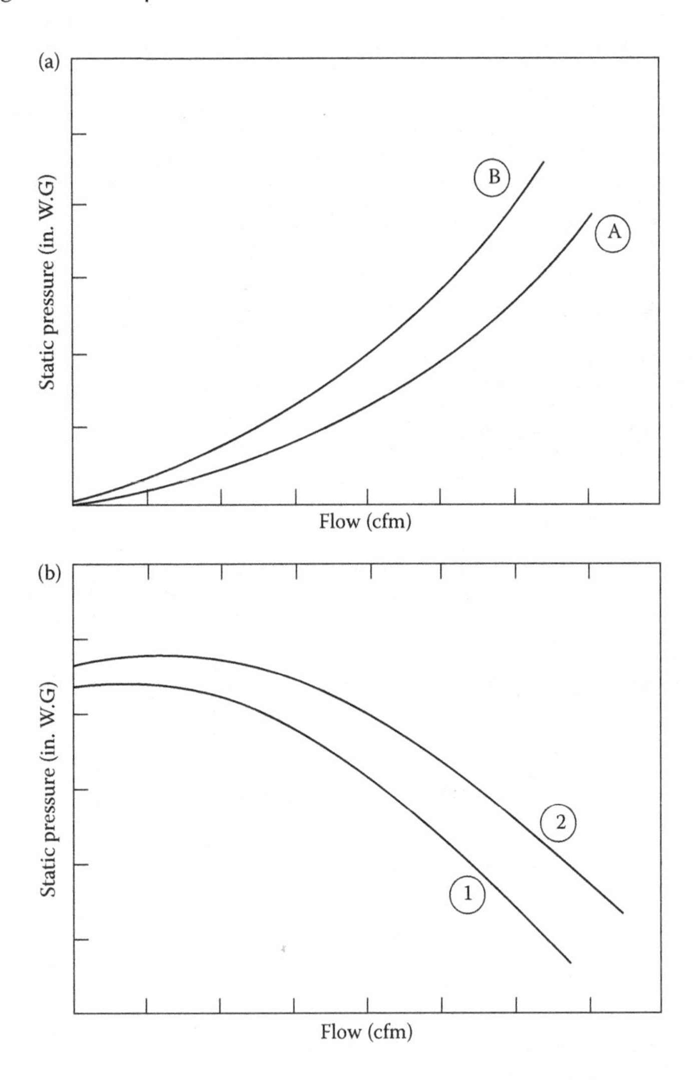


**FIGURE 10.6** (a) Change in tower system curve with increased resistance, Curve A to Curve B. (b) Change in fan performance curve with increased speed, Curve 1 to Curve 2.

point is the one that satisfies both the fan and the system characteristic curves, that is, at the intersection of the two curves.

Approximate relationships for assessing the effects of changes in fan speed, airflow rate, and fan power are known as the *fan laws*:

1. *Volumetric airflow rate (CFM) is directly proportional to fan speed (RPM).*
2. *Static pressure (SP) is proportional to the square of the fan speed.*
3. *Brake horsepower (BHP) is proportional to the cube of the fan speed.*

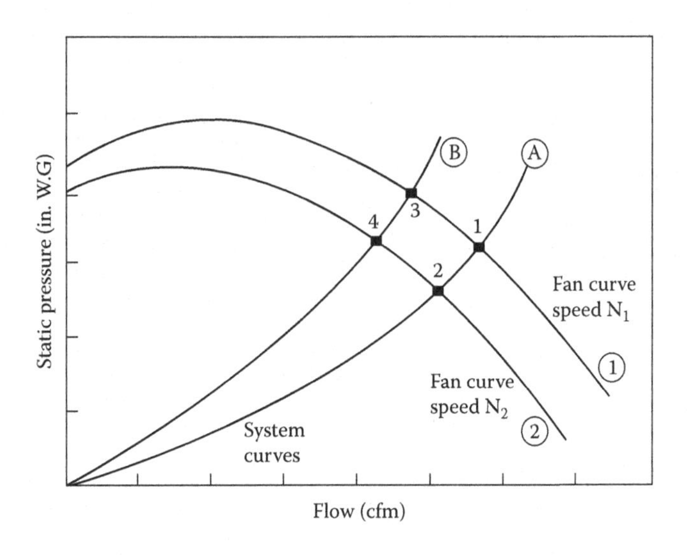


**FIGURE 10.7** Fan and tower curves imposed to show the determination of the operating point.

The mathematical expressions for the fan laws for changing conditions from state "1" and state "2" are shown by the following relationships:

$$RPM2 = RPM1\left(\frac{CFM2}{CFM1}\right)$$

$$SP2 = SP1\left(\frac{CFM2}{CFM1}\right)^2$$

$$BHP2 = BHP1\left(\frac{CFM2}{CFM1}\right)^3$$

At any airflow and pressure condition, the theoretical fan brake horsepower can be computed as follows:

$$BHP = CFM \times \left\{0.0003 + \left[\frac{SP}{6346 \times EFF}\right]\right\}$$

where

BHP = total theoretical brake horsepower
CFM = total airflow through the tower
SP = static pressure drop through the tower (wg)
EFF = fan efficiency (typical):
Backward-inclined centrifugal 0.70

Forward-curved centrifugal 0.60
Propeller 0.70

Cooling towers may be equipped with a single fan or with multiple fans. If multiple fans are provided, the tower design must be such that it can continue to operate when one or more fans are off. The most common approach to this requirement is to provide internal baffles to form individual fan plenums and prevent air discharged by one fan from entering the discharge plenum of an adjacent fan. Another approach is to use backdraft dampers on each fan discharge to prevent reverse airflow through the fan when it is off.

## Motors

There are three basic types of motors used in HVAC applications:

*Open drip proof (ODP).* ODP motors are constructed with openings in the motor casing that provide a cooling air path to the motor windings. ODP motors are designed for dry applications and, if installed outdoors, must be weather protected.

*Totally enclosed fan-cooled (TEFC).* The TEFC motor shaft extends from the motor casing to drive a cooling fan installed within a shroud on the nondrive end of the motor.

*Totally enclosed air-over (TEAO).* TEAO motor windings are enclosed in a moisture-tight enclosure. The casing is provided with large fins designed to radiate the motor heat to an air stream passing over them.

ODP motors are never used in cooling towers simply because they are not designed for outdoor, wet locations and will quickly fail in this environment. TEFC motors are typically applied in cooling towers where the motor is located outside of the tower airflow path. TEAO motors are used where the motor is located within the tower airflow path. Both TEFC and TEAO motors have the advantage of better cooling than ODP motors and can generally be operated continuously at up to 15% overload without failure, while an ODP motor overheats and fails under almost *any* overload condition.

Using an "energy-efficient" rated motor in lieu of a "standard" motor will reduce fan energy use by 4.6–6.8%, depending on motor size, and given the small price premium is an excellent choice. A "premium efficiency" rated motor can reduce energy use by an additional 1–3%, depending on motor size, but the cost increase for these motors is rarely justified for cooling tower applications.

Because cooling tower fans must cycle off and on for capacity control (see Chapter 12), it is important that tower motors be selected for frequent start/stops. Table 10.1 provides motor ratings for intermittent duty. In Table 10.1, "A" represents the maximum number of starts per hour; "B" is the product of the maximum number of starts per hour and the motor load $Wk^2$; and "C" is the minimum rest

**TABLE 10.1**
**Approximate Motor Allowable Starts and Starting Intervals (NEMA Design A and B Motors, 1800 rpm, 60 Hz)**

| Motor HP | A | B | C |
|---|---|---|---|
| 1 | 30 | 6 | 38 |
| $1\frac{1}{2}$ | 26 | 9 | 38 |
| 2 | 23 | 11 | 39 |
| 3 | 20 | 17 | 40 |
| 5 | 16 | 27 | 42 |
| $7\frac{1}{2}$ | 14 | 39 | 44 |
| 10 | 12 | 51 | 46 |
| 15 | 11 | 75 | 50 |
| 20 | 10 | 99 | 55 |
| 25 | 9 | 122 | 58 |
| 30 | 8 | 144 | 60 |
| 40 | 7 | 189 | 65 |
| 50 | 7 | 232 | 72 |
| 60 | 6 | 275 | 85 |
| 75 | 6 | 338 | 90 |
| 100 | 5 | 441 | 110 |

period or *off time*, in seconds, between starts. *Thus, the maximum allowable number of starts per hour is the lesser of "A" or "B" divided by the load* $Wk^2$.

To reduce the wear and tear of fan cycling capacity control, two-speed motors are often used on belt-drive cooling towers. These motors are available in either single-winding or dual-winding configuration.

*Single-winding motors* utilize their entire winding whether on high speed or low speed, but the number of active poles is reduced from 8 at low speed to 4 at high speed. Since the motor speed is a function of the number of poles, low speed is always exactly half of the high speed (typically 1800/900 rpm).

*Dual-winding motors* have two sets of windings, one that is used on high speed and the other on low speed. The advantage of this configuration is that the ratio between high and low speed can be something other than 2:1.

As an alternative to a single two-speed motor, the tower can be equipped with two separate motors, one to operate at high speed and the second to operate at low speed, again at a 2:1 speed ratio. The low-speed second motor, often called a "pony" motor, will (based on the fan laws) be about 1/8th the horsepower of the high-speed motor. During tower operation, when operating at high speed, the pony motor will "free wheel." Then, when switched to low speed, the pony motor drives the fan(s) and the larger high-speed motor will free wheel. (Contrary to some engineer's thinking, the idling motor does not act as a generator, simply because there is no excitation current available to it. There are, however, additional motor windage and drive losses that must be carried by the operating motor.)

When variable frequency speed drives are applied to cooling tower capacity control (see below and Chapter 3), since the output of a VFD is not a true sine wave, the fan motor must be rated for the potential of increased heat buildup.

## Mechanical Drives

While small centrifugal fans may be designed to operate at motor speeds (1750–1800 rpm) and be directly connected to a motor, most cooling tower fans are designed to operate at slower speeds since the static pressure losses through towers are relative low (1″ water guage (wg) or less) and high-speed fan operation is simply not needed. Two types of mechanical drives are applied to HVAC cooling towers:

*Belt drives.* Belt drives, belt and pulley assemblies, are used in a wide range of HVAC cooling tower configurations. These drives perform two functions: they transfer the motor power to the fans while reducing the motor speed to the required fan speed. The power transfer dictates the belt(s) load rating, while the sheaves (one on the motor and one of the fan) dictate the speed change in direct proportion to their diameters (a 5″ motor sheave and a 10″ fan sheave will reduce the speed by half).

Historically, V-belt drives, with multiple belts selected to meet the required load rating, have been applied. Efficiency of power transfer can be improved by about 3% if "cogged" V-belts are utilized. However, because of higher maintenance requirements due to alignment and stretch problems with any type of V-belt and to improve power transfer even more, many cooling towers today utilize the single "power" belt. This single belt is much wider than the V-belt for the same load and may have teeth or ridges that interlock with its sheaves to prevent slippage (and "squeal") on start-up.

*Gear drives.* Gear drives (or gear "boxes"), illustrated in Figure 10.8, are used by some manufacturers on larger towers. Here, a shaft connects the motor to a gearbox that performs three functions: it transfers drive power from the motor to the fan while changing the rotational direction by 90° (from horizontal to vertical) and reducing the motor speed to the required fan speed.

Gear drives are typically applied only to larger towers with 50-hp or larger fan motors. These drives typically use spiral bevel and helical gear sets and, depending on the speed reduction ratio and the motor horsepower, may have a single-stage or multistage gear set.

The service life of a gear drive is directly related to the "surface durability" or "wear hardness" of the gears and the number of hours the drive is operated. The American Gear Manufacturers Association (AGMA) has published Standard 420 to establish requirements for geared speed

**FIGURE 10.8** Typical cooling tower gear drive assembly. (Courtesy of The Marley Cooling Tower Company, Overland Park, KS.)

reducers. Cooling tower gear drives should be selected in accordance with AGMA Standard 420. CTI has published Standard 111 that recommends "service factors" specifically for cooling towers that should be applied to gear drives.

High-quality bearings are critical to gear drive life and bearings should be selected on the basis of a 100,000-h L-10 life. (L-10 life is defined as the life expectancy, in hours, during which 90% or more of the bearings will still be in service.) The incremental cost for this bearing life is small and, for HVAC cooling towers, generally ensures that bearings will need replacement only every 15–25 years.

The gear drive lubrication system must be designed to provide reliable lubrication in both forward and reverse operation of the drive.

*Direct-drive*: An alternative to gear- or belt-drive-induced draft cooling tower fan drives is the *direct-drive variable speed motor and controller* that has been introduced by Baldor Electric Co. This TEAO motor is designed specifically for cooling tower applications and is mounted vertically within the discharge cone and direct-connected to the fan shaft via a flexible coupling (much like base-mounted pump and motor assemblies). A companion VFD speed control system allows for soft starting and a wide speed range for tower capacity control. The direct-drive motor and control can be applied to retrofit existing towers or be specified for new towers.

# *Part II*

## *Cooling Tower Design and Application*

# 11 Tower Configuration and Application

## TYPES OF COOLING TOWERS

Almost all HVAC cooling towers are film or splash fill, mechanical draft type. Basic HVAC tower configurations are dictated by (1) the direction of the air versus water flow through the tower fill, and (2) the location of the tower fan(s), as shown in Figure 9.9.

*Air/water flow arrangement.* In *counterflow* towers, water and air flow in opposite directions: the water flows vertically downward and the air flows vertically upward. In *crossflow* towers, the two flow streams are oriented 90° to each other: the water flows vertically downward while the air flows horizontally.

*Fan location.* The tower is called *forced draft* when its fan(s) is arranged to blow air through the tower. In this configuration, the fan(s) is located at the air intake(s) of the tower and the fill is under positive pressure. In *induced draft* towers, the fan(s) is located at the air exit(s) of the tower and the fill is under negative pressure.

Figures 11.1 through 11.4 illustrate four common configurations typical for HVAC cooling towers: forced and induced draft counterflow and forced and induced draft crossflow.

### COUNTERFLOW VERSUS CROSSFLOW

There are waterside and airside advantages and disadvantages of both counterflow and crossflow cooling towers, as summarized in Table 11.1.

However, the largest differences between these types of cooling towers are as follows:

*Size/arrangement.* For any given capacity, counterflow towers will typically have a smaller footprint than crossflow towers, thus requiring less space for their installation. However, this size advantage may disappear when multiple tower cells are required to meet the imposed cooling load:

1. Small counterflow towers may require air intake on only one side and it is relatively simple to arrange multiple towers in a line, spaced just far-enough apart for maintenance (about 4′).

**FIGURE 11.1** Typical forced draft crossflow cooling tower. (Courtesy of the Baltimore Aircoil Company, Baltimore, MD.)

**FIGURE 11.2** Typical induced draft crossflow cooling tower. (Courtesy of The Marley Cooling Tower Company, Overland Park, KS.)

**FIGURE 11.3** Typical forced draft counterflow cooling tower. (Courtesy of Tower Tech, Inc., Oklahoma City, OK.)

**FIGURE 11.4** Typical induced draft counterflow cooling tower. (Courtesy of Evapco, Inc., Westminster, MD.)

**TABLE 11.1**
**Counterflow versus Crossflow Cooling Towers**

| Configuration | Flow | Advantages | Disadvantages |
|---|---|---|---|
| Crossflow | Waterside | Lower pump head, pump power requirement, and pumping energy<br>Easier access to wet deck for maintenance<br>Better acceptance of variation in water flow with economizer | Potential orifice clogging and poor water distribution over fill<br>Wet deck basin may house biological fouling<br>Larger tower footprint |
| | Airside | Lower static pressure loss with lower fan power requirement and energy consumption<br>Reduced drift<br>Reduced recirculation<br>Requires fewer cells for larger capacities | Large inlet louver surface area makes icing control more difficult |
| Counterflow | Waterside | Spray distribution improves water droplet size<br>Tower is taller and increased height accommodates closer approach | Increased pump head due to spray nozzles, pump power requirement, and pumping energy<br>Spray nozzles are difficult to access and clean |
| | Airside | Counterflow improves heat transfer | Higher static pressure losses, fan power requirement, and energy consumption<br>High inlet velocities may pull debris into basin<br>Tendency for uneven airflow across fill, reducing tower efficiency |

2. For larger counterflow towers, two cells can be joined, which effectively reduces the air entry condition for each cell from four sides to three sides. If more than two cells are required, the next one or two must be spaced at least one tower width away to ensure proper air entry.
3. Multiple crossflow tower cells can be joined side by side to form a tower as long as desired. However, it is better to limit this side-by-side arrangement to only three or four cells and to provide an airflow and maintenance separation of at least one tower width between each set of cells. When more than four cells are joined, there are often adverse airflow patterns at the middle cell(s).

*Fan horsepower/energy.* Counterflow cooling towers generally have higher airflow static pressure losses than crossflow cooling towers, requiring more fan horsepower and resulting in greater fan energy consumption. These higher losses are primarily due to counterflow towers being taller, with a greater fill height, and the water spray wet deck.

*Pump horsepower/energy.* The nozzles used in the counterflow cooling tower spray wet deck require, typically, 5 psig residual water pressure to provide proper atomization. This, coupled with the fact that counterflow towers tend to be taller than crossflow towers, results in increased pump head, pump motor horsepower, and pump operating energy.

In general, crossflow towers are the better selection when it is desirable to minimize tower fan energy consumption, minimize pump size and pumping energy, and provide ease of maintenance. The crossflow tower is better with waterside economizer applications.

The counterflow tower is the better selection where the available space (footprint) is limited and/or where icing during winter operation is a concern. The counterflow tower may also be the better selection when very close approach is needed (<5°F).

## Mechanical Draft

As discussed in Chapter 9, almost all HVAC cooling towers utilize mechanical draft since the tall draft chimney and large size required by natural draft towers would be almost impossible to incorporate in most building designs.

Crossflow towers may utilize forced draft or induced draft. *Single-flow* forced draft crossflow towers (one-cell) utilizing propeller fans are available in smaller capacities (from 30 to about 250 tons). The fan(s) is mounted vertically in the tower air inlet and forces the air through the fill and out of the exit on the opposite side of the tower.

As cooling load increases, more heat transfer surface is needed and the *double-flow* crossflow tower, with up to about 1100 tons capacity in one factory-assembled cell, is typically utilized since it is considerably less costly to use one large cell than to use two smaller cells. In this configuration, it is more efficient to use one large fan to induce airflow through the tower rather than multiple fans to force the airflow. The induced draft fan is mounted at the top of the tower and induces airflow from each end, through the fill, and into a common exit air plenum.

Counterflow towers may also utilize forced draft or induced draft up to about 1300 tons capacity in one factory-assembled cell. Both propeller and centrifugal fans are used for forced draft, but only propeller fans are used for induced draft.

In forced draft counterflow configurations with propeller fans, the fan(s) is located at the bottom of the tower, which requires a very special basin design since the air must flow through the basin. In these towers, a water collection system, consisting of sloped channels above the fans, moves the water from the bottom of the fill to an adjacent enclosed basin or tank.

Induced draft counterflow towers have a far simpler configuration. The fan is located at the top of the tower and the air intakes are located between the top of the basin and the bottom of the fill, typically on all four sides of the tower. As for crossflow towers, the largest single tower cell that is available is about 1100 tons.

Generally, there is no advantage of forced draft over induced draft or vice versa. However, in applications where cooling towers are used in the winter and there is the potential for *icing*, forced draft towers, particularly those with centrifugal fans, should be avoided. The forced draft fans are located in the entering cold air stream rather than in the warm leaving air stream, making them highly susceptible to icing and potential mechanical failure. Thus, for cold weather operation, the induced draft tower is the better choice since both the potential for icing is reduced and, as discussed in the section "Basin and Outdoor Piping Freeze Protection" in Chapter 14, measures to control any icing that does occur are much easier to implement.

## CAPACITY AND PERFORMANCE PARAMETERS

In Chapter 9, two factors were defined as establishing the HVAC cooling tower performance requirements:

1. The cooling load imposed on the tower, which is defined by the condenser water flow rate and the selected range
2. The approach, which is dictated by the design outdoor wet bulb temperature and the required condenser water supply temperature

### TEMPERATURE RANGE AND APPROACH

As defined in the section "Nomenclature" in Chapter 9, *range* is the temperature difference between the condenser water entering (condenser water return) and leaving the tower (condenser water supply), while *approach* is the temperature difference between the condenser water supply and the entering air wet bulb. These temperature differences are illustrated in Figure 11.5.

The vast majority of HVAC condenser water systems are designed for a 10°F range. This range was established many years ago based on desired flow velocities in condenser tubes and maintaining approach at a reasonably high level. The "standard" selection criteria for water chillers are 85°F condenser water supply temperature and a 10°F range, resulting in a 95°F condenser water return temperature.

Since the design ambient wet bulb temperature is fixed, the approach is changed by changing the range (and resulting flow rate) and/or changing the required condenser water supply temperature. From Figures 9.5 and 9.7, it is obvious that there is a trade-off between approach and range—increasing range reduces the required tower characteristic, but if increased range reduces the condenser water supply temperature, the resulting reduced approach quickly increases the required tower characteristic.

If the range, and the resulting condenser water flow rate, is a design option, it is best to evaluate a number of cooling towers, each at a different range from 8°F to 14°F, to determine the optimum selection. However, *if the resulting approach for any range is <5°F, the size of the tower will increase exponentially for the*

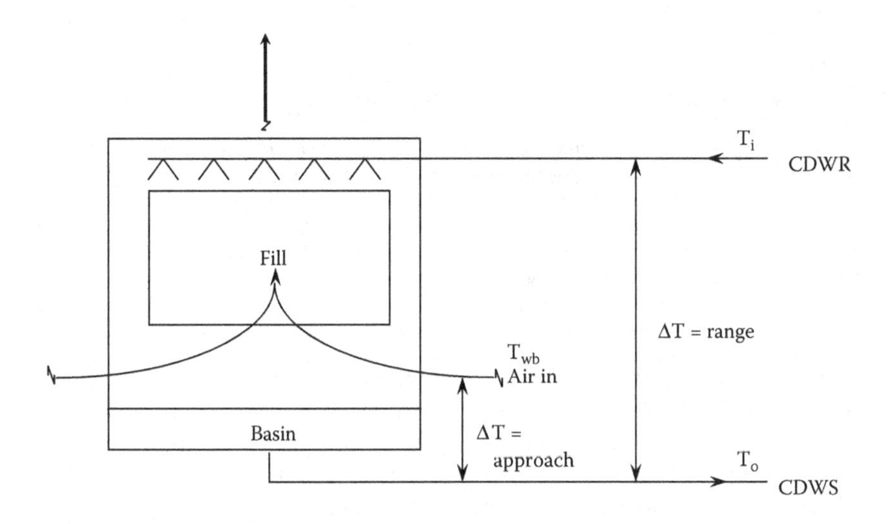


**FIGURE 11.5** Definition of "range" and "approach" for condenser water systems.

*same capacity and, at 3°F, no further reduction to approach is possible with current cooling tower technology.*

## Ambient Wet Bulb Temperature

Appendix A lists the recommended design outdoor entering wet bulb temperature for locations in the United States for cooling tower sizing and selection. The recommended temperature is the 0.4% criteria ambient wet bulb temperature from ASHRAE weather data, plus 1°F to provide a "safety factor" against local variations and the effects of climate change that are now well documented. These values range from 75°F to 82°F and, with the standard 85°F condenser water supply temperature, the approach will vary from 3°F to 10°F (with 6–8°F being most common).

The actual wet bulb temperature of air entering a cooling tower may be higher due to recirculation. Normally, there is always some recirculation, but it is usually small enough to ignore, for example, the ambient and entering wet bulb temperatures are essentially the same. However, poor tower placement, excess tower screening, or wind patterns can create excess recirculation and the entering wet bulb temperature may be several degrees above the ambient level. The section "Tower Placement and Installation" addresses measures required to minimize this condition.

## Condenser Water Heat Rejection

The total heat rejection to the condenser water system is determined, as shown in Chapter 9, from the design condenser water flow rate and its design temperature range. Typically, the range for an HVAC condenser water system is selected to be

**TABLE 11.2**
**Required Condenser Water Flow Rate for 10°F Range**

| Compressor Input Power (kW/ton) | Total Heat Rejection (Btu/ton) | $F_{cdw}$ (gpm/ton of Cooling Load) | $F_{cdw}$ Reduction (%) |
|---|---|---|---|
| 0.8 | 15,000 | 3.00 | 0 |
| 0.7 | 14,640 | 2.93 | 2.0 |
| 0.6 | 14,280 | 2.86 | 4.7 |
| 0.5 | 13,920 | 2.78 | 7.3 |
| 0.4 | 13,680 | 2.74 | 8.7 |

10°F. At this range, the required condenser water flow rate, in terms of "gpm per ton of cooling load" can be computed as shown in Table 11.2.

For many years, "standard" condenser water conditions were based on electric-drive chillers with a power input of 0.8 kW/ton. Older literature will sometimes refer to a *cooling tower ton* as 15,000 Btu/h and current manufacturer ratings are still based on 3.0 gpm/ton as the "standard" flow rate. *None of these old data are valid today.* Chillers can be purchased with initial peak power input requirements as low as 0.4 kW/ton, and so the actual condenser water flow rate for any application must be computed based on the actual compressor power input, as shown in Table 11.2. This, in effect, minimizes pumping energy and the cooling tower size.

## CHILLER/COOLING TOWER CONFIGURATION

When there is only one water-cooled chiller, the only option is to select one cooling tower to serve that chiller's condenser. However, when there are multiple chillers, there are three options for designing the condenser water system as shown in Figure 11.6.

1. Each chiller can have an individual, dedicated cooling tower.
2. Multiple cooling towers or a large multicell tower can be connected to a common condenser water system to serve all of the chillers. Each chiller and cooling, however, must have two-position isolation control valves to isolate each component that is "off." The advantage of this configuration is versatility—any tower and any pump can serve any chiller.
3. Multiple cooling towers or large multicell towers are connected to a common condenser water system with an individual pump serving each chiller. This configuration eliminates the need for chiller isolation valves and *is the recommended configuration when the chillers have different capacities.* (The downside of this configuration is, of course, loss of the versatility provided by the common parallel pumping arrangement.)

With individual, dedicated towers, individual condenser water pumps and supply and return condenser water piping between each chiller and its tower is

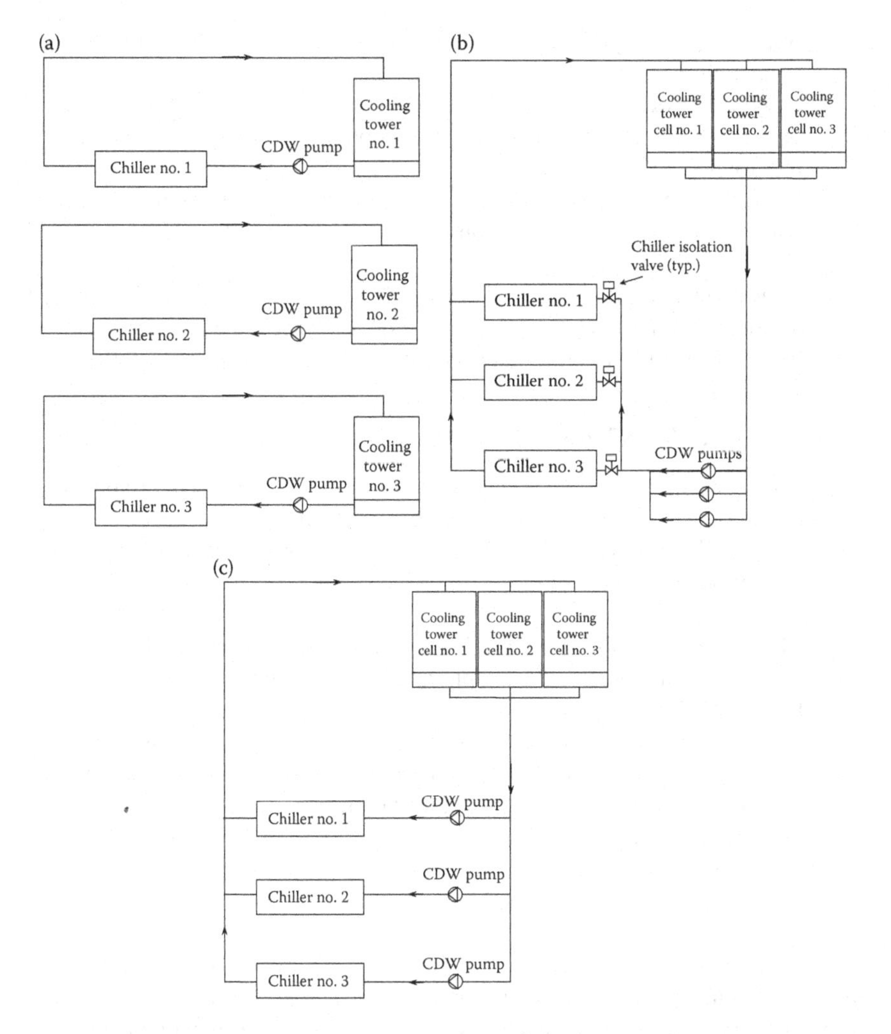


**FIGURE 11.6** Multiple chiller/tower configuration: (a) option 1, (b) option 2, and (c) option 3.

required. The cost of this piping is typically higher than the cost of a single (but larger) condenser water supply loop serving all chillers from all towers.

The multicell approach has a number of operating advantages:

1. More than one cell can be used to serve a single chiller, providing a lower condenser water supply temperature and reducing chiller energy consumption.
2. The failure of a tower cell does not necessarily mean that a specific chiller cannot be utilized, since all chillers are served by all of the tower cells.
3. An additional, redundant tower cell can be included to provide “back-up” in the event of a tower cell failure or to maintain the required condenser

water supply temperature in periods of higher-than-expected wet bulb temperature.

4. Only one tower makeup water control system is required, rather than one for each tower under the first option, reducing maintenance requirements.
5. One cell (or more) can be shut down for maintenance without interfering with a specific chiller's operation.

A disadvantage of the multicell approach is that automatically controlled two-position isolation valves are required on both the inlet and outlet of each tower cell. These isolation valves eliminate water flow through cells that are not active.

## TOWER PLACEMENT AND INSTALLATION

To maximize performance and reduce the amount of dirt, leaves, and so on entering a cooling tower, *the best place to locate any HVAC cooling tower is on the roof of the building it serves.* This placement satisfies two conditions that are critical to proper tower performance:

1. The elevation of the tower's operating water level must be above the condenser water pump (and any piping between the pump suction and the tower basin).
2. Towers must have sufficient clearance from walls and other obstructions around them to prevent *recirculation* of discharge air back into the inlet. If there are multiple towers, they must be located to avoid the discharge of one tower entering any adjacent tower. Generally, a building roof is clear of obstructions that can result in recirculation, *provided that obstruction by aesthetic screening is not added.*

The section "Cooling Tower Piping" addresses condenser water pump placement and piping options. This section addresses tower placement and installation to avoid recirculation of tower airflow if the tower must be located on the ground.

The airflow requirement of an HVAC cooling tower is typically 270–300 cfm/ton, which is equivalent to 90–100 cfm/gpm. Thus, (1) sufficient free area around the tower for the unobstructed flow of large quantities of air must be provided, and (2) the tower must be located sufficiently clear of obstructions to prevent discharge air from reentering the tower (or entering an adjacent tower).

Ground-level placement considerations for a tower depend, initially, on the type of tower, crossflow or counterflow. Crossflow towers have an air intake on either one (*single-flow* towers) or two sides (*double-flow* towers), while counterflow towers have air intakes on all four sides. However, for both types of towers, the following recommendations apply:

*Adjacency.* Building walls, screen walls, trees, and so on form obstructions for cooling towers if the tower is located too close to them. While most tower manufacturers provide information about layout for their cooling towers, most of the recommended clearances are minimal and are not sufficient to always ensure proper tower performance. Therefore, the recommended clearance from adjacent obstructions is shown in Figure 11.7.

The cooling tower should be located at least 30 ft away from adjacent building walls. As buildings get taller, the obstruction formed increases and, thus, the tower setback should be increased to at least half the building height to prevent recirculation.

Cooling towers are rarely pretty; thus, architects will often specify that screening (wood, masonry, or metal) be installed around the tower. To prevent tower performance problems due to screening, the following design criteria for screening must be followed:

1. The screening must provide at least 50% open area on the air intake side(s) of the cooling tower and the open area must be sufficient to allow for 600 fpm or lower airflow velocity through the screen.
2. The top of the screening must be no higher than the tower air discharge (see "Elevation" below).
3. A clearance must be provided between the screening and tower air intake side(s) of >10′ or the width of the tower. If there are multiple towers or multiple tower cells within a screened enclosure, these clearances must be increased by at least 50% (to 15′). Maintenance clearance between the screening and the other (nonactive) sides of the tower must be at least 4′ to provide adequate passage and maintenance space.

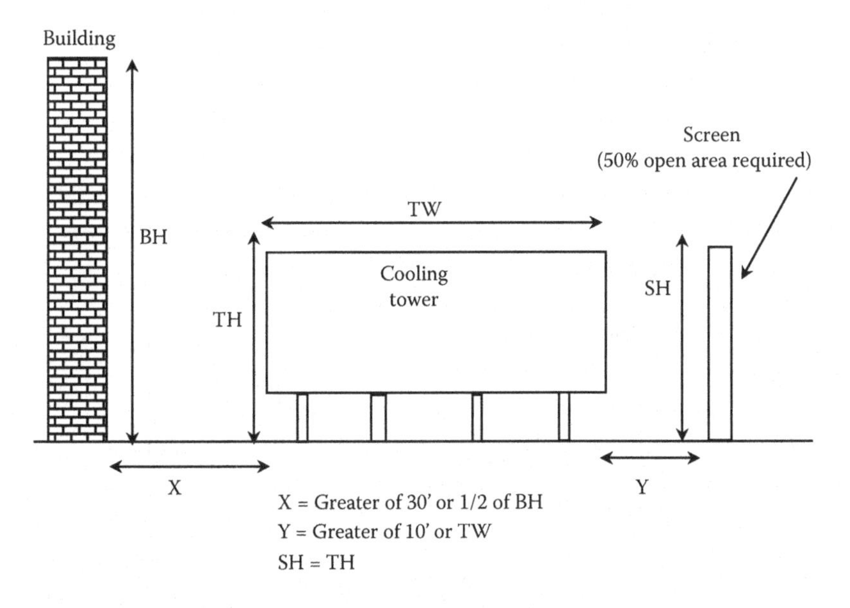


**FIGURE 11.7** Recommended cooling tower siting parameters.

*Elevation*. In every HVAC cooling tower installation, the operating water level must be higher than the condenser water pump and all of the condenser water supply piping between the tower basin and the pump. Thus, the pump is maintained at a "flooded" condition and cannot lose prime.

Where the tower is installed close to a building (i.e., closer than the half the building height or 30′, whichever is greater), the fan discharge elevation must be at or above the adjacent building roofline or the top of any architectural screening that is provided. As shown in Figure 11.8, this can be accomplished by (1) providing a discharge cowl or plenum on the tower and/or (2) providing a structural support system to elevate the tower to the proper height.

An elevating structural support system or "grillage" can be easily fabricated from structural steel elements, pipe columns, and so on. The discharge cowl, however, must be engineered to minimize static pressure loss and avoid the need for a larger fan motor and increased fan energy consumption. The cowl must have a discharge taper (called an *evase discharge*) in order to reduce the air discharge velocity efficiently and regain the available velocity pressure to offset friction-based static pressure losses.

Table 11.3 tabulates the static pressure regain factor for a cowl of length $L$ (ft), a fan discharge diameter D1 (ft), and a final discharge diameter

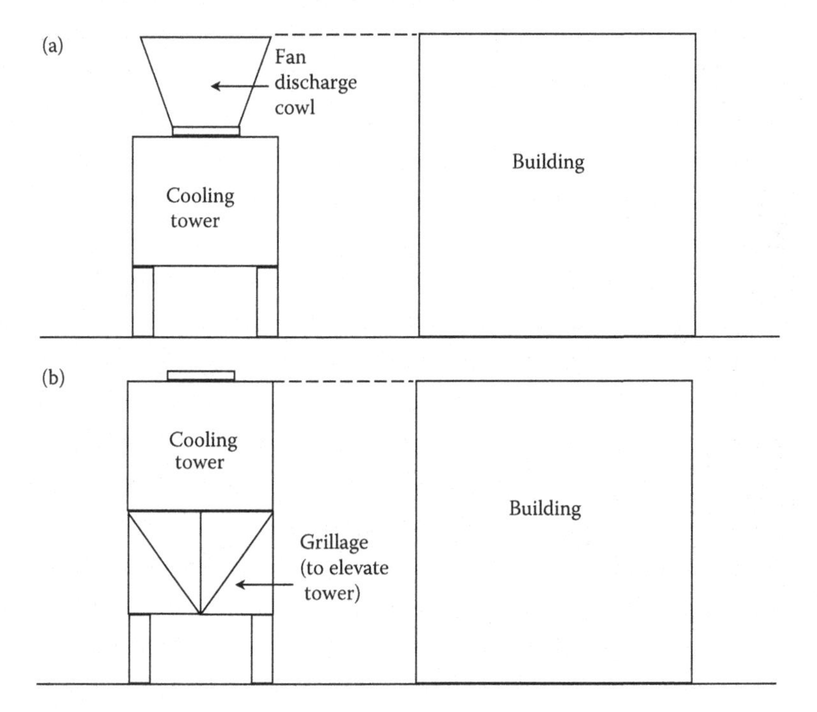


**FIGURE 11.8** Tower elevation: (a) option 1 and (b) option 2.

**TABLE 11.3**
**Discharge Cowl Static Pressure Regain Factors (CRF)**

| | Static Pressure Regain Factor of D2/D1 Ratio | | |
|---|---|---|---|
| ***L*/D1** | **1.2:1** | **1.4:1** | **1.6:1** |
| 1.0:1 | 0.26 | 0.27 | 0.22 |
| 1.5:1 | 0.27 | 0.33 | 0.31 |
| 2.0:1 | 0.29 | 0.36 | 0.36 |
| 3.0:1 | 0.31 | 0.40 | 0.42 |
| 4.0:1 | 0.32 | 0.42 | 0.44 |
| 5.0:1 | 0.33 | 0.43 | 0.46 |
| 7.5:1 | 0.34 | 0.44 | 0.49 |

D2 (ft). There is little improvement to the static pressure regain factor for an *L*/D1 greater than 7.5:1 or D2/D1 greater than 1.6:1.

The amount of regained static pressure is computed as follows:

$$SP = (V/4005)^2 \times CRF$$

where
SP = regained static pressure in wg
$V$ = fan discharge velocity (fpm)
CRF = cowl regain factor from Table 11.3

The amount of static pressure regain can be subtracted from the static pressure loss through the cowl, which can be determined from any duct sizing nomograph using the average cowl diameter [(D1 + D2)/2], the average airflow velocity through the cowl, and the cowl length. Designed correctly, the discharge cowl should impose little or no additional static pressure for the tower fan to overcome.

*Prevailing wind.* Since counterflow towers typically have air intakes on all four sides, tower orientation relative to the prevailing summer wind is not important. However, always attempt to orient a crossflow cooling tower as shown in Figure 11.9. Wind can influence the performance of a double inlet tower by creating an unbalanced airflow through the two sides. If a crossflow tower must be installed with one inlet in the direction of the prevailing wind, the tower manufacturer must be apprised of this condition so that the proper sizing adjustments can be made.

*Multiple towers or tower cells.* Figure 11.10 illustrates the recommended layout for multiple towers or tower cells for both crossflow and counterflow cooling towers. As discussed earlier, no more than four crossflow cooling tower cells should be close coupled. For counterflow towers, usually only two cells can be close coupled and even then the tower

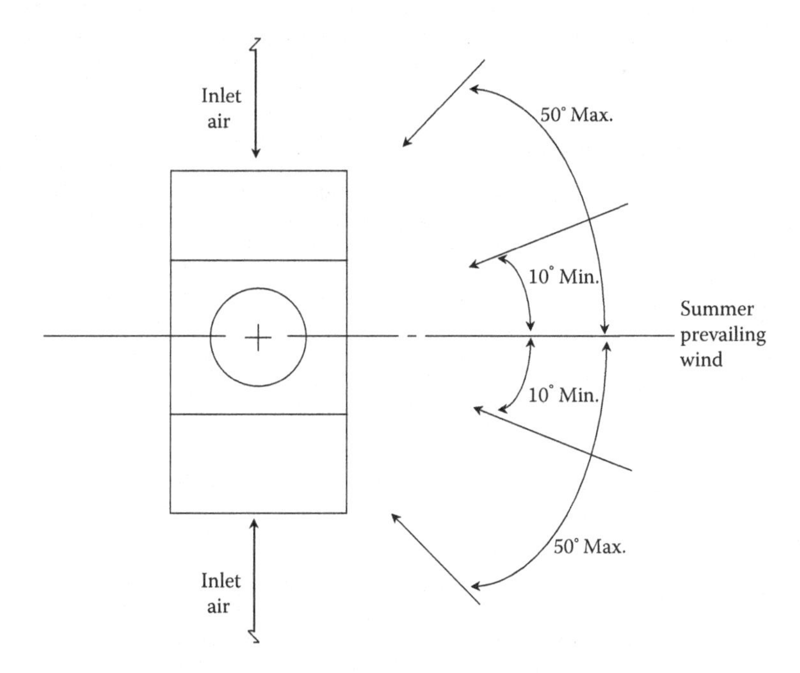


**FIGURE 11.9** Crossflow tower placement relative to the prevailing wind.

manufacturer must guarantee that the individual cell performance will not be impaired since each cell is losing one intake side.

Again, wind direction is relatively unimportant to counterflow towers. However, for crossflow towers, recirculation from one tower bank to an adjacent one can be significantly increased by the prevailing wind. Figure 11.11 shows recommended tower layouts relative to the prevailing winds for crossflow tower bank installations.

*Dunnage and grillage. Dunnage* is the steel and/or concrete structural foundation upon which the tower support *grillage* rests. The tower grillage routinely consists of structural steel members arranged to support the tower at the required bearing points.

On the roof of a building, dunnage should consist of structural column extensions so that the tower weight is transferred directly to the building foundation elements. *Never attempt to install a tower directly on the roof structural system!* Roofs are never designed for the types of loads introduced by a cooling tower and its water load. Structural columns should be extended 3–4 ft above the roof to provide clearance for roofing maintenance.

On the ground, dunnage many consist of concrete piers or columns, each with properly designed footings and reinforcement.

Table 11.4 provides approximate operating and shipping weights for typical cooling towers.

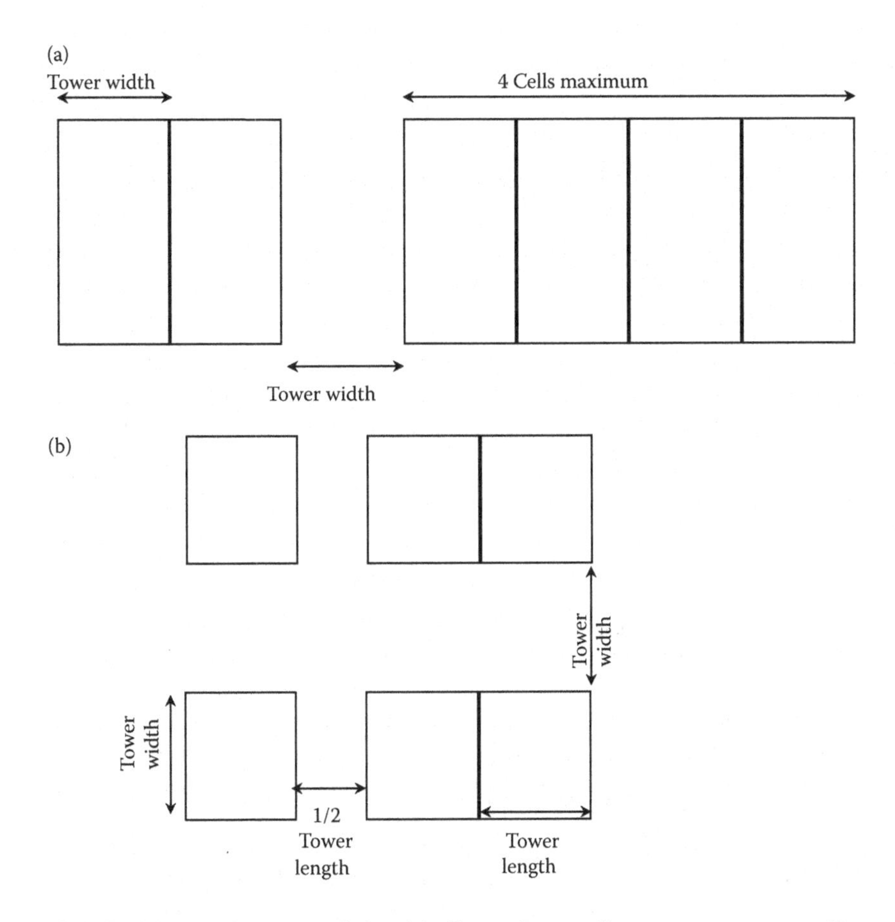


**FIGURE 11.10** (a) Recommended multicell crossflow cooling tower arrangement (four or more cells). (b) Recommended multicell counterflow cooling tower arrangement (two or more cells).

Care must be taken to ensure that dunnage/grillage height is sufficient to satisfy the tower operating water level elevation requirement.

## COOLING TOWER PIPING

Three separate piping systems are required for each cooling tower, as shown in Figure 11.12 and described in the following subsections.

### Condenser Water Piping

Condenser water return (CDWR) piping delivers warm water from the condenser to the tower, while condenser water supply (CDWS) piping delivers cool water from the tower to the condenser. A manual isolation valve should be provided at both the supply and return connections to the tower to facilitate tower maintenance.

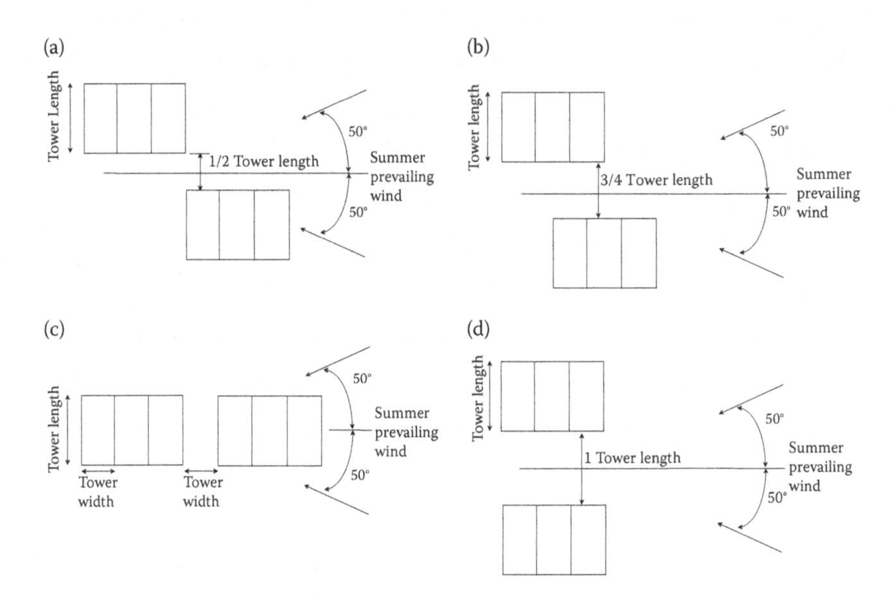


**FIGURE 11.11** Recommended multicell crossflow cooling tower arrangement relative to the prevailing wind: (a) option 1, (b) option 2, (c) option 3, and (d) option 4.

Electric heat tape (or other type of heat tracing) may be required to protect the condenser water from freezing during off periods in winter (see Chapter 14).

Black steel or PVC piping materials are normally used for condenser water service. In the past, galvanized steel piping was often used because it was thought that this piping was more resistant to corrosion and the higher cost for galvanized over black steel was justified. However, since galvanized pipe cannot be welded without destroying the galvanizing at each joint, screwed connections are required. In the larger pipe sizes required for condenser water systems, increased labor costs have made screwed joints prohibitively expensive and galvanized piping is rarely used today.

CDWS piping between the tower and the condensers should be insulated with at least 2″-thick nonhydroscopic insulation (cellular glass indoors; cellular glass or polyisocyanurate outdoors) to reduce heat gain by the condenser water to the chiller. At normal condenser water temperatures, a vapor barrier is not required, but, if a waterside economizer cycle is incorporated or CDWS temperatures below 70°F are anticipated, a vapor barrier becomes necessary to prevent surface condensation.

*Vortexing* is term used to describe the whirlpool effect that sometimes occurs at the CDWS connection to the tower basin, particularly with a vertical discharge connection. Vortexing causes air to be entrained in the water and can, in extreme cases, cause pump cavitation and result in mechanical damage. Most tower manufacturers offer an option for a side outlet discharge sump that can be added to the basin. This type of sump will minimize air entrainment and *should always be used.*

**TABLE 11.4**
**Approximate Cooling Tower Weight**

| Nominal Capacity (tons) | Approximate Operating Weight (lb) | | | |
|---|---|---|---|---|
| | Forced Draft Crossflow[a] | Induced Draft Crossflow[a] | Forced Draft Counterflow[b] | Induced Draft Counterflow[a] |
| 100 | 4300 | 9000 | 12,000 | 5000 |
| 200 | 8200 | 14,000 | 12,000 | 6500 |
| 250 | 9500 | 14,000 | 13,000 | 7500 |
| 300 | 14,000 | 14,000 | 13,000 | 8500 |
| 350 | 16,000 | 18,000 | 16,000 | 10,000 |
| 400 | 16,000 | 18,000 | 16,000 | 11,000 |
| 450 | 19,000 | 22,000 | 19,500 | 14,000 |
| 500 | 19,000 | 28,000 | 19,500 | 14,500 |
| 600 | — | 29,000 | 19,500 | 17,000 |
| 700 | — | 35,000 | 22,600 | 19,000 |
| 800 | — | 37,000 | 22,600 | 22,000 |
| 900 | — | 44,000 | 23,600 | 28,000 |
| 1000 | — | 44,000 | 23,600 | 28,000 |
| 1500 | — | 75,000 | 47,200 | 43,500 |
| 2000 | — | 90,000 | 47,200 | 63,000 |

[a] Shipping weight is ~30–35% of operating weight.
[b] Shipping weight is ~55–65% of operating weight.

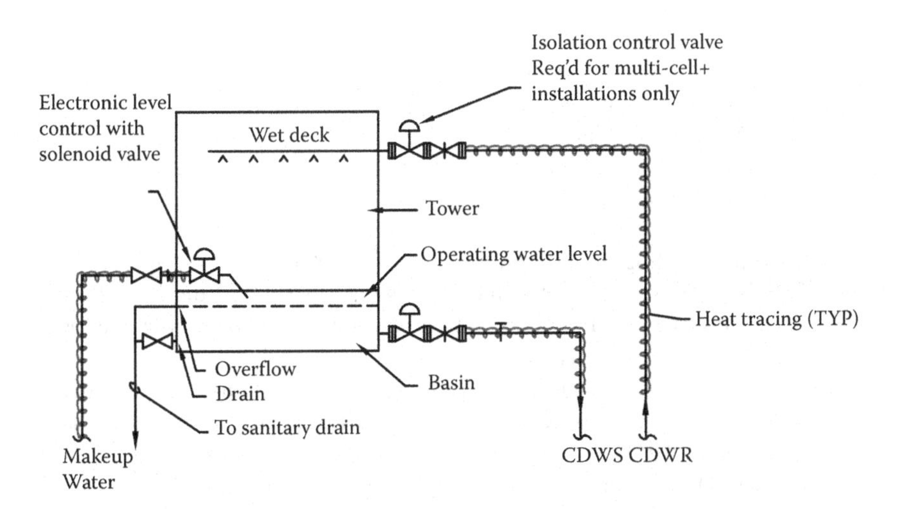


**FIGURE 11.12** Typical cooling tower piping installation schematic.

## Makeup Water Piping

Makeup water is potable water from municipal water systems and/or wells that is added to the condenser water flow to offset the water losses due to evaporation, drift, and blowdown. Again, a manual isolation valve is required at the tower connection point for maintenance purposes. Copper tubing or PVC piping is normally used for makeup water piping and all piping should be insulated with at least 1″-thick glass fiber insulation with an integral vapor barrier.

The makeup piping terminates at a makeup water control valve, which may be either mechanically or electronically actuated in response to the water level in the basin. Electric heat tape (or other type of heat tracing) is normally required to protect the makeup water from freezing during off periods in winter (see Chapter 14).

Water losses from an HVAC cooling tower result from evaporation, drift, and blowdown. Each of these must be determined to properly size the makeup water system.

*Evaporation loss*. The heat of vaporization of water is 1045 Btu/lb at 85°F. Therefore, to determine the evaporation rate required for a given heat rejection (condenser water cooling) rate, it is necessary to know the flow rate, the range, and the heat rejection required. In the general case (a flow rate of 1 gpm with a range of 1°F), the evaporation requirement can be computed as follows:

Flow: 1 gal/min (equivalent to 500 lb/h)
Range: 1°F
Heat load:

$$Q = 1 \text{ gal/min} \times 8.34 \text{ lb/gal} \times 60 \text{ min/h} \times 1 \text{ Btu/lb°F} \times 1\text{°F} = 500 \text{ Btu/h}$$

Heat of vaporization:

$$Q_v = 1045 \text{ Btu/lb}$$

Evaporation rate:

$$E = \frac{500 \text{ Btu/h}}{1045 \text{ Btu/1b}} = 0.478 \text{ lb/h}$$

Thus, dividing the evaporation rate of 0.478 lb/h by the flow rate of 500 lb/h yields an evaporation rate of essentially 0.1% of the condenser water flow rate per °F of range, ~1% of the condenser water flow rate for the 10°F range typically used in HVAC applications.

*Drift loss*. Drift loss data can be provided by the manufacturer for the specific cooling tower selection being considered. However, drift loss from modern HVAC cooling towers is typically within the range of 0.0001–0.005% and most engineers simply ignore drift loss when evaluating makeup water requirements.

*Blowdown loss*. Water treatment programs for deposition control routinely require *blowdown*, the intentional wasting of condenser water to increase the amount of makeup water to maintain the ratio of dissolved solids in the condenser water at a desirable level (see Chapter 13). The water treatment program will establish the required number of *cycles of concentration*, which is the ratio of the amount of dissolved solids in the condenser water to the amount in the makeup water. Once the cycles are determined, the amount of required blowdown can be computed as

$$\mathrm{BD} = \frac{E}{(\mathrm{cycles} - 1)}$$

where
BD = blowdown flow (gpm)
$E$ = evaporation rate (gpm)
cycles = cycles of concentration established by water treatment program
Typically, the cycles of concentration will be between 5 and 10, and the lower value, which requires greater blowdown to achieve, is often used to establish makeup water flow rates.

Thus, the maximum required makeup water flow rate, per gpm of condenser water flow, can be conservatively computed as follows:

| | | |
|---|---|---|
| Evaporation: | (0.01 × 1 gpm) | 0.0100 gpm |
| Blowdown: | [0.01 gpm/(5–1)] | 0.0025 gpm |
| Total makeup: | | 0.0125 gpm |

A 500-ton chiller with a rated condenser water flow rate of 3 gpm/ton would have a condenser water flow rate of 1500 gpm and the maximum amount of makeup water required would be (1500 gpm × 0.0125 gpm make-up/gpm CDW) or 18.75 gpm. For an "optimum" chiller with a condenser water flow rate of 2.8 gpm/ton, the make-up water requirement would be 17.5 gpm at the peak load of 500 tons.

A water meter, of a type satisfactory to the local municipality, should be installed in the makeup water supply line so that the amount of water consumed by the tower can be accurately determined. Most municipal water and sewer agencies will provide a credit to eliminate sewer charges for makeup water lost through a cooling tower.

Normal operation of cooling coils produces condensate water that typically drains to the sewer. But, condensate is clean water that can be captured and reused for nonpotable water applications. Typical applications for recovered condensate include cooling tower makeup.

Design considerations for condensate recovery systems include the following:

1. Condensate recovery works by gravity flow. A drain line runs from each air-handling unit to a central connection point in the penthouse, and from there a single line runs to the cooling towers.
2. Collected condensate water is at temperatures between 50°F and 60°F.
3. To use condensate as cooling tower makeup, a three-way valve in the line feeding makeup water to the cooling towers is required to allow the system to draw from reclaimed condensate or service water as needed for level control. A cooling tower may require more makeup water than can be recovered from the condensate, in which case the system uses supplemental domestic water.
4. During periods when there is some excess condensate, it can be diverted to waste or a cistern can be used to store the excess for later use.

Cooling coils produce condensate at a rate approximately as follows:

$$\text{GPH} = 0.055 \times \text{CFM}_{\text{OA}} \times (\text{H} - 0.0078)$$

where

$\text{CFM}_{\text{OA}}$ = Outdoor ventilation airflow rate

H = Average outdoor humidity ratio (lb of water/lb of dry air) during the summer cooling season (normally May through September). (This data can be computed from monthly weather data for the project location available from NOAA.)

Where possible, recovery of this water for cooling tower make-up or other use should be part of the design.

## Drain and Overflow Piping

Every cooling tower must have drain and overflow piping connected to a sanitary waste water system. The Clean Water Act makes it *illegal to drain chemically treated condenser water into storm or site drainage systems.*

The overflow portion of the system is simply an open drain at the same level as, or slightly above, the operating water level in the basin. If excess water enters the basin, it automatically flows into the overflow drain and is wasted. A manual isolation valve controls the basin drain connection, which allows the basin water to be emptied for periodic maintenance.

## Multiple Towers or Cells Piping

When two or more cells are piped together to act as one cooling tower, each cell is typically provided with the piping systems outlined. However, there are some modifications and additions that must be considered.

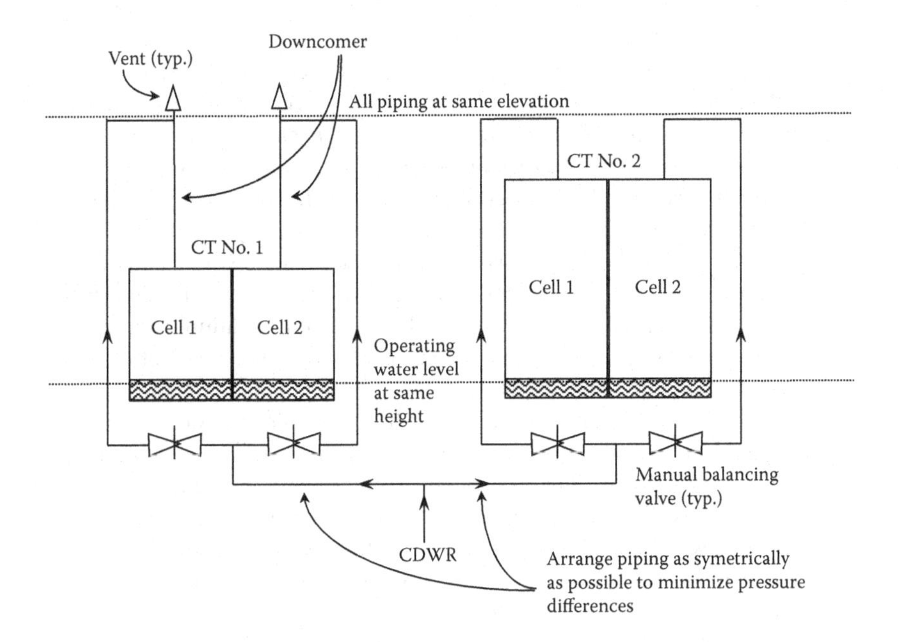


**FIGURE 11.13** Condenser water supply piping schematic for multiple towers or cells with different inlet elevations.

*Common condenser water piping system.* If each cell is piped from a common condenser water piping system, automatic isolation valves on both supply and return are required to control flow through the cell (see Chapter 12). Without the isolation valves, part of the flow may go through inactive cells, destroying the ability to control CDWS temperature.

The elevation and inlet pressure requirements for multiple cells must be considered if different types of cooling towers or tower cells (or similar towers by different manufacturers) are to be utilized together. First, the operating water level for each tower or cell must be at the same elevation. For crossflow towers, the wet deck basins must have the same elevation or flow will tend to favor the lower basin, which will routinely overflow while the higher basin tends to run dry. To avoid this problem, it is necessary to install the supply piping to each cell at the same elevation above the higher inlet and let the water fall into the individual inlets. To prevent siphoning, a vent, as shown in Figure 11.13, is required for the lower cell and the *downcomer* pipe must be sized on the basis of gravity flow. Also, as shown in Figure 11.13, piping to each cell should be kept as symmetrical as possible to reduce flow imbalance resulting from different pressure losses through different flow paths.

Even when multiple gravity flow inlets are at the same elevation, equal flow distribution is, for all practical purposes, impossible to achieve and there will always be slight flow imbalance from cell to cell.

For counterflow towers, water spray connections should be at essentially the same elevation and fitted with nozzles requiring the same water pressure to equalize static head requirements for the condenser water pump.

*Avoid mixing crossflow and counterflow tower cells on the same condenser water piping system.* The different elevations and inlet pressure requirements for the two types of towers make it essentially impossible to balance flow between them well enough to maintain satisfactory operation.

*Common basin.* Where the cells share a common basin, or where individual cell basins are connected together, a single makeup water valve and level control is required, not one for each basin. With multiple make-up connections to a single basin, the controllers will "fight" each other and level control will be impaired.

Multiple, individual cell basins can be connected together to form a single basin by installing manufacturer-fabricated *basin connectors* between each basin. These connections create large openings between each cell basin, effectively creating one common basin. An alternative approach would be to operate the tower with "dry" individual cell basins and drain all supply condenser water to a common remote sump (see the section "Basin and Outdoor Piping Freeze Protection" in Chapter 14).

Where each cell has an individual basin, the basins can be connected together with an *equalizer line*, as shown in Figure 11.14. No matter how closely matched the piping is between cells, flow to each cell is never exactly balanced. Without the equalizer line, one cell basin would probably overflow while another was pumped dry. The equalizer line allows water to flow between tower basins and compensate for the minor flow imbalances between the cells that will exist.

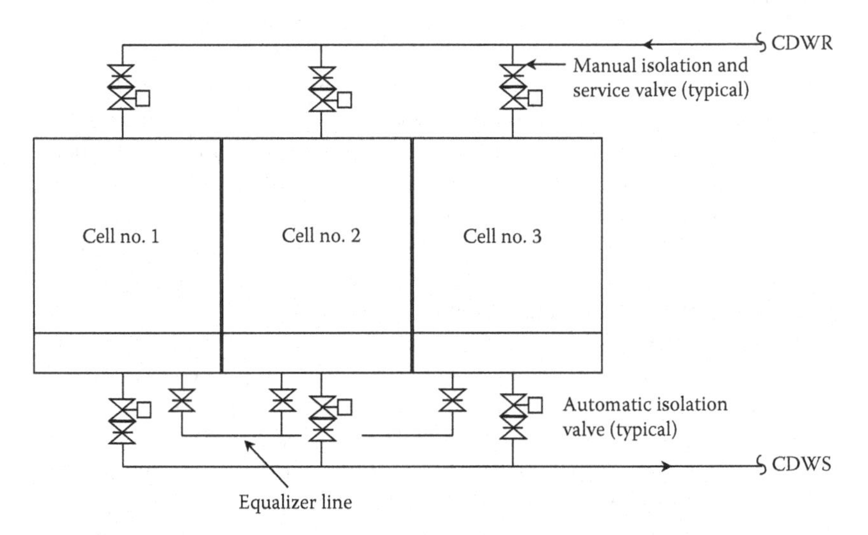


**FIGURE 11.14** Recommended multiple tower or cell isolation and equalizer piping schematic.

**TABLE 11.5**
**Multicell Tower Equalizer Line Size**

| Largest Cell Flow Rate (gpm) | Equalizer Line Size (in.) |
|---|---|
| Up to 240 | 4 |
| 241–627 | 6 |
| 628–1167 | 8 |
| 1168–1925 | 10 |
| 1926–2820 | 12 |
| 2821–3465 | 14 |
| 2336–3850 | 2 @10 |
| 3851–5640 | 2 @12 |

Based on field experience, the equalizer line should be sized to accommodate at least 15% of the flow rate of the largest cell. On this basis, Table 11.5 summarizes recommended equalizer line sizes relative to tower cell flow rate.

With an equalizer line, the water operating level will be essentially the same in each cell basin. Therefore, it is critical that each cell be installed so that its recommended operating water level is at the same elevation as adjacent cells.

## PUMP SELECTION, PLACEMENT, AND PIPING

For HVAC condenser water systems, centrifugal pumps (as discussed in the section "Pump Selection and Piping" in Chapter 3) are the most commonly used type of pump and inline, end-suction, horizontal split case, and vertical split case pumps can all be used for condenser water service. However, for some applications, vertical turbine pumps are preferred. Vertical turbine pumps, as shown in Figure 11.15, have two or more impellers mounted in *stages* along a common shaft, with the discharge of each stage serving as input to the next. Both the motor and pump shafts are arranged vertically. This pump is designed with a piped discharge connection, but the intake is designed to be simply submersed in a tank or pool of water (such as a cooling tower basin).

Part of the head that must be produced by a condenser water pump consists of the piping friction losses in the condenser water system (including piping, fittings, valves, etc. and the chiller condenser) and these losses can be computed in the same way as for chilled water systems (see the section "Pump Head and Horsepower" in Chapter 3). However, since the condenser water system is an open system, two additional pressure conditions must be evaluated:

1. *Static head* is the height, in terms of feet of water column, that the condenser water must be lifted and is normally simply the elevation difference between the operating water level in the basin and the wet deck, as shown

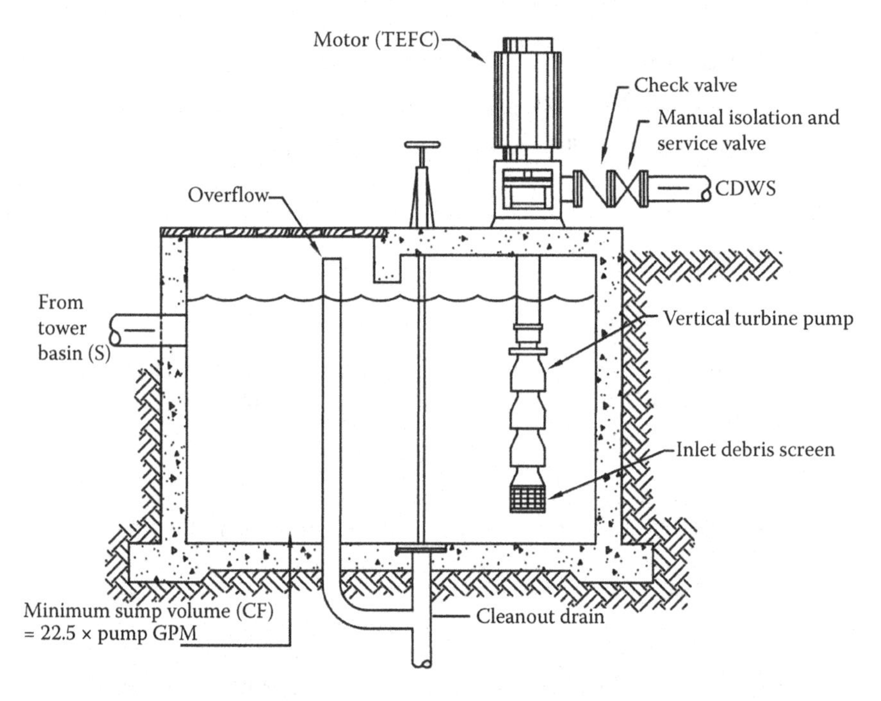


**FIGURE 11.15** Recommended installation for remote condenser water sump with a vertical turbine pump.

in Figure 11.16. It makes no difference if the tower is on the roof of a tall building or sitting on the ground outside the chiller room; the static head is represented only by the height of the "open" portion of the condenser water system, which is always the tower itself.

2. Required residual pressure at the wet deck is the final component of the condenser water head requirement. For crossflow towers, the gravity wet deck system imposes no residual pressure requirement. But, for counterflow towers, the required spray nozzle pressure (typically 2–5 psig) must be included. (The residual head requirement is computed by multiplying the psig pressure requirement by 2.31 ft of water/psig.)

*A check valve must always be installed at the pump discharge.* The check valve prevents condenser water in the piping on the discharge side of the pump from "draining back" through the pump and into the basin while seeking an equilibrium level (Bernoulli's law) when the pump is not operating.

With base-mounted end-suction or split case pumps, the pump may be located near the cooling tower or may be located remotely in a mechanical equipment room. In both cases, two conditions must be satisfied:

1. The pump and all condenser water suction piping (piping between the cooling tower basin and the pump inlet) must be installed below the base

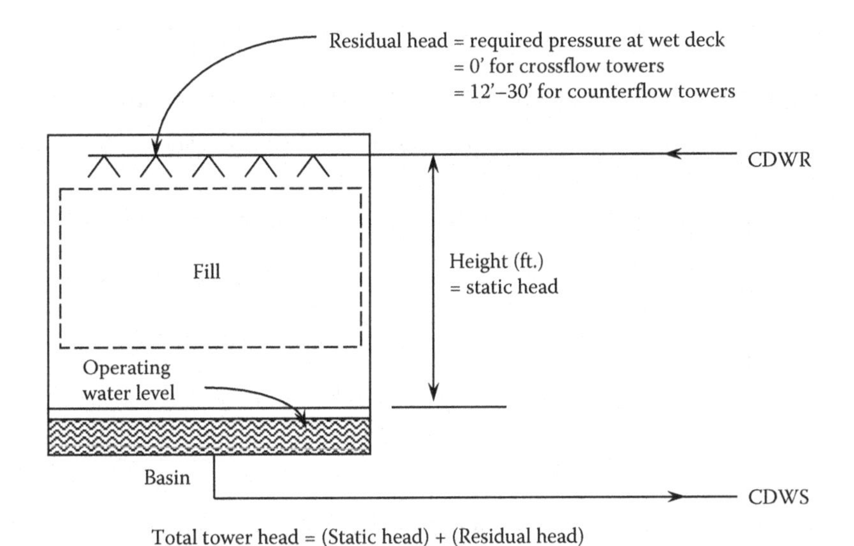


**FIGURE 11.16** Condenser water pressure loss elements at the cooling tower.

tower basin operating water level. If any section of the suction piping is installed above the basin water level, a portion of the water will drain back anytime the pump is shut down. Unfortunately, the next time the pump is started, there will be an air gap in the piping and there will be no condenser water flow once air reaches the impeller.

2. The available net positive suction head (NPSH) must be equal to or greater than the pump's required NPSH in order to prevent pump cavitation.

Cavitation is a condition that results when the pressure at the pump impeller inlet is below the vapor pressure of the water, resulting in "flash" vaporization and the formation of water vapor bubbles within the flow stream. As the pump impeller increases the pressure, these bubbles collapse and the resulting noise is very similar to rattling marbles in a can. The pump can be damaged by cavitation (localized pitting on both the impeller and the casing) and, thus this condition must be avoided.

Available NPSH is determined as follows:

$$\text{NPSH} = \text{BP} - H - \text{SH} - \text{VP}$$

where

NPSH = available net positive suction head (ft of water column)
BP = barometric pressure (ft of water, absolute)
$H$ = pressure loss due to friction in the suction piping (ft of water)
SH = static head on the suction side of the pump ("lift") (ft of water)
VP = water vapor pressure (ft of water, at water temperature)

**TABLE 11.6**
**Water Vapor Pressure (VP) as a Function of Temperature**

| Temperature (°F) | Vapor Pressure (ft of Water, Absolute) |
|---|---|
| 40 | 0.28 |
| 45 | 0.35 |
| 50 | 0.42 |
| 60 | 0.59 |
| 70 | 0.89 |
| 80 | 1.17 |
| 85 | 1.38 |
| 90 | 1.61 |
| 100 | 2.19 |

BP and VP for typical HVAC conditions are summarized in Tables 11.6 and 11.7.

Vertical turbine pumps are always located at the tower since the suction side of the pump must be installed in the basin or in a sump or basin extension specifically constructed for the pump. Vertical turbine pumps are a good option when it is necessary to locate the pump at the tower due to one or both of the following conditions that would result in an NPSH problem for a base-mounted centrifugal pump that is located remotely from the tower:

1. Long suction piping lengths to a remote pump.
2. The tower elevation is significantly below the elevation of a remote pump.

Side outlet sumps on HVAC cooling tower basins are designed to minimize vortexing and associated air entrainment in the supply condenser water. But, as water flows though the cooling tower, it always becomes aerated and leaves the tower with some entrained air. In the pump suction piping, that is, piping between

**TABLE 11.7**
**Barometric Pressure (BP) as a Function of Elevation**

| Elevation (ft) | Atmospheric Pressure (ft of Water) |
|---|---|
| 0 (sea level) | 33.9 |
| 500 | 33.3 |
| 1000 | 32.8 |
| 1500 | 32.1 |
| 2000 | 31.5 |
| 3000 | 30.4 |
| 4000 | 29.2 |
| 5000 | 28.2 |

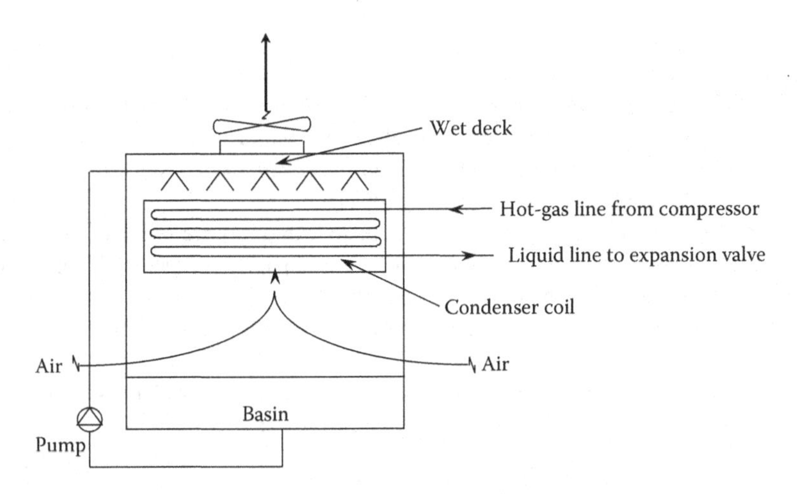


**FIGURE 11.17** Evaporative condenser schematic.

the tower basin and the pump inlet, some of this air will tend to separate. At the pump, if an eccentric reducer is installed in a *flat-on-bottom* configuration, released air will be "trapped" in the top of the suction piping, reducing the effective pipe area and impeding flow. Additionally, free air can enter the pump, causing a noise similar to cavitation. This condition can simply be avoided by using concentric reducers or installing an eccentric reducer in a *flat-on-top* configuration.

## EVAPORATIVE CONDENSERS AND COOLERS

Evaporative condensers and coolers are "first cousins" of the cooling tower, the difference being the use of a closed coil in lieu of open fill.

The vast majority of direct expansion (DX) cooling systems use ambient air as their heat sink for condensing the refrigerant during the vapor compression refrigeration cycle. For comfort cooling, the refrigerant condensing temperature is typically 110–140°F, resulting in 75–100°F refrigerant *lift* or energy input. With evaporative cooling, the condensing temperature can be reduced to 95–105°F, saving 40–60% peak energy input.

The *evaporative condenser*, as illustrated schematically in Figure 11.17, replaces the tower fill with a refrigerant-to-air cooling coil. Water is introduced through a spray-type wet deck in a counterflow configuration to the airflow. By wetting the coil surface and allowing evaporative cooling to occur, the effective air temperature is reduced by 10–15°F, which results in significantly improved heat transfer and a lower condensing temperature compared to dry air cooling. Evaporative condensers are usually designed for saturated condensing temperatures between 95°F and 105°F with an entering wet bulb temperature of 78–80°F.

Water flow to an evaporative condenser is typically between 1.0 and 1.25 gpm/ton simply because their range is higher (usually 25–30°F). The evaporation rate, however, is the same as for a cooling tower, 0.1% per degree of range, and water treatment is required (see Chapter 13).

The *evaporative cooler* is configured in the same way as the evaporative condenser except that a liquid, rather than refrigerant vapor, flows through the coil. This device is sometimes referred to as a *closed-circuit cooler* since the liquid is not circulated through an "open" cooling tower.

Evaporative coolers are larger, more expensive, and require greater fan energy input than a cooling tower sized for the same capacity, range, and approach. However, they allow for the cooling of liquids other than water, such as the water–glycol mixes that are often used in data processing cooling and water source heat pump systems.

The use of both evaporative condensers and evaporative coolers is somewhat limited in HVAC applications and a detailed discussion is beyond the scope of this text.

# 12 Cooling Tower Controls

HVAC cooling tower control falls into four areas: start-up control, capacity or condenser water temperature control, makeup water control, and operating safety control.

## START/STOP CONTROL

Obviously, the first step in controlling the operation of a cooling tower is to simply turn it on and off when needed. For a "dedicated" tower, starting and stopping are controlled by an electrical "interlock" with the chiller so that the tower is allowed to operate anytime the chiller is in operation. This interlock first starts the condenser water pump(s) to establish flow and, then, the actual tower operation is controlled by the capacity control scheme applied (see the section "Capacity Control"). In a multitower or multicell common condenser water system, the tower start/stop control is an integral part of the capacity control scheme for the system.

Figure 12.1 illustrates a common hardwired dedicated chiller, condenser water pump, and cooling tower interlock scheme.

For any condenser water system, either single tower or multitower/cell, the initial system start-up can create a problem. If the system has been shut down overnight, as often happens in spring and fall, the condenser water temperature in the tower basin may be relatively cold (below about 65°F). If this cold water is circulated to the condenser, the chiller may shut down by its safety operating controls due to "low condenser water temperature" or "low refrigerant temperature." In either case, the chiller will simply not start until the condenser water temperature is increased to above 65°F. *To prevent this problem, a modulating three-way control valve should always be included with HVAC cooling tower systems.*

The control valve is normally open to the bypass (normally closed to the tower). Thus, upon start-up, all condenser water flow bypasses the tower and simply circulates through the condenser. As the chiller loads, the condenser begins to reject heat to the water, raising its temperature. As the temperature rises, to setpoint (usually 70–75°F), the control valve modulates the bypass port closed, diverting the water to the tower, upon which the tower capacity control scheme takes over.

The bypass control valve should be placed as close to the condenser as possible, but it must also be at an elevation at or below the tower basin operating water level if the bypass line is connected directly to the condenser water supply line. If not, when the chiller and its condenser water pump shut down at night, the bypass port reopens and the water in the bypass line will drain to the same level as the tower

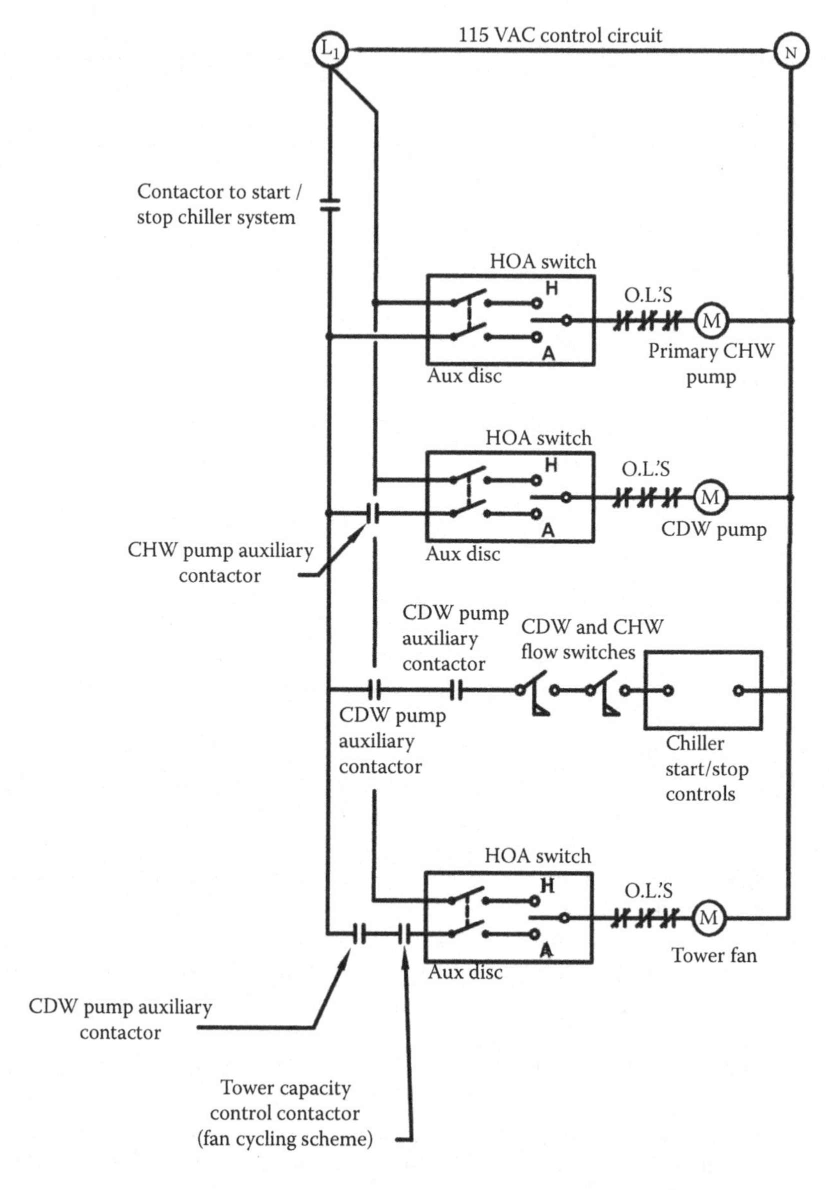


**FIGURE 12.1** Dedicated chiller and cooling tower interlock wiring schematic.

operating water level. This, in turn, creates an "air gap" or "bubble" in the bypass. Upon start-up the next morning, this air bubble will enter the pump, causing it to lose prime and all flow will stop, immediately causing the chiller to shut down!

If the valve cannot be installed below the tower operating water level, the bypass line must be piped to the tower basin. Thus, while the drain down will still occur, the resulting air bubble cannot enter the pump. (An alternative would be to install a check valve in the suction piping between the bypass connection and the tower basin. But, over time, the check valve will probably leak and the problem will return.) In this configuration, however, the basin water, at a low temperature, is supplied to the condenser, so chiller start-up problems may result. Figure 12.2 illustrates the correct bypass valve locations in the condenser water piping.

For proper bypass valve installation and operation, the valve should be located and piped as follows:

1. Locate the valve as close to the condenser as possible.
2. The valve and all piping should be below the tower(s) water operating level.
3. The valve bypass should be piped directly to the condenser water supply.

The bypass control valve, ideally, would have linear flow characteristics. However, butterfly valves, which are normally used for bypass duty, have equal

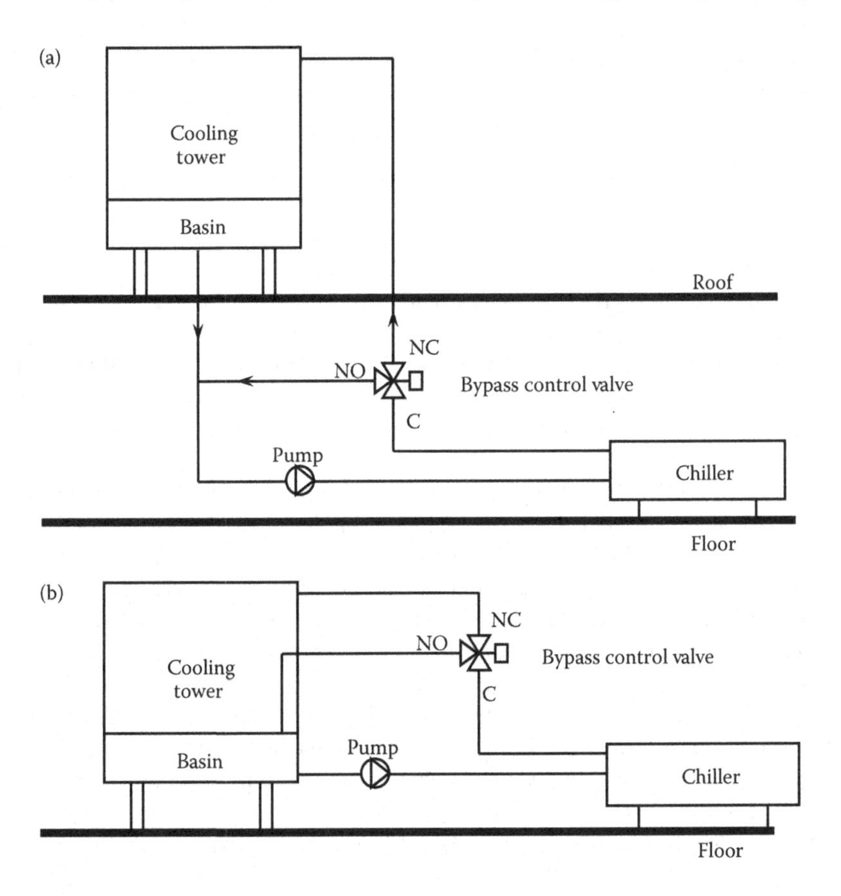


**FIGURE 12.2** (a) Cooling tower bypass valve installation: chiller, pump, and piping located below the tower basin level. (b) Cooling tower bypass valve installation: chiller, pump, and piping located at the same level as the tower basin.

percentage flow characteristics. However, a butterfly valve will have a nearly linear relationship between "% open" and "% flow" if it is selected to operate with a maximum 70° rotation and a 33% authority setting.

Additionally, to aid maintenance, the butterfly valves should be the full-lug type, with operators sized for tight shutoff. Thus, these valves can be used as isolation valves, also.

To select the bypass valve on this basis, compute the control valve flow coefficient $C_v$ as follows:

$$C_v = \frac{PD^{0.5}}{F_{cdw}}$$

where

PD = required pressure drop across a valve at design flow (psig)

$F_{cdw}$ = design condenser water flow rate (gpm)

PD must be equal to or greater than the sum of the static head and residual head required by the tower at design flow.

Based on the computed flow coefficient, the correct valve size can be selected from Table 12.1.

## CAPACITY CONTROL

Tower capacity control is synonymous with condenser water temperature control via manipulation of the airflow through the tower. Condenser water supply temperature will decrease if the imposed cooling load falls, as indicated by a reduced condenser water return temperature, and/or the outside air WB temperature lowers. Thus, to prevent the condenser water from becoming too cold, the tower capacity must be reduced to maintain the condenser water supply temperature setpoint.

Vapor compression cycle systems have a 1.5–2.5% improvement in operating efficiency and resulting energy consumption for each 1°F decrease in condenser water supply temperature. Most manufacturers have set the minimum condenser water supply temperature requirements at 70–75°F for centrifugal chillers, although at least one manufacturer guarantees operation down to 55°F condenser water supply temperature. Rotary screw compressors can operate with a condenser water supply temperature lower than for centrifugal compressors, but the minimum temperature will vary from manufacturer to manufacturer. Thus, it is better to establish the condenser water temperature control setpoint as 70–75°F.

While a lower condenser water supply temperature will result in lower chiller energy consumption, the cooling tower fan energy consumption will increase somewhat for this mode of control. But, studies have shown that chiller energy savings usually far exceed the additional cooling tower energy consumed within the temperature ranges discussed here.

Figure 12.3 illustrates a simple condenser water temperature control sequence, including both tower bypass control and tower capacity control.

**TABLE 12.1**
**Tower Bypass Butterfly Control Valve Size**

| Approximate $C_v$ at 70° Rotation | Pipe Size (in.) | Required Butterfly Valve Size (in.) |
|---|---|---|
| 69 | 3 | 2 |
| 136 | 3 | $2\frac{1}{2}$ |
| 201 | 4 | 3 |
| 496 | 4 | 4 |
| 348 | 6 | 4 |
| 611 | 6 | 5 |
| 722 | 6 | 6 |
| 533 | 8 | 5 |
| 784 | 8 | 6 |
| 913 | 8 | 8 |
| 729 | 10 | 6 |
| 1462 | 10 | 8 |
| 2948 | 10 | 10 |
| 1347 | 12 | 8 |
| 2367 | 12 | 10 |
| 4393 | 12 | 12 |
| 2174 | 14 | 10 |
| 3565 | 14 | 12 |
| 5939 | 14 | 14 |
| 3261 | 16 | 12 |
| 4923 | 16 | 14 |
| 7867 | 16 | 16 |
| 4509 | 18 | 14 |
| 6597 | 18 | 16 |
| 10,065 | 18 | 18 |
| 6042 | 20 | 16 |
| 8629 | 20 | 18 |
| 12,535 | 20 | 20 |
| 8789 | 24 | 18 |
| 11,666 | 24 | 20 |

## Fan Cycling

The simplest method of condenser water temperature control is to vary airflow by cycling the cooling tower fan(s) on and off. It is necessary to establish both a condenser water supply temperature setpoint and a reasonable throttling range (or control differential), usually 2–3°F. Thus, the fan is started when the supply temperature rises to setpoint plus half of the differential and stopped when the

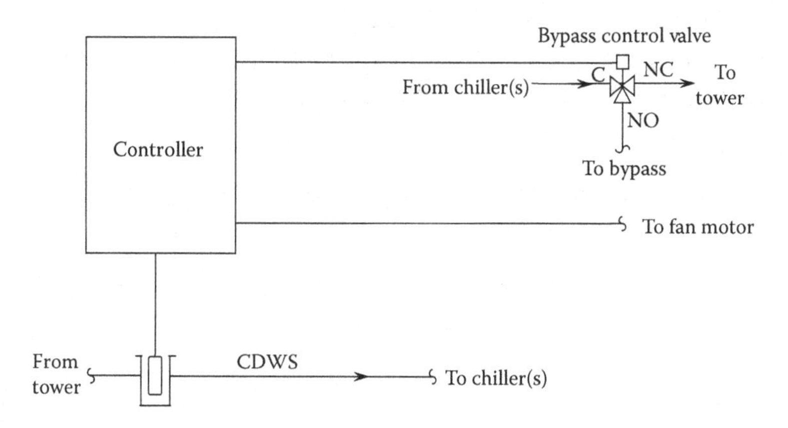


**FIGURE 12.3** Simplified condenser water temperature control system.

supply temperature is reduced to setpoint minus half of the differential. Thus, with a 75°F setpoint and a 3°F differential, the fan starts at 76.5°F water temperature and stops at 73.5°F.

Figure 12.4 illustrates this control process. Note that the "actual" differential will be greater than the controller differential due to thermal mass in the system and time delay in initiating and terminating the heat transfer process in the tower and condenser.

When selecting the controller differential, it is important to pick one that is reasonable (typically 2–3°F). Too great a differential will cause erratic control by allowing too much temperature variation, while too small a differential will result in rapid or "short" cycling of the tower fan(s) and can result in rapid motor failure, drive damage, or even fan or shaft failure.

A drawback of fan cycling control is that under very light-load conditions, the tower capacity is so great compared to the imposed load that rapid fan cycling

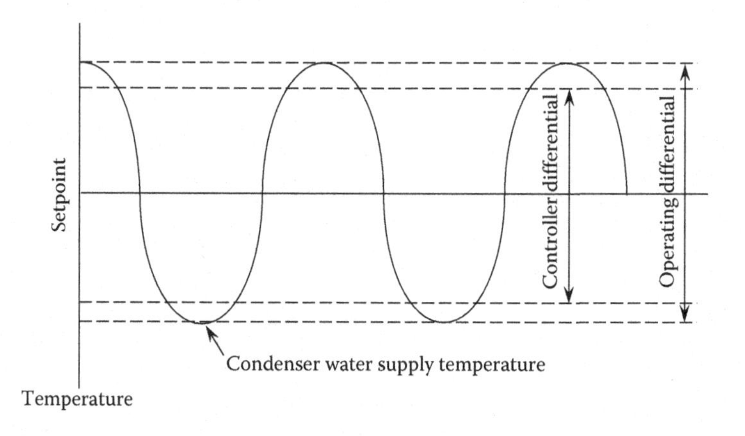


**FIGURE 12.4** Condenser water supply temperature control based on tower fan cycling.

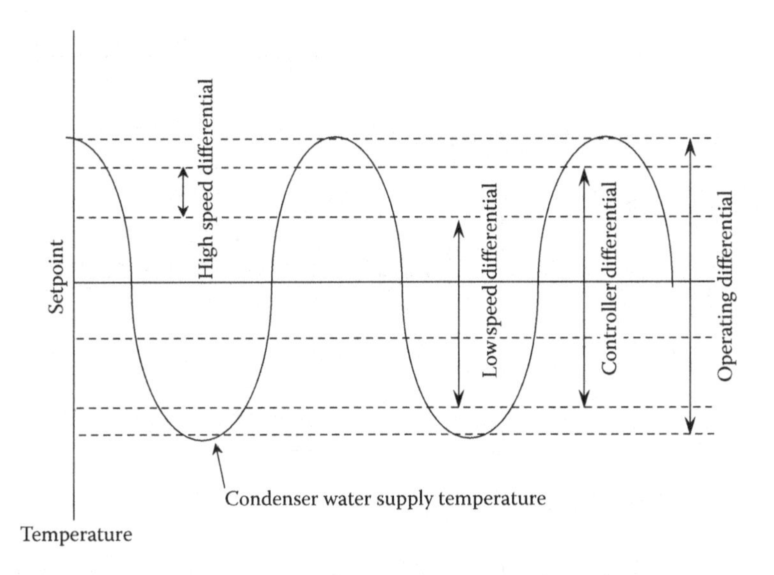


**FIGURE 12.5** Condenser water supply temperature control based on tower two-speed fan cycling.

could occur. Thus, as discussed in the section "Motors" in Chapter 10, two-speed fan motors or tandem pony motors are recommended for cooling towers since they allow an improvement beyond on–off fan cycling control by adding an intermediate capacity level: half speed. The fan cycling control concept is the same, except that a second stage of control is added, as shown in Figure 12.5.

Figure 12.6 illustrates a typical condenser water interlock wiring scheme for a two-speed tower fan motor or a main motor and pony motor.

With two-speed or pony motor control, a short time delay (about 15 s) must be included when gear drives are used to prevent drive damage that may result from rapid speed change and the resulting "braking" effect.

## Fan Speed Control

In lieu of the "step control" of tower fan(s) cycling, modulating or "proportional" control can be achieved through the use of tower inlet dampers, variable-pitch axial fans, or fan speed control.

Dampers, applied to the inlet side(s) of a cooling tower, have proved to be serious maintenance headaches while, due to their nonlinear control characteristics at low velocities, providing relative poor airflow control. Thus, dampers are generally no longer used in HVAC cooling towers.

The controllable-pitch axial fan concept is widely applied in propeller-driven aircraft, but has a poor operational history in cooling tower applications. The mechanical linkages utilized have proved to be unreliable and a constant maintenance problem. This concept is rarely used in HVAC cooling towers.

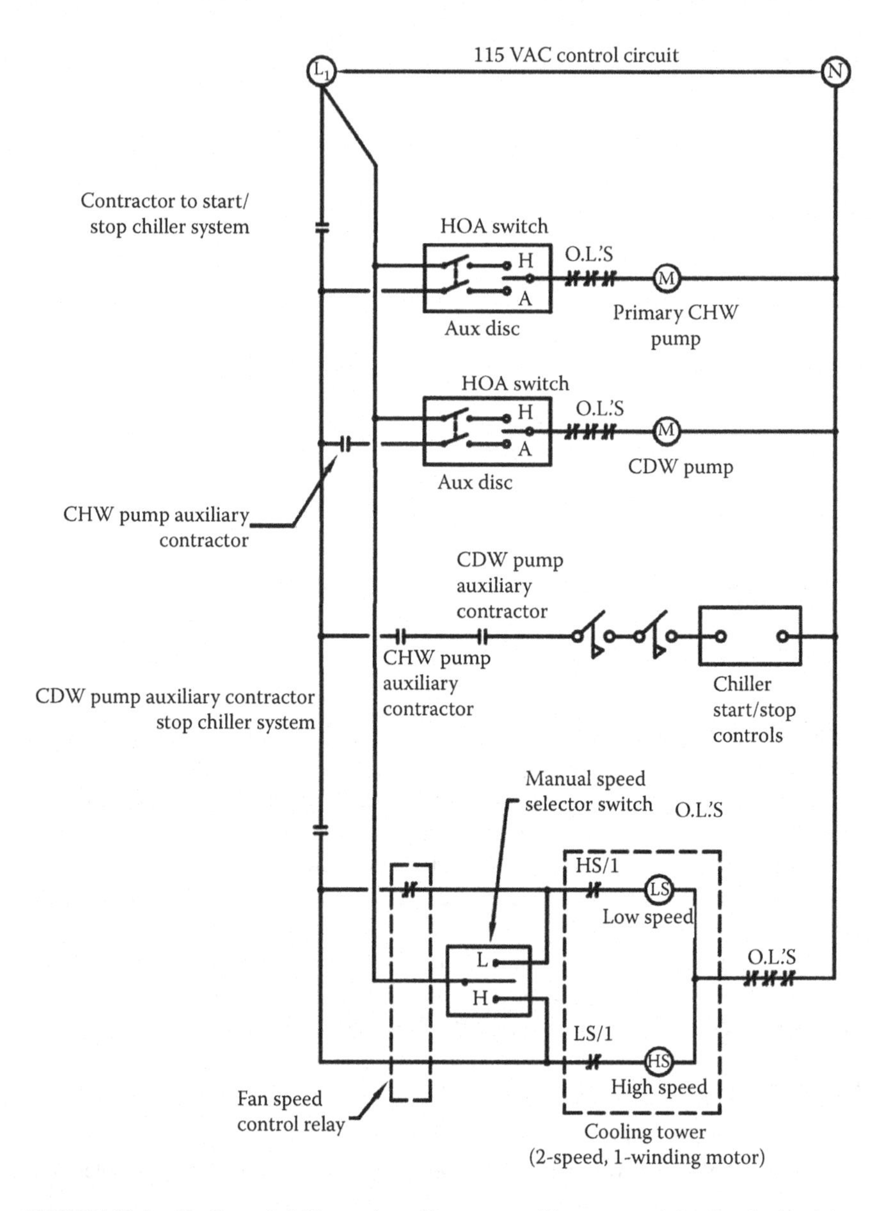


**FIGURE 12.6** Dedicated chiller and cooling tower with two-speed fan interlock wiring schematic.

Variable frequency drives have been successfully applied to cooling tower fan motors so that condenser water supply temperature could be controlled by changing the fan speed in direct proportion to the change in condenser water supply temperature. Thus, a 10% increase above setpoint of the condenser water supply temperature would result in a 10% increase in fan speed, that is, *proportional* control.

With proportional control, there is a linear relationship between the final control element (the VFD) and the controlled variable (condenser water supply temperature). The setpoint is typically in the middle of the throttling range, but the final control element is seldom in the middle of its range, resulting in a control point that is "offset" from the setpoint. Typically, the control point and setpoint will be the same at 50% load, but the condenser water temperature will tend to be lower at light-load conditions (<50% load) and higher at heavier-load conditions (>50% load). To eliminate offset, an *integral function* should be introduced to the proportional control loop. The integral function acts to "reset" the control point anytime it drifts off setpoint. This control mode is called *proportional + integral* (PI) and is the recommended mode for fan speed control. Often, control vendors will attempt to add a *derivative* element to the control mode to create a *proportional + integral + derivative* (PID) mode for operation. PID controls should be avoided since adding the derivative function tends to create more problems than it solves.

The application and cost effectiveness of VFDs have been widely discussed in the industry and the following guidelines for their application have developed:

1. Avoid operating the cooling tower fan at its "critical speed," or at a multiple thereof.
2. Be very careful when applying VFDs to gear drives. The braking action created by the motor as the speed is slowed can cause damage to the gear drive and the VFD. Also, at low speeds, gear drives may cause vibration, noise, or gear "chatter." A separate, electrically driven oil pump may also be required to provide lubrication at low speeds. (Coordinate these items carefully with the tower manufacturer.)
3. Figure 12.7 illustrates the relative energy use by an HVAC cooling with a two-speed fan motor versus the same tower with a VFD. Thus, unless the tower is very large (40 hp or greater motor size), the savings produced by the VFD will rarely offset the additional cost for the VFD.
4. Using VFDs on each tower in a multitower system, no matter what the motor size, is a waste of money since the operating cost savings over tower staging control will not offset the additional costs for multiple VFDs. In a case where operation under very light loads is anticipated, for example, below 50% of a single tower's or cell's capacity, then a VFD on one tower may be cost effective if (a) the tower is large, and (b) this tower is always the first and last stage in the tower operation sequence.
5. VFDs must have bypass switching (automatic or manual) so that the tower can be operated even if the drive fails. The fan speed control system, then, must include provisions to revert to fan cycling control in this event since the fan can only operate at full speed.

Even with VFDs, there is a "minimum" speed required to keep the fan motor from overheating. If the tower capacity at the minimum fan speed exceeds the required capacity, the fan control mode must revert to fan cycling, albeit at the minimum speed.

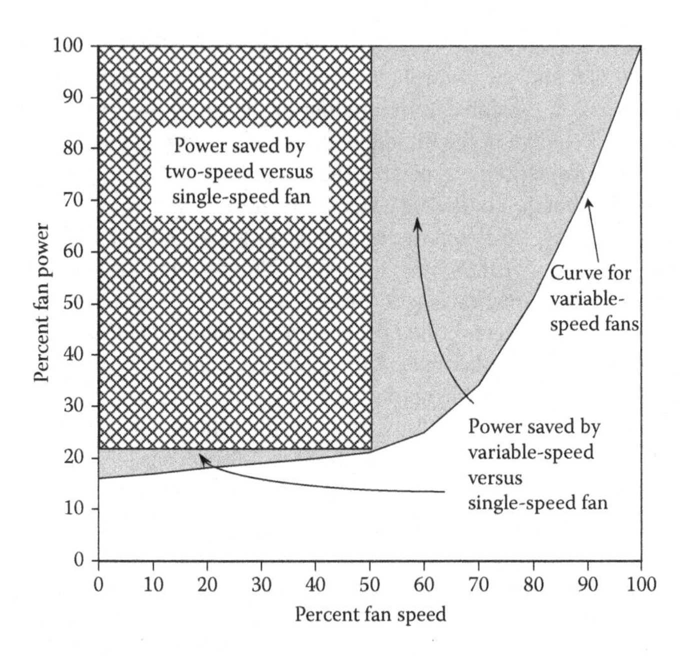


**FIGURE 12.7** Cooling tower fan power requirement: two-speed fan cycling control versus variable-speed control.

The problems with gear drives with fan speed control can be eliminated through the use of a cooling tower direct-drive motor and speed control system introduced by Baldor Electric Company in 2009. This motor allows gear drive systems to be replaced with a directly coupled motor that is designed especially for cooling tower application. This motor is a TEAO permanent magnet-type, synchronous motor with no slip. Since a permanent magnet motor cannot run "across the line" as does a conventional motor, the VFD provided by Baldor is designed specifically to start and control the speed of this motor. The Baldor cooling tower motor and speed controller package can be specified for a new tower or used in retrofit applications with existing towers.

With retrofits, drive losses of 2–5% are eliminated with application of the Baldor direct-drive motor, effectively improving the efficiency of the tower in addition to the energy savings associated with fan speed control (especially for single-speed motor installations).

## Tower Staging

In a multiple tower or cell common condenser water system, capacity control is accomplished by a combination of water flow control and fan cycling.

With increasing load, the automatic isolation valves at each tower or tower cell should be sequenced open so that flow is provided through all towers before any tower fan is started. At light cooling loads, maximizing the available heat transfer surface before running a fan is more energy efficient. Once flow is established to each tower or cell, the fans can be staged on and off in sequence to control capacity.

When multiple towers or tower cells utilize two-speed or pony motors, it is more energy efficient to operate all cells on low speed before switching any cell to high speed with increasing load. The incremental power increase associated with starting an additional fan on low speed is less than 79% of the power increase from switching a fan from low speed to high speed. (Likewise, on a falling load, all fans should be switched to low speed before any fan is cycled off.) The use of additional towers or tower cells also increases the heat transfer surface that is available, decreasing approach and the condenser water supply temperature.

Where multiple towers or tower cells are used, with proper sizing and staging control, only one tower or cell requires fan speed control, beyond two-speed motors. Once the load on the tower system has been reduced to less than the capacity of one tower or cell operating at low speed, the application of fan speed control over fan cycling may be advantageous.

## MAKEUP WATER CONTROL

The makeup water to an HVAC cooling tower is controlled by monitoring the tower basin water level. Two types of devices are used for this duty:

1. Mechanical float valves are simple float-operated valves that modulate open to add water as the float falls with the basin water level and modulate close as the float rises. However, these valves are not reliable or accurate and are routine maintenance headaches. Thus, they are not recommended except for the smallest of HVAC cooling towers.
2. Electronic-level sensors and controllers are commonly applied to HVAC cooling towers. The level sensor has four probes that detect water level in the basin:
   a. High-water-level alarm
   b. High operating water level
   c. Low operating water level
   d. Low water level alarm

   When the water level drops to the low operating water level, a signal opens a makeup water solenoid valve to add water to the system. When the water level rises to the high operating water level, a signal closes the makeup valve.

## OPERATING SAFETY CONTROLS

The high- and low-water-level alarms from the tower electronic-level controller(s) should be wired to the facility management system or, at a minimum, to both visual and audible alarm annunciators. (A tower that floods can be annoying but a tower that runs dry puts the cooling system out of commission.)

For towers with propeller fans, there is always a potential, however small, that a fan may be damaged or it may lose a fan blade. If the fan, now badly out of balance, continues to operate, vibration damage to the fan, the drive, and the tower structure can result. *To prevent this condition, it is recommended that all propeller fan towers be equipped with a vibration switch that deenergizes power to the fan motor in the event of tower vibration.* The switch is normally installed outside the fan cylinder, near the motor, in a weatherproof enclosure (NEMA 3R or 12 type).

# 13 Condenser Water Treatment

Condenser water treatment is required to prevent or control deposition (scaling), corrosion, and microbiological fouling of cooling towers, condensers, and piping.

## DEPOSITION CONTROL

*Deposition* is the term that is used to describe scale and other deposits that may form on the heat transfer surfaces of a condenser water system. Water has been called the "universal solvent" and makeup water will always contain dissolved mineral solids to a greater or lesser extent. The purpose of the deposition control portion of a condenser water treatment program is to prevent the level of dissolved solids from becoming so high that they will begin to be deposited on the wetted surfaces of the system.

The amount of dissolved solids in makeup water depends on the source of the water and the geology of the area. However, most regions of the United States have some sedimentary rock, such as limestone, with which water comes into contact. The water readily dissolves the calcium and magnesium carbonate, the major components of limestone, which, in solution, separate into ions of positively charged calcium and magnesium and negatively charged carbonate and bicarbonate.

Obviously, the more limestone there is in the local geology, the greater the amount of dissolved solids, and the "harder" the water. Areas with little limestone will have lower concentrations of dissolved solids and will be "soft" by comparison. Table 13.1 defines the levels of total dissolved solids relative to various water hardness levels. Since calcium, magnesium, and carbonate are found all together and they are not easily distinguishable by chemical test, all three are expressed as $CaCO_3$ (calcium carbonate) for water treatment applications.

Deposits of calcium carbonate on wetted surfaces in a condenser water system act as insulation and seriously reduce the heat transfer effectiveness. Deposit thickness can increase to a level where it may reduce flow areas and, thus, significantly reduce flow rates as associated flow velocity and pressure loss increase. In effect, significant calcium carbonate deposits will destroy the effectiveness of the condenser water system and the refrigeration system it supports.

As a portion of the condenser water flow is lost by evaporation through the cooling tower, the concentration of dissolved solids increases since solids are left behind as the liquid evaporates. Makeup water, which is added to the condenser water system to offset the evaporation losses, will add dissolved

**TABLE 13.1**
**Water Hardness Scale**

| Water Hardness | Concentration of $CaCO_3$ (ppm) |
|---|---|
| Soft | 0–75 |
| Moderately hard | 76–150 |
| Hard | 151–300 |
| Very hard | Over 300 |

minerals at a lower concentration level than the condenser water and, thus, some equilibrium concentration level will be maintained. However, if this equilibrium concentration level is high enough that deposition occurs, a chemical-based water treatment program must be added to control solids concentration at a lower level.

The concentration of dissolved solids can be reduced by adding more makeup water, which has a lower solids concentration, to the condenser water, which has a higher concentration. However, to add water to the system, an equal amount must be removed from the system by *blowdown*, the intentional "dumping" of condenser water to drain.

The tendency to form calcium carbonate deposits (scale) is a function of several factors, including temperature, pH, calcium hardness, total alkalinity, and total dissolved solids of the water.

*Temperature.* The condenser water supply temperature for HVAC systems will range from 70°F at low ambient wet bulb and light load to 85°F at design wet bulb temperature and load. Most scale-forming dissolved solids have the unusual property of becoming less soluble as the water temperature increases. Thus, the greatest potential for scaling occurs at design conditions.

*pH.* Water is made up of hydrogen and oxygen as follows:

$H^+$ protonated water molecule
$OH^-$ hydroxyl radical or hydroxide

Combining $H^+$ with $OH^-$ yields $H_2O$, water.

Chemically, pH is a negative logarithmic scale defining the relative concentration of $H^+$, as shown in Table 13.2. Each increase or decrease of pH by 1 represents a 100-fold increase or decrease in the amount of acidity or alkalinity (base).

*Calcium hardness.* It is the amount of calcium carbonate in the water (ppm).

*Total alkalinity.* It is the amount of negatively charged ions of hydroxide ($OH^-$), bicarbonate ($HCO_3^-$), and carbonate ($CO_2^{2-}$) in the water (ppm) and represents the ability of the water to neutralize acid. Alkalinity occurs in all water with a pH above 4.4.

*Total dissolved solids.* It is the amount of all dissolved solids (calcium, magnesium, phosphate, iron, etc.) in the water.

**TABLE 13.2**
**pH Scale**

| pH | H⁺ Concentration (g mol/L) | Relative Acid | Relative Base |
|---|---|---|---|
| 14 | $10^{-14}$ | | 10,000,000 |
| 13 | $10^{-13}$ | | 1,000,000 |
| 12 | $10^{-12}$ | | 100,000 |
| 11 | $10^{-11}$ | | 10,000 |
| 10 | $10^{-10}$ | | 1000 |
| 9 | $10^{-9}$ | | 100 |
| 8 | $10^{-8}$ | | 10 |
| 7 | $10^{-7}$ | Neutral | |
| 6 | $10^{-6}$ | 10 | |
| 5 | $10^{-5}$ | 100 | |
| 4 | $10^{-4}$ | 1000 | |
| 3 | $10^{-3}$ | 10,000 | |
| 2 | $10^{-2}$ | 100,000 | |
| 1 | $10^{-1}$ | 1,000,000 | |
| 0 | $10^{0}$ | 10,000,000 | |

Since 1936, with the introduction of the Langelier Saturation Index (LSI), water treatment specialists have attempted to "predict" the potential for deposition based on water chemistry and, thus, establish the required "cycles" to maintain deposition control.

The LSI is computed from the following relationship:

$$LSI = pH_a - pH_s$$

where

$pH_s$ = pH at saturation

$pH_a$ = actual pH

The $pH_s$ is computed on the basis of the water analysis data by the following equation:

$$pH_s = 12.3 - (\log_{10}Ca + \log_{10}TA + 0.025T - 0.011TDS^{0.5})$$

where

Ca = calcium hardness (ppm)

TA = total alkalinity (ppm)

$T$ = temperature (°C)

TDS = total dissolved solids (ppm)

LSI generally has a value from +3.0, which indicates high deposition potential, to –3.0, which is often used to indicate a high corrosion potential.

However, the LSI prediction for deposition or corrosion is, particularly for cooling tower systems, often wrong. This value is a function of the properties of the makeup water, the actual cycles of concentration, the quality of mixing of the condenser water and makeup water streams, airborne contaminants in the water, and so on and is almost impossible to predict accurately. In the water treatment industry, the chemistry of the make-up water is used as the starting point and, as cycles are increased, the expected value of the actual pH is then computed. While the result may not be always accurate due to a limited number of factors that are addressed, it does provide a basis upon which to begin.

In 1944, an attempt to improve the prediction of deposition potential was introduced by the Ryzner Stability Index (RSI):

$$RSI = 2(pH_s) - pH_a$$

Typically, the RSI will have a value of 3.5, indicating high deposition potential, to 9.0, indicating high corrosion potential. However, the limitation on the accuracy of predicting the $pH_a$ is not improved by the RSI, and the RSI is no more or no less valid as a predictor than the LSI.

Because of the unreliability of both LSI and RSI to accurately predict deposition or corrosion conditions in cooling tower systems, in the 1970s, a new index called the Puckorius Scaling Index (PSI) was developed. The PSI is based on correcting the actual pH to match the total alkalinity of the water and is computed as follows:

$$PSI = 2(pH_s) - pH_{eq}$$

where $pH_{eq}$ = the adjusted or equilibrium pH computed from the following relationship:

$$pH_{eq} = 1.465 \log_{10}TA + 4.54$$

The range of values of PSI is essentially the same as for RSI.

For corrosion control, condenser water pH should be maintained between 4 and 10, as shown in Figure 13.1. In most cooling tower water treatment programs, the desirable range for pH is between 8 and 9 in order to maintain water alkalinity at a reasonable level (400 ppm or less). However, in metal towers, a pH of 7.0–8.0 is preferred to help prevent white rust corrosion (see the section "White Rust"). Therefore, at least for most metal cooling towers, the ideal pH range is 7.5–8.5.

Water hardness and alkalinity is a function of the hardness and alkalinity of makeup water, the amount of evaporation and drift loss from the cooling tower

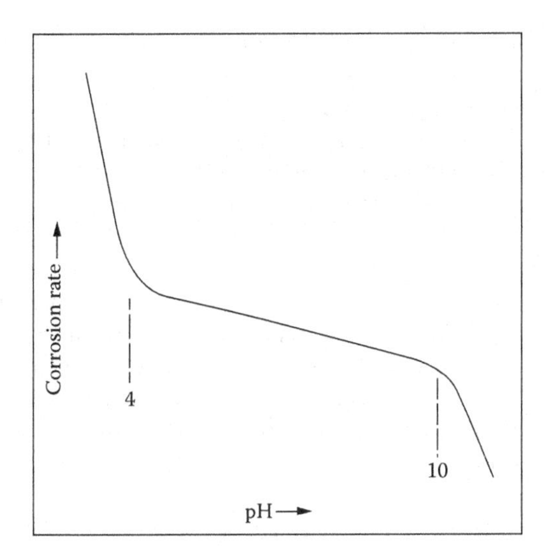


**FIGURE 13.1** Corrosion rate as a function of water pH. (Courtesy of BetzDearborn, Inc., Trevose, PA.)

operation, and the blowdown to proposed to yield the desirable pH to prevent both deposition and corrosion. The term *cycles of concentration* defines the ratio of the desired concentration of dissolved solids in the condenser water to the concentration of dissolved solids in the makeup water as follows:

$$\text{Cycles (of concentration)} = \frac{\text{Dissolved solids (ppm) in blowdown}}{\text{Dissolved solids (ppm) in makeup}}$$

This relation can be express in terms of water flow as follows:

$$\text{Cycles(of concentration)} = \frac{\text{MU}}{\text{BD}}$$

where

MU = total makeup water flow, which is the sum of evaporation + blowdown (gpm)

BD = blowdown flow (gpm)

Since the amount of evaporation, $E$, is typically 0.1% of the condenser water flow rate per degree of range, MU in the above relationship can be replaced with the value ($E$ + BD) to yield the following:

$$\text{BD} = \frac{E}{(\text{cycles} - 1)}$$

Thus, after the number of cycles is determined based on the makeup and desired condenser water (blowdown) concentration of dissolved solids, the actual blowdown requirement can be calculated.

Since drift water loss is not included in these calculations, the actual required BD flow can be reduced by the amount of drift loss from the tower.

There are two ways of controlling blowdown in an HVAC cooling tower system:

1. Constant blowdown with manual adjustment based on periodic water hardness analysis is the simplest method. However, since the amount of blowdown is constant, the loss in water and water treatment chemicals is high and this really represents the most expensive approach.
2. Controlled blowdown based on continuous monitoring of the water hardness as indicated by its conductivity is another method. Automatic control minimizes the waste of water and water treatment chemicals and is the preferred method, as described in the section "Water Treatment Control Systems."

Table 13.3 summarizes the water flows associated with a cooling tower for various cycles of concentration. From Table 13.3, it is clear that the amount of makeup water is reduced significantly as the number of cycles is increased from two to six. However, there is only a further 5% reduction as the number of cycles is increased from 6 to 10, and only a further 2% reduction as the number of cycles is increased to 20. *Therefore, in most cooling tower applications, the number of cycles of concentration is maintained between 5 and 10 (as indicated by the LSI, RSI, and/or PSI) and chemical deposition inhibitors are added as necessary.* While lower number of cycles represents loss of more water and treatment chemicals, the amount of treatment chemicals required tends to go down with the number of cycles, and 5–10 cycles usually represent a good balance point.

**TABLE 13.3**
**Makeup Water Requirement as Function of Cycles**

| Cycles | Evaporation (gpm/ton) | Blowdown (gpm/ton) | Makeup (gpm/ton) | % Makeup |
|---|---|---|---|---|
| 2 | 0.0300 | 0.0300 | 0.0600 | 100 |
| 3 | 0.0300 | 0.0150 | 0.0450 | 75 |
| 4 | 0.0300 | 0.0100 | 0.0400 | 67 |
| 5 | 0.0300 | 0.0075 | 0.0375 | 63 |
| 6 | 0.0300 | 0.0060 | 0.0360 | 60 |
| 10 | 0.0300 | 0.0033 | 0.0333 | 55 |
| 15 | 0.0300 | 0.0023 | 0.0323 | 54 |
| 20 | 0.0300 | 0.0015 | 0.0315 | 53 |

Reduction of hardness by blowdown or even by water softening pretreatment may not eliminate the potential for deposition in the condenser water system. In these cases, chemical treatment, that is, the addition of deposition inhibitors to the water system, becomes necessary.

For a dissolved salt to precipitate and deposit on the wetted metal surfaces of the condenser water system, it forms a crystal growth that attaches itself to metal surfaces as it comes out of solution. The most common scale inhibitors used in condenser water systems are *phosphonates*, which are organic phosphate compounds, such as hydroxyethylidene-1,1-diphophonic acid (HEDP), which function by adsorption on the mineral crystals as they form and prevent them from attaching to the metal. Thus, these crystals precipitate out of solution, usually into the tower basin.

## CORROSION CONTROL

### GALVANIC CORROSION

Metal corrosion occurs as a result of *galvanic action* at a negatively charged "pole" or site on the metal surface. Both *anodes*, negatively charged sites, and *cathodes*, positively charged sites, can be created on the metal due to impurities in the metal, localized stress, metal grain size or composition differences, or even scratches on the metal surface. Due to the differences in charges, there is an electrical potential between the anode and cathode and an electrical current (electrons) flows from anodes to cathodes, using the surrounding water as a conductor or an *electrolyte*, as shown in Figure 13.2.

For steel, the anodic reaction is for the iron to give off two free electrons, thus becoming positively charged. This positively charged iron then combines two hydroxyl radicals from water to form ferrous hydroxide, which combines further with water to form ferric hydroxide that, when dehydrated, is iron oxide or rust.

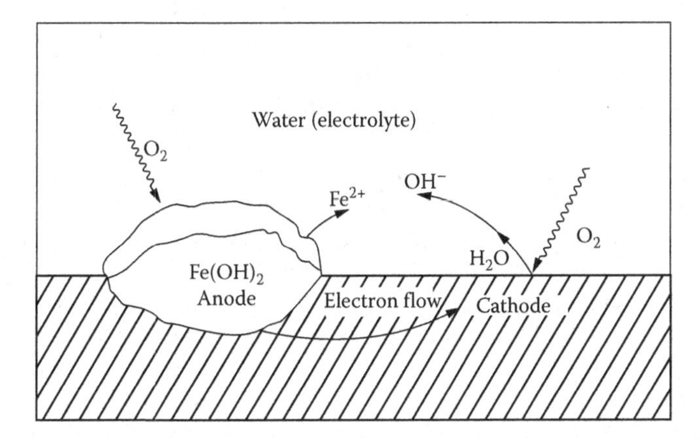


**FIGURE 13.2** Typical steel corrosion chemistry. (Courtesy of BetzDearborn, Inc., Trevose, PA.)

At cathodes, the electrons given off at the anode combine with water to yield hydroxyl radicals, to balance the reactions.

Corrosion then is the loss of metal; it literally "dissolves." Corrosion can exhibit two characteristics depending on the underlying reason for the anodic and cathodic sites. *General corrosion* is widespread and is caused, usually, by impurities in the metal or characteristics of the metal or its environment that results in an overall fouling of the metal surface. *Localized corrosion* results, mostly, from scratches, stress, or localized environment and is the most common reason for "metal failure."

If dissimilar metals with different electrical potentials are used in a condenser water system, galvanic corrosion is enhanced and the metals simply corrode faster, particularly at and near the point(s) of contact between the metals.

The first step in corrosion control is to minimize the contact between water and mild steel materials. All primary wetted surfaces—wet decks for induced draft towers and basins—should be constructed of stainless steel. This typically increases tower costs by 20–40% and is generally a worthwhile investment. Use plastics or fiberglass for the tower casing, wet deck covers, intake louvers, drift eliminators, and fill. Finally, if mild steel is used for the tower structural frame, it should be galvanized and, as discussed in the section "White Rust," coated with an epoxy or polymer final protective coating.

However, the piping in most condenser water systems will be steel and must be protected from corrosion. This is accomplished by using one or more treatment programs as follows:

*Passivating (anodic) inhibitors.* These chemicals form a protective oxide film on the metal surface that is not only tough, but, when damaged, quickly repairs itself. Typical chemicals that act as passivating inhibitors include molybdate, polyphosphates, and orthophosphate. These chemicals are oxidizers that promote passivation by increasing the electrical potential of the iron. The drawback of the use of molybdate is its high cost, and it is used only when blowdown levels are kept as low as possible. *Orthophosphates should not be used in condenser water systems containing stainless steel since it will make the metal brittle over time.*

*Precipitating inhibitors.* At cathodic sites, the localized pH at the site is increased due to the higher concentration of hydroxide ions that are being produced. Precipitating inhibitors form complexes that are insoluble at higher pH and, thus, precipitate out of the water. Zinc is a good precipitating inhibitor. Molybdate will also act as a precipitating inhibitor and, thus, can serve as a corrosion inhibitor using two mechanisms.

*Adsorption inhibitors.* These are organic compounds containing nitrogen, such as amines, or sulfur or hydroxyl groups. Due to the shape, size, orientation, and electrical charge of the molecule, they will attach to the

**TABLE 13.4**
**Water pH and Chemistry for Corrosion Programs**

| Program | Water pH and Chemistry |
|---|---|
| Zinc | 7.5–8.5 |
| Molybdate or molybdate/zinc | 7.5–9.5 |
| Orthophosphate | 7.5–8.5, with a phosphate deposition inhibitor |
| Organic adsorption | 7.5–9.5, with 300–500 ppm alkalinity |

surface of the metal, preventing corrosion. Their drawback is that they form thick, oily surface films that reduce heat transfer capability.

Each program is designed for particular condenser water pH range and water chemistry, as summarized in Table 13.4.

To protect copper in heat exchanger tubes and piping from corrosion, aromatic triazoles, such as benzotrizole (BZT) and tolyltriazone (TTA), are used in most condenser water treatment systems. These compounds bond with the cuprous oxide on the metal surface and protect it.

## White Rust

Since the late 1970s, there has been a significant increase in the use of galvanized steel cooling towers, particularly in the HVAC market. With the advent of new cooling tower water treatment polymers and corrosion inhibitors, most condenser water treatment programs now utilize little or no acid addition, and operating pH levels have increased from 6.5–7.5 to as high as 9.5 (with the new alkaline polymer approach). As the technology of the water treatment industry has changed, the corrosion of galvanized steel has now become a major concern.

The term *white rust* refers to the premature, rapid loss of galvanized coating on cooling tower metal surfaces. White rust is evidenced by a white, waxy, nonprotective zinc corrosion deposit on wetted galvanized surfaces. This rapid loss of the galvanizing results in the corrosion of the underlying steel and, instead of tower systems that will last 20–25 years, equipment will have drastically shortened lifespans.

Many believe that the change to the dry kettle method of galvanizing, with continuous sheet steel (see Chapter 10), has had an effect on the increased formation of white rust. Initial research has shown that the levels of aluminum and lead in the galvanizing have changed. With the wet kettle method, lead levels were 0.60–1.0% and aluminum levels were 0.005%. Levels of lead have dropped to 0.05% and levels of aluminum have increased to 0.40% since going to the continuous sheet galvanizing. This increase in aluminum and decrease in lead is believed to help increase the brightness of the metal surfaces, making a better-looking product. The higher aluminum percentage is also necessary for better bonding of the zinc coating to the steel.

The cooling tower manufacturers do not believe that changes in the galvanizing process or in the lead and aluminum levels deposited are responsible for the increase of white rust. The industry has stated that aluminum levels have not changed and that the galvanized coating is actually more than twice as thick (2.35 oz of zinc per square foot of steel sheet) as they were 30 years ago.

However, it can be documented that cooling water programs operated with a pH range of 8.0 to 8.5 for many years without having white rust problems. It can also be documented that new towers added to existing systems, using the same makeup water, chemical treatment, and controls, have developed white rust, while the original units have not. Therefore, white rust does not appear to be a water quality and/or water treatment problem.

So, while the argument continues, and based on current information, it is well established that white rust may form if the following conditions exist:

1. The galvanized coating is not properly "passivated" when the tower is placed in service. Passivation is a process that allows the zinc coating to develop a natural nonporous surface of basic zinc carbonate. This chemical barrier prevents rapid corrosion of the zinc coating from the environment, as well as from normal cooling tower operation. The basic zinc carbonate barrier will form on galvanized surfaces within 8 weeks of tower operation with water of neutral pH (6.5–8.0), calcium hardness of 100–300 ppm, and alkalinity of 100–300 ppm.
2. Condenser water is maintained at pH above 8.0.
3. High condenser water alkalinity (above 300 ppm).
4. Low condenser water calcium hardness level (below 100 ppm).
5. The lack of phosphate-based corrosion inhibitor in the condenser water treatment program.

For most galvanized metal HVAC cooling towers, white rust will occur if not prevented by the following steps:

1. Provide a secondary barrier coating on all wetted surfaces, such as epoxy or polymer finish for a new tower or coal tar (bitumen) on an existing tower. An even better approach is to specify new towers to have wetted surfaces such as basins and wet decks to be constructed of stainless steel. This option is normally available for only a 15–20% cost increase.
2. Run the cooling water treatment program at a pH between 7.0 and 8.0, which may require pH control.
3. Make sure the galvanized tower is properly passivated upon system start-up. Where white rust has occurred, the metal can be "repassivated" by

treating the surface with a 5% sodium dichromate 0.1% sulfuric acid, brushing with a stiff wire brush for at least 30 s, then rinsing thoroughly with clean water.

4. Incorporate a phosphate-based product into the water treatment program, along with proper dispersants.

## BIOLOGICAL FOULING CONTROL

### Biological Fouling

Biological fouling results from bacteria, fungi, zooplankton, and phytoplankton or algae introduced through makeup water or filtered from the air passing through an HVAC cooling tower. "Fouling" results when these micoorganisms grow in open systems rich in oxygen (an aerobic process) and form slime on the surfaces of the tower, piping, and heat transfer surfaces of the condenser water system. Slime is an aggregate of both biological and nonbiological materials. The biological component, called the *biofilm*, consists of microbial cells and their by-products. The nonbiological components consist of organic and/or inorganic debris in the water that has become adsorbed or imbedded in the biofilm layer.

The impact of biological fouling is twofold: the slime acts as an insulator and reduces heat transfer efficiency in the system and microbial activity within the slime can accelerate corrosion by creating a localized oxygen-rich environment that accelerates oxidation.

A very special case of biological fouling in cooling towers is the bacterium *Legionella*, which is discussed in Chapter 14.

The primary method of controlling biological fouling is to keep cooling towers clean. At least twice during the cooling season, the tower should be drained, scrub cleaned, and allowed to fully dry before refilling. Then, the use of chemical treatment will complete the control chore.

There are two kinds of antimicrobial chemicals or *biocides* used in cooling tower water treatment programs to control biological fouling: oxidizing and nonoxidizing.

*Oxidizing chemicals*. These chemicals include chlorine, bromine, and ozone that *oxidize* or accept electrons from other chemical compounds. Used as antimicrobials, these chemicals reacted directly with the microbes and degrade cellular structure and/or deactivate internal enzyme systems. They penetrate the cell wall and disrupt the cell metabolic system to "kill" it. (*Warning! Oxidizing chemicals, particularly chlorine, can react with steel, including stainless steel, and cause rapid corrosion. To prevent this, concentrations of these chemicals must be kept low, ideally to less than 0.7 ppm. Oxidizing chemicals must be introduced into the condenser water system in such a way as to be rapidly dispersed to prevent localized high concentrations.)*

*Nonoxidizing antimicrobials.* These attack cells and damage the cell membrane or the biochemical production or the use of energy by the cell, resulting in its death, and are sometimes referred to as "surface-active" biocides. Typical nonoxidizing biocides include isothiazolines, gluteraldehyde, mercapto benzothiazole MBT, and polyquat.

Microbials in condenser water systems can become resistant to a single method of attack, or some microbials may be more or less immune to one type of attack. Therefore, it is recommended that both types of treatment chemicals be used (oxidizing and nonoxidizing), either blended together or in alternating treatment patterns, as indicated by periodic water testing results. The key to a successful biological treatment program is maintaining adequate chemical treatment levels at all times via continuous feed of antimicrobials into the condenser water system.

## Microbiologically Induced Corrosion

Microbiologically induced corrosion (MIC) is caused by sulfate-reducing bacteria (SRB) in the water and is usually evidenced by reddish or yellowish nodules on metal surfaces. When these nodules are broken, black corrosion by-products are exposed and a bright silver pit is left in the metal. A "rotten egg" smell when the nodule is broken is also evidence of SRB corrosion.

SRBs obtain their energy from the anaerobic reduction of sulfates that are available in most water. The bacteria contain an enzyme that enables it to use hydrogen generated at a cathodic site to reduce sulfate to hydrogen sulfate and act like a cathodic "depolarizing agent." *Iron corrosion by this process is very rapid and, unlike rusting, is not self-limiting.*

Once SRBs begin to grow in a system, they are very difficult to eliminate. Thus, a *preventative program* is far more effective than a clean-up program.

1. Keep the system clean through sidestream filtration and regular cleaning.
2. Prevent contamination by oils and/or grease. Even very small amounts can cause problems.
3. Eliminate potential bacteria sources (such as bathroom and kitchen vents, diesel exhaust, etc.) near cooling towers.
4. Run the condenser water pump as much as possible. Do not allow stagnant conditions to exist.

The condenser water system should be regularly tested for SRBs in accordance with ASTM Standard D4412-84, *Standard Test Methods for Sulfate-Reducing Bacteria in Water and Water-Formed Deposits.* Normally, control of SRBs can be accomplished with the housekeeping measures outlined above, coupled with the use of an oxidizing antimicrobial. *However, if SRBs do become established in the system, biocides are no longer effective and a special clean-out program must be designed for each system.*

### Foam Control

Some condenser water systems have surface foam that occurs due to contamination from ambient air, oxidation of organics by oxidizing antimicrobial chemicals, or overfeed of surfactants that may be added to reduce filming in cooling tower basins. Foam can interfere with heat transfer and can become airborne, contaminating areas within drift range of the tower. There are *antifoam chemicals* that may be added to control the tendency to foam, including certain oils and water-soluble organic silicone compounds.

## WATER TREATMENT CONTROL SYSTEMS

Automatic condenser water treatment, as opposed to manual treatment, is highly recommended. An automatic condenser water treatment control system must control four elements:

1. Blowdown cycles of concentration
2. Deposition (scale) inhibitor
3. Corrosion inhibitor
4. Antimicrobials (usually two of which are required)

Figure 13.3 illustrates a typical cooling tower water treatment system and its components.

The amount (cycles) of blowdown is controlled by monitoring and maintaining a specific conductivity setpoint for the condenser water. Conductivity in water is a function of the amount of dissolved solids in the water, since the dissociated salts serve as an electrolyte. Therefore, conductivity can be measured and a desirable setpoint established to maintain the desired amount of dissolved solids.

Blowdown is initiated by opening a blowdown solenoid valve on the condenser water return line (water returning to the tower). Blowdown can be controlled by "continuous bleed" that modulates the control valve and amount of blowdown in direct proportion to the level of conductivity or by "intermittent bleed" by opening and closing a two-position valve to maintain conductivity within established high and low limits.

The injection of deposition and corrosion inhibitors and dispersants are fed in proportion to the makeup water or blowdown water flow rate, measured by flow meters in both flow streams, by modulating metering pump speeds.

Antimicrobials are fed, typically, on the basis of a constant rate of injection established from periodic water testing. The two types of antimicrobials (oxidizing and nonoxidizing) can be alternately fed on a fixed time schedule, fed continuously together, or one type can be fed continuously and the other added manually to "shock" the system.

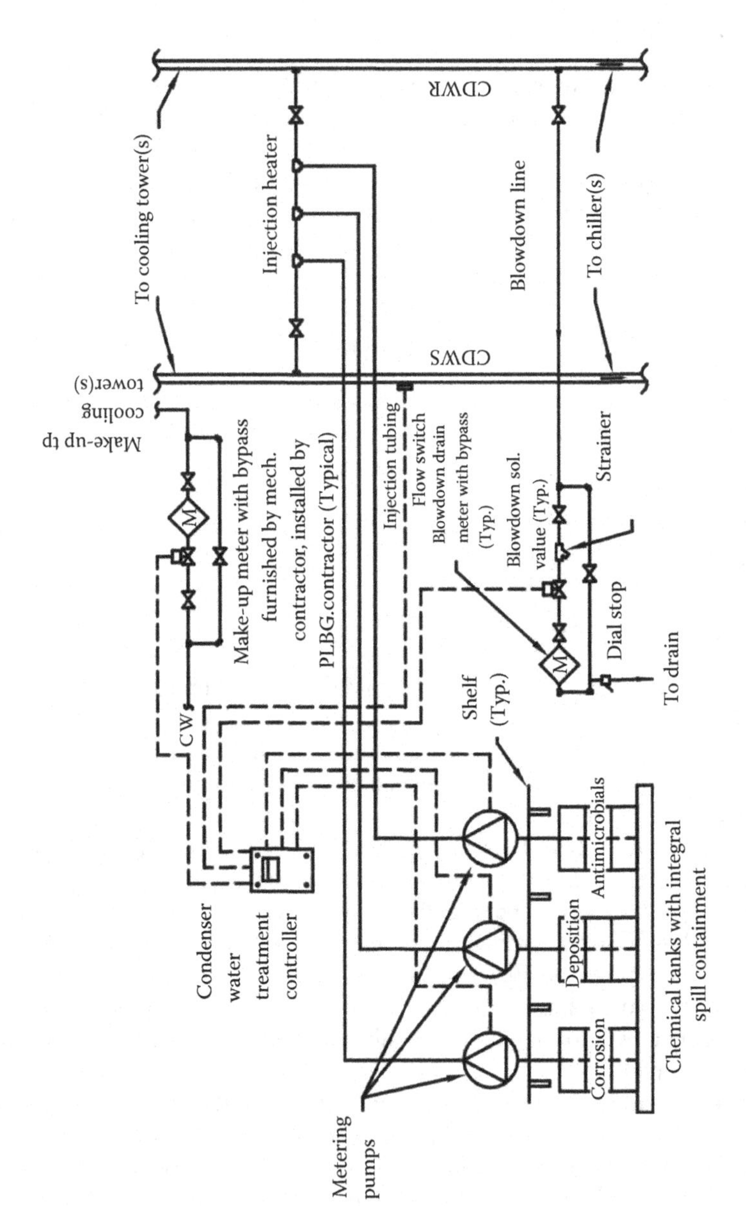


**FIGURE 13.3** Typical automatic condenser water treatment control system schematic.

## ALTERNATIVE WATER TREATMENT METHODS

### Sidestream Filtration

HVAC cooling towers are excellent "air washers." Pollen, dust, microbes, leaves, and other debris in the air are readily trapped and removed by the water and deposited in the basin to form "sludge," which can foul heat transfer surfaces. In the majority of cases, regular basin cleaning will control the problem, but in many urban and/or industrial areas, the amount of sludge formed during normal tower operation may be so great that specific control measures must be installed.

The most common general fouling control method is the use of *sidestream filtration*. Here, a portion of the condenser water flow is diverted through a filter for removal of dirt and suspended solids. Typically, filters are sized so that the entire water volume is filtered each hour. Thus, it is necessary to determine the total system water volume contained in the tower basin and wet deck, condenser water piping, condenser(s), etc. and divide that volume, in gallons, by 60 min to establish the required filter flow rate, in gpm.

The most common (and economical) type of sidestream filter is the sand filter with backwash, much as used for swimming pool applications. These filters remove suspended solids with 50 μm and larger particle size.

### Ozone Treatment

Ozone is a condenser water antimicrobial that eliminates the need for chemical treatment for microbiological fouling.

Ozone ($O_3$) is an unstable form of oxygen ($O_2$) that has a relatively short half-life, usually less than 10 min. Ozone is a powerful biocide and virus deactivant and will oxidize many organic and inorganic compounds.

When reacting with another molecule, the ozone molecule is destroyed, producing carbon dioxide, water, and a partially oxidized form of the original reactant molecule. Any residual ozone will decompose and recombine as oxygen, producing no toxic or carcinogenic by-products.

Ozone can be produced by UV radiation or by high-voltage corona discharge. However, the UV radiation method is very inefficient and the high-voltage corona discharge method is normally applied. This is accomplished by passing a high-voltage AC current (6–20 kV) across a dielectric gap of 10–15 mm, through which dry, compressed air is injected. As the compressed air passes through the voltage field, the oxygen molecules dissociate and form single oxygen atoms, some of which combine to form ozone (about 1–4%). The compressed air and ozone mixture is then injected into the condenser water, typically at the tower basin.

Filtration downstream of the ozone injection point is required since the ozone generation process will produce mineral deposits that must be removed to prevent fouling. As the cycles of concentration increase, calcium and carbonate ions in solution will precipitate out as calcium carbonate and settle in low water velocity areas of the system, typically in the tower basin.

Ozone treatment cannot be used with water with excessive hardness (500 ppm or higher calcium carbonate) or with sulfates greater than 100 ppm. Also, because of its short life, ozone should not be used in large systems or systems that have long piping runs that would require long residence times to get complete coverage. Water temperatures above 110°F eliminate the effectiveness of ozone treatment.

Ozone exposure for workers is also a consideration. USOSHA has established an exposure limit of 0.1 ppm in the air over an 8-h exposure period. Monitoring of ozone concentration is required if towers are located at ground level, adjacent to personnel.

The potential benefits from ozone treatment are twofold:

1. Effective antimicrobial treatment without the use of expensive chemicals.
2. Reduced blowdown since the ozone generation process precipitates out dissolved solids.

## UV Treatment

Ultraviolet light can be used in lieu of chemicals as antimicrobial treatment of condenser water to prevent microbiological fouling. Typically, a portion of the condenser water flow (3.5–10%) is diverted through the UV treatment equipment.

UV systems are offered in several configurations. The UV lamps that are used have two different wavelengths, 185 and 254 nm. For cooling tower applications, the 254-nm lamp has been found to be most effective. This wavelength is effective for germicidal applications and does not break down the other chemicals used for deposition and corrosion control.

Aside from wavelength, lamp intensity is also important. Lamps are typically available at both 39 and 65 W. Based on the system size and the flow rate, the smallest number of lamps will result in the lowest maintenance (replacement) costs.

Each UV lamp is installed with a quartz sleeve, which must include a "wiper system" to keep the sleeve clean so that the maximum output of UV light is provided.

The UV system is typically piped in a "sidestream" arrangement on the condenser water pump discharge. A filter must be installed ahead of the UV system to minimize the fouling of the lamp quartz tubes. Filtration can be cartridge, bag, sand, or cleanable mesh type, but they must be rated to stop particles 50 μm and larger in size.

## Magnetic Treatment

Every few years, salesmen make the rounds to tower owners and designers and talk about the "wonders" of magnetic treatment for deposition control in condenser water systems. Their product consists of "strap-on" permanent magnets installed around condenser water piping. Promoters of these devices claim that dissolved solids, which dissociate into charged ionic salts in solution, can be easily removed by allowing them to pass through a magnetic field.

*Tests by independent sources have proved conclusively that magnetic treatment has no effect on deposition rates in condenser water systems.* Magnetic treatment is simply a scam and, when encountered, should be immediately rejected as a viable water treatment method.

## TREATMENT FOR WOODEN TOWERS

For very large cooling towers, for example, 1000–1500 tons and larger, wood is still widely used for cooling towers. Wood is composed of three major elements:

1. Cellulose, which forms long fibers in wood and give it its strength
2. Lignin, the soft material that acts as a cementing agent for the cellulose
3. Extractives, the natural compounds that enable wood to resist decay

Wood deterioration can result from chemical attack, biological attack, and/or physical decay due to (elevated) temperature degradation or rupture of wood cells by crystallization of dissolved solids in the condenser water. Physical decay is relatively uncommon in HVAC applications.

Chemical attack commonly results in *delignification* of the wood, resulting in wood that has a white or bleached appearance and a fibrillated surface. Delignification is usually caused by oxidizing agents and alkaline materials and is particularly severe when high-chlorine residuals (more than 1 ppm free chlorine) and high alkalinity (pH of 8 or higher) occur simultaneously.

The biological organisms that attack cooling tower wood are those that use cellulose as a source of carbon for growth. These organisms degrade the cellulose by secreting enzymes that convert the cellulose into compounds they can then absorb as food. This type of attack deletes the cellulose and leaves a residue of lignin. The wood appears dark, loses much of its strength, and becomes soft and spongy.

Biological attack is of two basic types: surface rot and internal decay. Surface rot is easily detected and can be repaired, but internal decay, usually found in the plenum area, cell partitions, decks, fan housing, supports, and other areas that are not continuously flooded, is more difficult to detect simply because the exterior of the wood will display little or no sign of the rot.

Water treatment and preventative maintenance is the only way to prevent wood deterioration:

1. For flooded areas of the tower, use nonoxidizing antimicrobials to control slime and prevent biological attack. (If chlorine is used, limit free chlorine to 0.7 ppm or less.)
2. Internal decay is the most common problem for nonflooded areas of a tower and these areas should be thoroughly inspected at least twice each year. Test for soundness with a blunt probe (long, thin screwdriver) and, when suspected, a sample of wood should be examined microscopically to detect internal microorganisms. *Any infected wood should be replaced immediately* with pretreated wood to prevent the spread of the

infection to healthy wood. (Periodic spraying with an antifungal treatment may be helpful in reducing biological attack, but it must be done very thoroughly and on a regular schedule, otherwise it is a waste of time and chemicals.)

## CHEMICAL STORAGE AND SAFETY

### Spill Control

Chemicals for water treatment are usually purchased in bulk and, thus, there is a potential for a chemical spill when the drums are moved or opened. To eliminate this potential, either of the two measures may be used:

1. Purchase chemicals in double-walled containers that have integral spill containment. While more expensive than simple drums, these containers provide protection from internal rupture and spillage while filling or emptying. The drawback of them is that there is no protection from a puncture that penetrates both the outer and inner walls of the container.
2. Construct a "spill containment" reservoir for use with conventional single-walled drums. This reservoir usually consists of a concrete curb enclosing the chemical storage area. The volume of containment must be greater than the single largest container and a "sump" or depressed area must be provided, with the floor sloped to it, for collection of any spilled liquid for pump-out. Grated flooring can be installed on top of the curb and the drums stored on it, or ramps can be installed on both sides of the curb and the drums sit on the floor inside the containment area.

### Safety Showers and Eyewash Stations

USOSHA Standard 29 CFR 191.151(c) requires eyewash and shower equipment for emergency use where the eyes or body of an employee may be exposed to injurious materials. *Typically, condenser water treatment chemicals would be considered to be potentially injurious.* This safety equipment must be designed, installed, tested, and maintained in accordance with ANSI Z358.1. This standard requires that eyewash and shower equipment be located based on the estimated time of travel of a person with compromised vision. Equipment must be in accessible locations that require no more than 10 s to reach. Combination units (shower plus eyewash) must be designed and connected to the water system so that both units can be operated simultaneously by the same user. Flow is required within 1 s of activation and stay-open valves are required so that the user is free to use his/her hands to remove clothing or hold the eyelids open.

ANSI Z358.1 mandates the delivery of *tepid* water—moderately warm or lukewarm—for 15 min at 20 gpm in emergency showers and 0.4 gpm in emergency eyewash units. The standard does not specifically define tepid, but the tepid range is generally considered to be 78–92°F, based somewhat on the normal surface temperature of the human eye. *There is no grandfathering provision in the standard; an existing cold water shower or eyewash is simply no longer compliant and must be changed.*

Tempering water for emergency showers and eyewash applications is a challenging problem. When selecting an automatic thermostatic mixing valve for a tepid emergency shower or eyewash system, consider the following:

1. Combination shower and eyewash units are frequently used. The shower has a high flow rate (about 20 gpm), while the eyewash has a low flow rate (about 0.6 gpm). The mixing valves must be capable of both high and low flow control.
2. A "temperature spike" can occur when large, equal quantities of hot and cold water are mixed instantaneously, creating a sharp rise in the outlet temperature. In domestic water applications, the piping system has time to flatten a temperature spike. In an emergency situation, however, valves are typically installed very close to the fixture and this time delay is lost. The mixing valve must respond very quickly to flatten any temperature spike as it begins.
3. The majority of domestic water mixing valves are designed to restrict flow in the event of hot- or cold-water failure. However, at an emergency shower, flow reduction is unacceptable. In an emergency, cold-water flow is better than no water flow (while in a domestic water situation that would be unacceptable).

# 14 Special Tower Considerations

## BASIN AND OUTDOOR PIPING FREEZE PROTECTION

Except in Northern climates, where towers are often drained and secured for the winter, towers will stay full of water and ready for use year-round in more moderate climates. However, even in these climates, there will be periods when the ambient temperature may drop below 32°F long enough for basin water freezing to occur.

To prevent that problem, basin heaters are normally installed. These heaters are most commonly electric resistance elements, but steam coils or steam ejectors can also be used.

The heat loss from a cooling tower basin, with a maintained water temperature of 40°F, can be estimated from the data provided in Table 14.1. Thus, a 12′-wide by 20′-long steel cooling tower basin with 12″ of water depth would have estimated heat losses for 20°F outdoor temperature computed as follows:

Surface loss: 12′ × 20′ × 80 Btu/h sf = 19,200 Btu/h

Sides loss: 1′ × [(2 × 12′) + (2 × 20′)] × 40 Btu/h sf = 2560 Btu/h

Bottom loss: 12′ × 20′ × 40 Btu/h sf = 9600 Btu/h

Total: 31,360 Btu/h

For an electric immersion heater, each kW produces 3413 Btu/h. In this case, a 9-kW heater would be required to maintain the tower basin at approximately 40°F with a 20°F outdoor air temperature.

Electric heaters are typically used to maintain the basin water temperature above freezing, since they are relatively inexpensive. But electric heaters suffer from two problems: first, they are costly to operate and, second, since the heating is localized, poor mixing and over- or underheating in different areas of the basin can result.

Steam injection heaters can be used if steam is available. These heaters are submerged in the basin and, when heating is required, a two-position automatic valve is opened and steam is injected through a Venturi or nozzle that sets up a convective current in the water, ensuring relatively good mixing. Steam coils, submerged in the basin water, can also be used.

**TABLE 14.1**
**Approximate Cooling Tower Basin Heat Loss**

| Outdoor Air Temperature (°F) | Approximate Water Surface Heat Loss (Btu/h sf) | Heat Loss Basin Sides and Bottom (Btu/h sf) | |
|---|---|---|---|
| | | Steel or Fiberglass | Concrete (8″ Thick) |
| 30 | 50 | 20 | 15 |
| 20 | 80 | 40 | 25 |
| 10 | 110 | 60 | 35 |
| 0 | 130 | 80 | 50 |
| −10 | 160 | 100 | 60 |
| −20 | 185 | 120 | 70 |
| −30 | 210 | 140 | 80 |

Outdoor condenser and makeup water piping must also be protected from freezing. Typically, this is accomplished by wrapping the pipe with electric heat tracing elements (though, again, steam tracing can be used) before the piping is insulated. For cooling tower piping applications, the criteria from Table 14.2 can be used to prevent pipe freezing. (In most of the southern United States, where outdoor winter design temperatures are 20°F or higher and there are no prolonged periods at these low temperatures, heat tracing can be safely eliminated for piping larger than 4″.)

**TABLE 14.2**
**Electric Heat Tracing for Outdoor Piping**

| Pipe Size (in.) | Watts/Linear Foot of Piping |
|---|---|
| $\frac{1}{2}$ | 1.6 |
| $\frac{3}{4}$ | 1.7 |
| 1 | 1.9 |
| $1\frac{1}{4}$ | 2.0 |
| $1\frac{1}{2}$ | 2.3 |
| 2 | 2.9 |
| $2\frac{1}{2}$ | 3.4 |
| 3 | 4.5 |
| 4 | 4.9 |
| 6 | 7.1 |
| 8+ | 0.68/in. of pipe diameter |

## WATERSIDE ECONOMIZER CYCLE

Commonly, HVAC systems are designed to utilize an *airside economizer cycle* to provide "free" cooling when the outdoor temperature is at or below the required supply air temperature. However, there are a number of secondary system types in which the airside economizer cannot be applied, such as with fan coil units. In these cases, the *waterside economizer cycle* may have application.

This cycle utilizes a cooling tower to produce cold water when the ambient wet bulb temperature is 50°F or lower. At 50°F WB, a typical cooling tower will produce 40–50% of its design capacity at an approach of about 5°F. Thus, a 500-ton cooling tower will deliver ~250 tons of cooling with a 45°F supply water temperature at 50°F ambient wet bulb. But, in most comfort cooling applications, the need for dehumidification is reduced during the heating season and a 5–10°F increase in the chilled water supply temperature can be tolerated. Thus, the number of hours of available free cooling can be significantly increased.

Obviously, this cold water can be used to provide cooling without operating a water chiller. However, to separate "dirty" condenser water from "clean" chilled water, a heat exchanger is typically used since only a 2–3°F increase in supply water temperature results.

Two types of heat exchangers may be used for waterside economizer applications:

1. Plate-and-frame heat exchangers, often simply called "frame heat exchanges," consist of thin corrugated plates separated by rubber or neoprene gaskets and mounted in a rigid frame. The two water flows are arranged counter to each other on opposite sides of each plate. Close temperature approach, 1–2°F, is obtained by having a very large heat transfer area in a relatively small package.
2. High-efficiency shell-and-tube heat exchangers provide high heat transfer area by (a) using a large number of small-diameter enhanced surface tubes, and (b) increasing the tube length with a U-shaped configuration. Figure 14.1 shows a typical high-efficiency shell-and-tube heat exchanger. An approach of 2°F can be obtained with this type of heat exchanger.

Shell-and-tube heat exchangers have lower cost, less weight, and more versatile configurations than plate-and-frame heat exchangers. They are also easier to insulate. But closer approach (higher efficiency) can be obtained with the plate-and-frame heat exchanger.

While a waterside economizer can be applied to a single-tower system, ASHRAE Standard 90.1 requires that it be an "indirect" system, as shown in Figure 14.2, which allows the chiller and the economizer to operate concurrently. Since chillers normally will not operate at a 45°F condenser water supply temperature, bypass temperature control becomes critical and often fails. However, where multiple chillers and cooling towers are available, one tower can be isolated for economizer duty, leaving at least one other chiller and

**FIGURE 14.1** Typical shell-and-tube heat exchanger. (Courtesy of the Baltimore Aircoil Company, Baltimore, MD.)

tower available to supplement the economizer cooling capacity, as shown in Figure 14.3.

The energy consumption by the waterside economizer for the pumps and cooling tower is only about 15–20% of the energy consumption required by mechanical cooling.

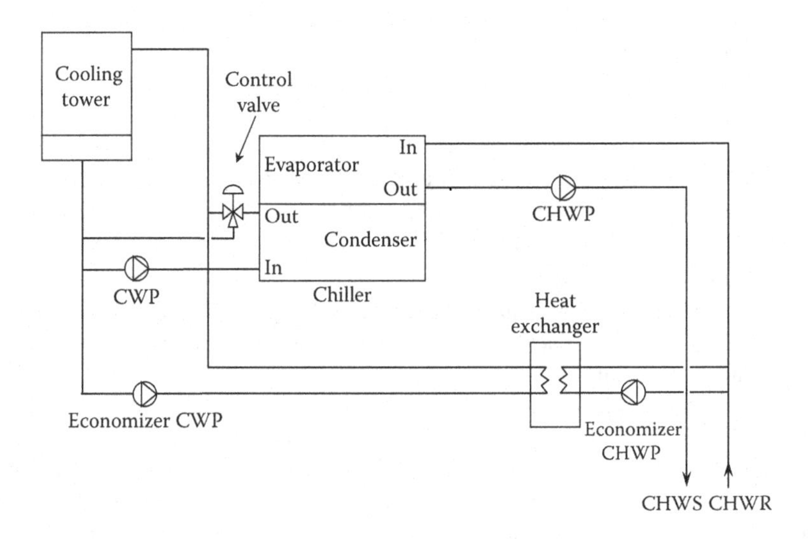


**FIGURE 14.2** Piping schematic for waterside economizer with a single chiller and cooling tower.

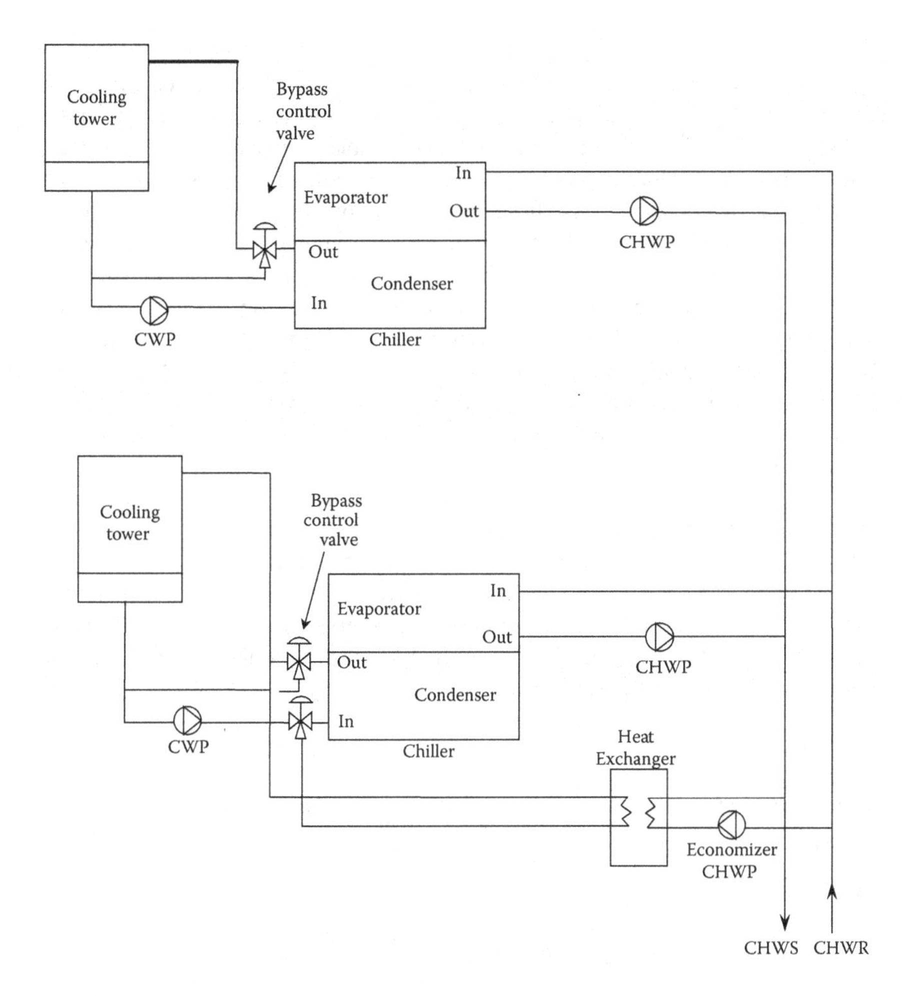


**FIGURE 14.3** Piping schematic for waterside economizer with multiple chillers and cooling towers.

## NOISE AND VIBRATION

*Noise* is simply objectionable sound and "community noise" and "noise pollution" are terms applied to sound generated on one property, but deemed objectionable by listeners on adjacent properties. Along with trucks and cars, trains, the neighbor's stereo, and leaf blowers, cooling towers are often the source of community noise problems.

In 1972, the U.S. Congress passed the Noise Control Act and authorized the EPA to begin a process of establishing emission standards for outdoor noise sources. However, funding for this effort was deleted in 1982 and, for all practical purposes, community noise control efforts have been delegated to the state and local governments.

Numerous states and municipalities have *property line sound limits* that must be met. Under this type of regulation, maximum sound pressure levels (dBA, see the section "Noise and Vibration" in Chapter 6), measured at adjacent property lines, are established. Typical maximum allowable sound pressure values are 50–60 dBA in residential areas and 60–70 dBA in industrial areas.

More stringent standards have been developed by the World Health Organization (WHO). The WHO standards are being promoted by numerous antinoise groups, and their adoption by at least some states and municipalities is considered but a matter of time. WHO makes the case that the effect on the health and safety of the listener by a combination of noise "events" is related to the combined sound energy of those events. Thus, WHO introduces the term "LAeq,T" to define A-scale sound pressure level (dBA) averaged over a time period *T*. As shown in Table 14.3, these standards establish maximum LAeq,T limits at or within various types of occupancies or environments.

Typical cooling tower noise emission is in the range of 65–85 dBA measured at (typically) 5 ft from the tower, and each tower manufacturer can provide specific sound emission data for the required operating conditions.

Fan noise is related to the fan design: centrifugal fans are inherently quieter than axial flow fans. For induced draft cooling towers that utilize axial fans, manufacturers have developed "low noise" options, such as adding extra fan blades so the fan can be operated at lower speed, and so on.

**TABLE 14.3**
**World Heath Organization Community Noise Limit Guidelines**

| Environment | Critical Health Effect(s) | LAeq (dBA) | *T* (h) |
|---|---|---|---|
| Outdoor living area | Serious annoyance | 55 | 16 |
| | Moderate annoyance | 50 | 16 |
| Dwelling (living areas, inside) | Speech intelligibility and moderate annoyance | 35 | 16 |
| Bedrooms (inside) | Sleep disturbance | 30 | 8 |
| Bedrooms (outside, window open) | Sleep disturbance | 45 | 8 |
| Schools (classrooms, inside) | Speech intelligibility, disturbance of information extraction, message communication | 35 | During class |
| Schools (playground, outdoors) | Annoyance | 55 | During play |
| Hospital (patient rooms, inside) | Sleep disturbance | 30 | 8 (night)<br>16 (day/evenings) |
| Industrial, commercial, shopping, and traffic areas (inside and outside) | Hearing impairment | 70 | 24 |
| Outdoors in parkland and conservation areas | Disruption of tranquility | Minimize disruption | |

There are two caveats relative to manufacturer-provided sound data:

1. The noise data provided are typically based on a single tower or tower cell. Where multiple towers or tower cells are utilized, the manufacturer must be queried for cumulative effect sound data. In the field, fan speeds on multiple towers or cells must be as nearly identical as possible to prevent annoying "out-of-phase" sound differences.
2. The noise emission from the tower may not be uniform in all directions. For crossflow towers, the inlet side(s) will be noisier than the enclosed side(s). The manufacturer data must define the tower orientation for each noise emission value.

The basic sources of cooling tower noise include the following:

1. Fans
2. Air movement and turbulence through the tower
3. Motor and/or drive
4. Structural or casing vibration

The tower location and orientation can have a pronounced effect on the sound condition. Always orient a quiet side of the tower toward any sound-reflecting wall or side of a building or toward adjacent property. Locate towers at least 20 ft away from any sound-reflecting surface.

Sound dissipates over distance. As shown in Table 14.4, the amount of sound attenuation varies with the multiples of distance from the point of sound measurement (distance from the rated noise measuring point). At a distance 10 times greater than the sound rating distance from the tower, ~20 dB attenuation is expected. Thus, a tower with a rated noise level of 86 dB at 5 ft from the tower will have an anticipated noise level of 65 dB at 50 ft from the tower, a 21 dB reduction. Note that the curve is not linear and an increase to 25 times the sound rating distance from the tower only reduces the sound level by ~25 dB.

*To avoid a community noise problem, locate each cooling tower as far from adjacent property lines as possible.* Since HVAC cooling towers have sound levels

**TABLE 14.4**
**Effect of Distance on Cooling Tower Sound Level**

| Multiple of Sound Rating Distance | Approximate Sound Reduction (dB) |
|---|---|
| 5× | 13 |
| 10× | 20 |
| 15× | 24 |
| 20× | 26 |
| 25× | 29 |

of 65–85 + dBA, the WHO criteria require that cooling towers not be located within 50–100 ft of property lines without taking additional attenuation measures:

1. Select/approve the quietest cooling tower possible for the application. Use centrifugal fans or select low rpm axial fans with the highest number of blades for the application.
2. Orient a quiet side of the equipment toward any sound-reflecting wall or side of a building and toward adjacent property. Locate cooling towers at least 20 ft away from any sound-reflecting surface, no matter how oriented.
3. Reduce radiated noise with a solid wall sound barrier between the equipment and the adjacent property. Sound barrier noise attenuation, however, is a function of the wall height and the distances between the tower, the barrier, and the property line. The maximum theoretical attenuation provided by a sound barrier is 24 dB, but the actual attenuation will always be less due to sound leakage around the ends of the barrier and sound reflected off buildings, trees, and so on. An acoustical consultant should be retained to design any sound barriers that may be required.
4. Only as a last resort should *sound attenuators* on the tower inlet and/or outlet be utilized. These devices are expensive, seriously impact on the performance of the tower (requiring, often, that a larger size be used), and create significant maintenance problems.

Cooling tower vibration results when fan balance and/or drive alignment is not correct. This is more often a problem with propeller fans and/or gear drives. Propeller fan blades, due to their relatively long length (large length-to-width ratio), tend to flex under air loading. Even if the blades are perfectly balanced by weight, dynamic balancing is required to eliminate unequal blade movement and resulting vibration. Shaft alignment into and out of gear drives is also critical to avoid vibration. Also, gear drives must be operated at speeds high enough to avoid "gear chatter."

Centrifugal fans, unless they have very long shafts, have less vibration potential because of their construction. However, many cooling towers using centrifugal fans will be constructed with two or more fan wheels on a long, common shaft driven by a single motor. To prevent vibration, the shaft must be carefully aligned and supported by bearing blocks at each end and *between each fan wheel.* The shaft must be rigid enough to avoid deformation, and the resulting vibration, due to the fan wheel(s) weight.

All propeller fan and/or gear drive cooling towers, and towers with multiple fan wheels on a common shaft, should be equipped with a vibration cutout switch (see Chapter 12).

## PLUME CONTROL

If HVAC cooling towers are operated during the winter, a significant and objectionable discharge *plume* can result. When warm air, at or near saturation conditions, leaving the cooling tower impacts with the cold ambient air, condensation occurs, resulting in a supersaturated air stream and visible "fog." This fog is trapped in the moving discharge from the cooling tower and is distributed upward and outward to form the plume.

Uninformed observers often equate visible cooling tower plume with "smoke," indicating a fire, or with "pollution." In either case, 911 phone calls can result.

Generally, plume problems can be handled with education. People who are routinely around the tower simply become used to this harmless condition. Fire and police dispatchers should be informed that this condition is not a problem so that they can handle public alarm accordingly.

In a small number of cases, cooling tower plume can become a more serious problem, since it can return to ground level to form a vision hazard or, in very cold weather, cause ice to form on downwind surfaces and structures. If plume abatement does become necessary, there are two basic approaches that can be applied:

1. The tower can be equipped with *bypass heating* as shown in Figure 14.4. In this configuration, additional ambient airflow is induced into the tower discharge. Enough heat is added to this induced air to raise its temperature high enough to allow it to absorb more moisture and prevent condensation until after the discharge plume has dissipated.
2. The tower can be equipped with *reheat* coils in the discharge air as shown in Figure 14.5. These coils are sized to provide enough heat to offset the cooling effect of the ambient air on the plume and, thus, prevent condensation until after the discharge plume has dissipated.

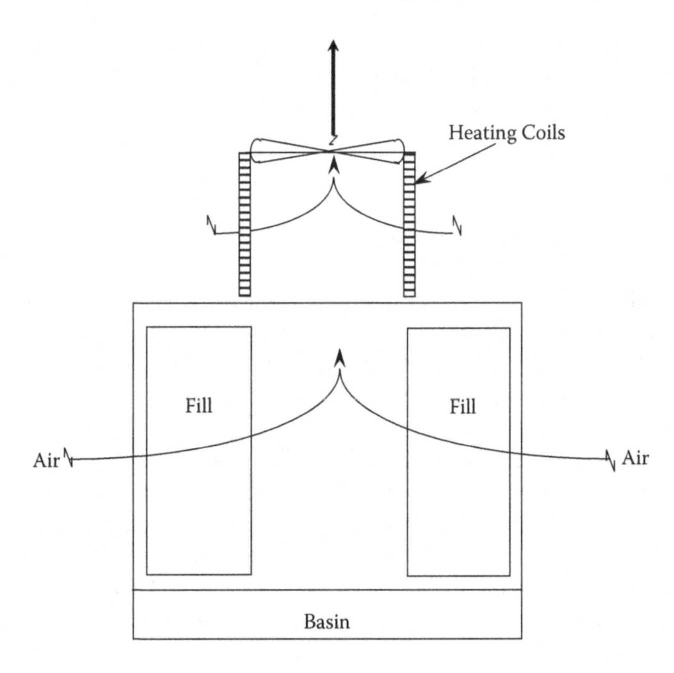


**FIGURE 14.4** Tower arrangement for plume control via bypass air heating.

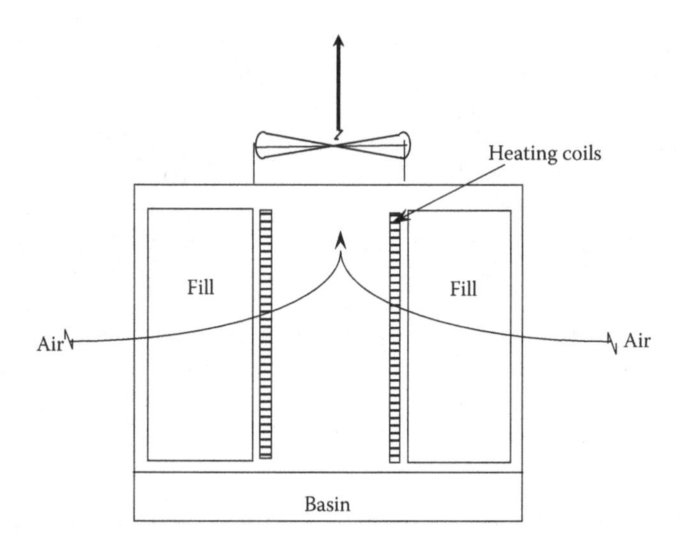


**FIGURE 14.5** Tower arrangement for plume control via discharge air reheating.

The configurations shown in Figures 14.4 and 14.5 represent a *series path wet/dry* (SPWD) approach to plume control. The drawbacks of this configuration are increased pressure drop and resulting energy consumption, for the cooling tower fan and high pump head, with increased energy consumption, for the hot-water pump. And, due to the highly corrosive nature of warm, moist air, the coil construction must be corrosion resistant, increasing first costs considerably.

In the 1970s, an alternative concept called the *parallel path wet/dry* (PPWD) was developed. With this concept, the height of the cooling tower above the fill was increased and air intakes, with bypass dampers, were installed. Thus, when plume control is needed, the dampers are modulated open and cold, dry air mixes with the warm, moist air leaving the tower fill. This has the effect of reducing both tower discharge air temperature and relative humidity, eliminating the potential for a discharge plume.

The disadvantage of the PPWD approach is that the tower fan must be sized for an increased airflow when the bypass dampers are open, increasing fan energy. But, since the coils and heating energy associated with the SPWD approach are avoided, this is a far more cost-effective method of dealing with the potential for discharge plume.

In recent years, a third approach has been developed. With this concept, crossflow air-to-air heat exchangers are installed in the tower above the fill discharge. With this concept, bypass air is introduced on one side of the heat exchanger, while warm, moist air leaving the tower fill enters on the other. The colder bypass (ambient) air then cools the warm, moist air, causing condensation to occur and reducing its dewpoint temperature to below the outdoor air temperature, eliminating plume potential. Since the condensate drains to the tower basin, both blowdown and make-up water requirements are reduced. Likewise, drift losses are reduced since the humidity ratio of the final tower discharge air is reduced. And, compared to PPWD designs, the dampers are smaller and less leaky, reducing fan energy requirements.

## FIRE PROTECTION

The NFPA Standard 214 defines *noncombustible* as any material that will not ignite, burn, support combustion, or release flammable vapors when subjected to fire or heat. When any HVAC cooling tower is constructed wholly or partly with materials that are not "noncombustible" (e.g., wood), a fire protection (sprinkler) system may be required. See local building codes for applicability of NFPA 214.

Any of the four types of sprinkler systems, applied and designed in accordance with NFPA Standard 13, may be used as follows:

1. Counterflow towers may be protected with a wet-pipe system, a dry-pipe system, a preaction system, or a deluge system.
2. Crossflow towers must be protected with a deluge system.

The minimum required rate of water application must be in accordance with Table 14.5.

When the fan deck is combustible and a deluge system is applied, the underside of the deck must be protected with a minimum water application rate of 0.15 gpm/sf. With crossflow towers, when wet deck covers are constructed of combustible materials, protection under these covers, at the same rate as for the fan deck, is also required.

All sprinkler piping, fittings, hangers, braces, attachment hardware, and so on must be galvanized to reduce corrosion. Exposed pipe threads and bolts must also be protected against corrosion.

Obviously, freeze protection for the sprinkler system is of paramount importance. While a wet-pipe system is allowed by NFPA, it really has only very limited geographic application in the United States and is rarely used. The most common systems applied are the (closed head) dry-pipe system, usually applied to smaller towers or where the water supply may be limited, and the (open head) deluge system.

**TABLE 14.5**
**Fire Protection Water Flow Rate Requirements**

| Location | Rate (gpm/sf) |
|---|---|
| Under fan decks (including fan opening), counterflow tower | 0.50 |
| Under fan decks (including fan opening), crossflow tower | 0.33 |
| Overfill area, crossflow tower | 0.50 |

Overall, the deluge system provides a higher degree of protection when there are adequate water supplies and this system has the advantage of minimizing the possibility of failure due to freezing.

## *LEGIONELLA* CONTROL

In 1976, 34 attendees at an America Legion Convention in Philadelphia died from a pneumonia-like disease that was later traced to the then-unknown bacteria that we now call *Legionella.* Since this initial outbreak (which was really preceded by earlier events in Austin, MN in 1957 and Washington, DC in 1964), the recognition of *Legionella* as a serious problem has grown significantly. According to the U.S. Centers for Disease Control (CDC), Legionnaires' disease infects ~25,000 people in the United States each year and 10–15% of these cases are typically fatal. *This equates to 3000–4000 deaths attributable each year to Legionella.*

*Legionella* is a bacterium that is common in surface waters, including lakes, rivers, and so on. The bacteria survive routine water treatment, and low concentrations are introduced into most potable water supplies. *Legionella* thrives in water temperatures between 68°F and 122°F, with optimal growth occurring between 95°F and 115°F. Low pH and high levels of aquatic growth (microbiota, amoebae, algal slime, etc.) enhance bacteria growth. Water temperatures above 135–140°F kill the bacteria.

At these temperatures, the ideal habitats for *Legionella* include cooling tower and evaporative condenser systems, where with temperatures typically range from 85°F to 100°F. Other potential breeding grounds for *Legionella* include domestic hot-water systems in schools and hospitals, humidifiers, spas or whirlpools, and even vegetable misters in supermarkets.

The major mechanism for infection by *Legionella* is via inhalation of aerosolized water droplets or particles containing the bacteria. Cooling tower or evaporative condenser sprays introduce aerosolized water droplets and, therefore, represent prime mechanisms for infecting humans. There is no evidence that drinking water with the bacteria in it will cause disease, nor can the disease be passed by human-to-human contact.

There are two types of disease caused by *Legionella*: (1) "Legionnaires' disease" is a severe form of pneumonia, and (2) "Pontiac Fever" is a nonfatal flu-like illness. Legionnaires' disease symptoms can vary from a cough and low fever to rapidly progressive pneumonia, coma, and death. Symptoms occur typically within 3–9 days after exposure.

To monitor and control *Legionella*, the following steps are recommended:

1. Test for *Legionella* in cooling tower water. While there is an academic debate over the cost versus benefit of routinely testing for *Legionella*, it is dumb (from both an ethical and a liability point of view) to ignore any potentially life-threatening condition in a facility. Testing must be specific for *Legionella*: "total bacteria" tests promoted by some water

treatment companies are inadequate since there is no correlation between total bacteria and *Legionella* concentrations.

The "gold standard" for detection and quantification of *Legionella* in cooling towers is the standard culture method defined by the CDC. The number of *Legionella* bacteria in cooling tower water is calculated as "colony forming units" (CFU) per milliliter.

Culture of *Legionella* from environmental samples is technically difficult and requires that the testing laboratory be experienced in the detection of this bacterium.

2. If *Legionella* is found, eliminate the bacteria concentration. Cooling towers (and evaporative condensers) can be decontaminated by slug chlorination with 50 ppm of free residual chlorine, along with a dispersant. Then, maintain 10 ppm of free residual chlorine for at least 24 h, while maintaining the water pH at 7.5–8.0. Finally, drain the system and repeat the process.
3. Cooling tower systems should be drained and flushed twice each year. All surfaces should be cleaned and allowed to air dry before reuse. Continuous-feed water treatment systems for biocides are required to maintain consistent concentration levels.
4. Keep up with the latest information on *Legionella*. ASHRAE has published Guideline 12-2000, *Minimizing the Risk of Legionellosis Associated with Building Water Systems*, and CTI has published Guideline WTB-148 (2008), *Legionellosis, Guideline: Best Practices for Control of Legionella*.

To reduce the potential for *Legionella* contamination, the tower water treatment program should provide continuous *halogenation* treatment (e.g., chlorine or bromine):

1. For relatively clean systems or where clean potable water makeup is used, feed a source of halogen (chlorine or bromine) continuously and maintain a free residual. Continuous free residuals of 0.5–1.0 ppm in the cooling tower hot return water are recommended by many agencies. Periodic monitoring of the residual at sample points throughout the cooling water system is needed to ensure adequate distribution. The effectiveness of either halogen decreases with increasing pH; bromine is relatively more effective at a higher pH (8.5–9.0).
2. Stabilized halogen products should be added according to the label instructions, and be sufficient enough to maintain a measurable halogen residual.
3. A biodispersant/biodetergent may aid in the penetration, removal, and dispersion of biofilm and often increases the efficacy of the biocide.
4. Continuous halogen programs may require the periodic use of nonoxidizing biocides. These may be required to control biofilm and planktonic organisms in systems that use make-up water from sources other than potable water sources, and those with process leaks or

contamination. The choice of nonoxidizing biocides should be based on the results of toxicant evaluations. Reapply as dictated by results of biomonitoring.

*Hyperhalogenation* for periodic online disinfection may be necessary for systems:

1. That have process leaks
2. That have heavy biofouling
3. That use reclaimed wastewater as makeup
4. That have been stagnant for a long time
5. When the total aerobic bacteria counts regularly exceed 100,000 CFU/mL
6. When *Legionella* test results show greater than 100 CFU/mL

Periodic hyperhalogenation will discourage development of large populations of *Legionella* and their host organisms. Consequently, periodic hyperhalogenation may eliminate the need for conducting more complicated and higher-risk off-line emergency disinfection procedures.

The following emergency disinfection procedure is based on OSHA and other governmental recommendations and is required when very high *Legionella* counts exist (i.e., >1000 CFU/mL), in cases where Legionnaires' disease are known or suspected and may be associated with the cooling tower, or when very high total microbial counts (>100,000 CFU/mL) reappear within 24 h of routine disinfection (hyperhalogenation):

1. Remove heat load from the cooling system, if possible.
2. Shut off fans associated with the cooling equipment.
3. Shut off the system blowdown. Keep makeup water valves open and operating.
4. Close building air intake vents in the vicinity of the cooling tower (especially those downwind) until after the cleaning procedure is complete.
5. Continue to operate the recirculating water pumps.
6. Add a biocide sufficient to achieve 25–50 ppm of free residual halogen.
7. Add an appropriate biodispersant (and antifoam if needed).
8. Maintain 10 ppm free residual halogen for 24 h. Add more biocide as needed to maintain the 10 ppm residual.
9. Monitor the system pH. Since the rate of halogen disinfection slows at higher pH values, acid may be added, and/or cycles reduced to achieve and maintain a pH of less than 8.0 (for chlorine-based biocides) or 8.5 (for bromine-based biocides).
10. Drain the system to a sanitary sewer. If the unit discharges to surface water under a permit, dehalogenation will be needed.
11. Refill the system and repeat steps 1 through 10, above.

12. Inspect after the second drain-off. If a biofilm is evident, repeat the procedure.
13. When no biofilm is obvious, mechanically clean the tower fill, tower supports, cell partitions, and sump. Workers engaged in tower cleaning must wear (as a minimum) eye protection and a 1/2 face respirator with High Efficiency Particulate Air (HEPA) filters, or other filter capable of removing >1 μm particles.
14. Refill and recharge the system to achieve a 10 ppm free halogen residual. Hold this residual for 1 h and then drain the system until free of turbidity.
15. Refill the system and charge with appropriate corrosion and deposit control chemicals, reestablish normal biocontrol residuals, and put the cooling tower back into service.

While the procedures described address the contamination issue of *Legionella* at the tower, there remains the important requirement that dispersion of *Legionella* via cooling tower drift be controlled to minimize the potential for community infection. Even though community-wide *Legionella* infection is rare, it has happened. And, unlike other source-based infections (grocery store misters, shower heads, etc.), cooling towers have been found as the source for *Legionella* infections miles away from the implicated cooling tower.

Exposure to *Legionella* bacteria in cooling tower drift is dependent on several factors:

1. Tower drift rate
2. Tower airflow volume
3. The dispersal or mixing of cooling tower discharge airflow by ambient air
4. The time the *Legionella* bacteria remains suspended in the airflow

As discussed previously, while older towers may have a drift rate of 0.02%, modern (since about 1990) induced draft, crossflow cooling towers will have drift rates of 0.0001–0.005%. Even these low drift rates, however, require that cooling tower operators routinely inspect and correct deficiencies in drift eliminator performance:

1. Repair/replace eliminators damaged by ultraviolet light degradation, hail or snow damage, and so on. Eliminators can be damaged during routine cleaning and other maintenance activities and these must also be repaired or replaced.
2. Eliminators clogged by the buildup of dissolved solids, tree debris, and so on causes increased air velocity in the open eliminator areas, decreasing the eliminator's effectiveness. Eliminators should be frequently inspected and cleaned.
3. Misaligned or missing drift eliminators are obvious signs of increased drift rates. Care must be taken after each maintenance procedure to make sure each eliminator section is properly replaced and all gaps

closed. After wind or hail storms, check to determine if eliminators are damaged (see above) or missing, and replace them immediately.

4. Damaged/clogged tower fill and/or tower inlet air obstructions will reduce the tower airflow, either overall or in localized areas. Either condition will negatively impact on the performance of the drift eliminators.
5. Unbalanced or uneven water distribution will result in high water flow rates in some portions of the tower, increasing the drift rate in that area. Water flows must be even across the entire airflow area.
6. Avoid the use of surfactants as part of the water treatment program. Surfactants, by lowering the surface tension of the water, cause the water to form very small droplets, increasing drift rates.

# Part III

## Cooling Tower Operations and Maintenance

# 15 Cooling Tower Operation and Maintenance

## TOWER COMMISSIONING

When the cooling system's installation has been completed, it is necessary to start the cooling tower and place it in service. Condenser water system commissioning can be broken down into the following basic elements, with numerous requirements associated with each element, outlined as follows:

A. Condenser Water Pump
   1. Check pump installation, including mountings, vibration isolators and connectors, and piping specialties (valves, strainer, pressure gauges, thermometers, etc.).
   2. Check pump shaft and coupling alignment.
   3. Lubricate pump shaft bearings as required by the manufacturer.
   4. Lubricate motor shaft bearings as required by the manufacturer.
   5. Turn shaft by hand to make sure that the pump and motor turn freely.
   6. "Bump" the motor on and check for proper rotation direction.

B. Cooling Tower
   1. Clean all tower surfaces; flush and clean the wet deck and basin.
   2. Clean the basin strainer.
   3. Lubricate fan shaft bearings as required by the manufacturer.
   4. Lubricate motor shaft bearings as required by the manufacturer.
   5. Test and adjust belt drive (if installed):
      a. Adjust belt(s) tension.
      b. Check and adjust belt(s) alignment.
   6. Test and adjust gear drive (if installed):
      a. Fill oil reservoir.
      b. Check shaft alignment.
      c. Check couplings for bolt tightness, excess play, and so on.
   7. Turn shaft by hand to make sure fan and drive turn freely.
   8. "Bump" motor(s) on and check fan(s) for correct rotation.

9. Run fan(s) for a short period and check for unusual noises and/or vibration. Verify that motor amps are in accordance with the manufacturer's data.
10. Confirm that the condenser water piping has been flushed and chemically cleaned.
11. Fill the basin and piping to the manufacturer's recommended basin operating level.
12. Run condenser water pump for a short period:
    a. Verify that the pump motor amps are within motor nameplate rating.
    b. Test the wet deck distribution, fill, and basin for proper water flow.
    c. Check the basin for vortexing.
13. Test the basin freeze protection thermostat and heater.
14. For multicell installations:
    a. Ensure that automatic isolation valves function properly.
    b. Balance the condenser water flow to and from each cell.
    c. Balance equalizer line to maintain a proper basin operating level in each cell.
15. Place the tower fan and condenser water motor starters in "automatic" position.

C. Controls
1. Check that all temperature sensors are properly installed.
2. Test operation of bypass control valve and adjust for 70° rotation.
3. Check controller setpoint.
4. Test fan control relays for proper functioning.
5. Test VFD (if installed) for proper operation.
6. Test the operation of vibration cutout switch.
7. Test the operation of basin level controls, including high- and low-level alarms.
8. Place controls in operation.

D. Water Treatment
1. Check that all pH and/or conductivity probes are properly installed.
2. Test operation of blowdown solenoid valve.
3. Place water treatment equipment in operation.

At this point, the condenser water system should operate and be controlled to maintain setpoint temperature under its automatic controls. However, during the first 24 h of operation, all aspects of condenser water pump cooling tower, controls, and water treatment should be checked and evaluated frequently. Unusual noises or vibration, erratic performance or operation, or other problems (see the section "Tower Performance Troubleshooting") may require that the system be shut down and repaired before being placed into full-time service.

Once the tower has been put into operation for several days and all start-up requirements have been met, test the tower for excess recirculation. This condition can be determined by measuring the ambient air wet bulb temperature and the wet bulb temperature entering the tower. An increase of more than about 1°F in the entering wet bulb over the ambient wet bulb indicates that tower excess discharge is reentering the tower. This condition must then be corrected based on the parameters outlined in the section "Capacity and Performance Parameters" in Chapter 11.

## COOLING TOWER MAINTENANCE

Once the cooling tower system is placed into full-time service, routine inspection and maintenance must be done to ensure proper tower operation and to obtain the expected service life of the equipment. The required maintenance can be broken into two areas: water treatment and mechanical maintenance.

### Water Treatment Management

Water treatment requirements and programs to prevent corrosion, deposition, and biological fouling in condenser water systems have been addressed in Chapter 13. However, as a routine matter, the tower owner must ensure that this requirement is being met effectively by the water treatment contractor. To this end, the following procedure is recommended:

1. Require and evaluate regular and frequent reports by the water treatment contractor, first, to ensure that regular water treatment is being done and, second, to "track" the various treatment parameters such as pH, TDS, chemical types and quantities used, and so on.
2. At least twice each year, send a water sample to an independent laboratory for analysis and compare the results with the most recent monthly report from the water service contractor.
3. During shutdown periods, the maintenance staff should inspect the tower and as much piping as possible for scaling or fouling that is being inadequately addressed by the water treatment program.
4. Track the chiller and tower performance on a routine basis to determine if the system is remaining free of deposition or fouling. The information shown in Figure 15.1 should be logged on a regular basis (at least once per day or once per shift) and the relationships between load, temperature difference, and power input tracked. If the relationships between these values change appreciably, this could indicate chiller fouling, tower fouling, or other performance problems.

COOLING TOWER LOG

Date ________ Shift ________ Data Collected By ________________

Outside Air Conditions

| Time | | | | | | | | |
|---|---|---|---|---|---|---|---|---|
| Dry Bulb Temperature | | | | | | | | |
| Wet Bulb Temperature | | | | | | | | |

Cooling Tower No. ________________

| Fan(s) On/Speed (H/L) | | | | | | | | |
|---|---|---|---|---|---|---|---|---|
| Water Level | | | | | | | | |

Cooling Tower No. ________________

| Fan(s) On/Speed (H/L) | | | | | | | | |
|---|---|---|---|---|---|---|---|---|
| Water Level | | | | | | | | |

Cooling Tower No. ________________

| Fan(s) On/Speed (H/L) | | | | | | | | |
|---|---|---|---|---|---|---|---|---|
| Water Level | | | | | | | | |

Cooling Tower No. ________________

| Fan(s) On/Speed (H/L) | | | | | | | | |
|---|---|---|---|---|---|---|---|---|
| Water Level | | | | | | | | |

**FIGURE 15.1** Chiller and cooling tower maintenance log.

## Mechanical Maintenance

Cooling towers require the following preventative maintenance measures:

*Start-up*: When the tower is to be started after seasonal shutdown, it must be thoroughly inspected and repaired as follows:

1. Check drift eliminators for proper position, cleanliness, and so on.
2. Check fans, bearings, motors, and drives for proper lubrication.
3. Rotate fan shaft(s) by hand to make sure they turn freely.
4. Check fan motors for proper rotation and adjust belt tension for belt drives.
5. Fill the basin with fresh water and check the operation of the level controller.
6. Start the condenser pump and check the wet deck for proper distribution.

7. Check fill for fouling and/or clogging and clean or replace if necessary.
8. Check access door gaskets and replace as necessary.
9. Thoroughly inspect all metal surfaces for corrosion, scale or fouling, or sludge. Clean as required. Any damaged metal should be cleaned down to the bare metal and refinished with a cold zinc coating.
10. Operate tower and look for and repair any water or air leaks from the basin, casing, or piping.

*Scheduled*:

1. Weekly
   a. Clean the basin strainer.
   b. Check the blowdown valve and makeup water valves to make sure they are working properly.
   c. Test water and adjust chemical treatment as necessary.
   d. Check/fill gear drive oil reservoir (on gear-drive towers).
2. Monthly
   a. Clean and flush basin. (This may be required more often for towers located adjacent to highways, industrial sites, and so on with high particulate emissions or in hot, humid climates with high biofouling potential.)
   b. Check the operating level in the basin and adjust as necessary.
   c. Check the water distribution system and sprays.
   d. Check drift eliminators for proper position.
   e. Check belts or gearbox and adjust as necessary.
   f. Check fans, inlet screens, and louvers for dirt and debris. Clean as necessary.
   g. Check keys and tightness of set screws.
3. Routinely
   a. Lubricate TEFC motor (TEAO motors are sealed and do not require lubrication) as follows:
      Relubricate on the basis of the following schedule:

| Motor RPM | Ambient Temperature (°F) | Regreasing Frequency |
|---|---|---|
| 1800 | 100 or less | 6 months |
| | 150 | 3 months |
| | 200 | 1 month |
| | 250 | 2 weeks |
| 3600 | 100 or less | 4 months |
| | 150 | 2 months |
| | 200 | 3 weeks |
| | 250 | 10 days |

Lubricant should comply with the following:

| Motor RPM | Temperature (°F) | | Grease Recommendation |
|---|---|---|---|
| | Ambient | Operating | |
| Up to 3600 | –20 | 250 | Lithium grease or polyurea grease |
| Over 3600 | 0 | 200 | Lithium complex grease with low-viscosity oil or nonsoap grease, like MIL-G-81422 grease |
| Below 3600, high temperature | — | 400 | Lithium complex grease, polyurea grease, or Mil-G-81322 grease |
| Vertical motors up to 3600 | –20 | 250 | Lithium base grease with heaver than NLGI 2 consistency |

Motor bearings must be flushed if new grease is not compatible with the old grease or if the old grease has become contaminated. When flushing, make sure that motor windings do not become contaminated.

Every 3 months, clean fan and "weep holes" at the bottom of the end housings that allows condensation or any other moisture accumulation to drain.

b. Lubricate fan shaft bearings. Lubrication of bearings must be done in strict accordance with the manufacturer's recommendations for the type of lubricant and intervals between relubrication.

However, when no manufacturer's data are available, the relubrication interval, at least initially, can be estimated on the basis of the following:

$$\text{Relubrication interval (operating hours)} = \frac{14{,}000{,}000}{\text{RPM} \times D^{0.5}} - (4 \times D)$$

where

RPM = shaft rotational speed (revolutions per minute)

$D$ = shaft diameter at bearing (mm) (where 1″ = 25.4 mm)

This relationship is valid for tapered or spherical roller bearings. For cylindrical or needle bearings, multiply the result by 5. For ball bearings, multiply by 10.

This relationship is valid for bearing operating temperatures of 160°F or less. If bearing operating temperatures exceed 160°F, multiply the results by the following factor:

| Bearing Operating Temperature (°F) | Multiplier Factor |
|---|---|
| ≤160 | 1.0 |
| 170 | 0.75 |
| 180 | 0.5 |
| 190 | 0.375 |
| 200 | 0.25 |

Note that bearing environment and position may require more frequent relubrication—dirty, dusty environments, vertical shafts, and so on will shorten relubrication periods.

Ultimately, the proper relubrication period must be determined by "trial and error." Using the manufacturer's data or the above formula as a starting point, visually examine purged lubricant at the end of the lubrication interval. If the lubricant is clean, lengthen the period between relubrication. If it is dirty or scorched, shorten the interval.

Overgreasing of bearing assemblies will result in seal failure—big globs of grease on the floor beneath the bearing are a sure sign that this had happened. In this case, bearing replacement is required (and care must be taken in the future to avoid failure from recurring).

To prevent overlubrication, there are two ways to determine the correct relubrication quantity. The first, and preferred, method is to contact the bearing manufacturer and request a recommendation based on the specific application. Failing that, the second option is to calculate the correct relubrication quantity using the following equation

$$\text{Grease quantity (oz)} = \frac{D \times B \times Q}{28.35}$$

where

$D$ = bearing outer diameter (mm)

$B$ = bearing width (mm)

$Q$ = relubrication interval factor (0.002 for a weekly relubrication interval, 0.003 for a monthly interval, 0.004 a yearly interval)

   c. Check the gear drive in accordance with the manufacturer's instructions. Change the gear drive oil every 2500 h of operation.
   d. Every 3 months, blow down (Y-type) or clean the (basket type) condenser water pump strainer.

4. Yearly
   a. Clean and touch up paint or other protective finish as necessary (including tower support steel).
   b. Dismantle and clean the condenser water pump strainer.
   c. Check motor control box for dirt, debris, and/or loose terminal connections. Clean and tighten connections as needed.
   d. Check motor contactors for pitting or other signs of damage. Repair or replace as needed.

## Induction/Venturi Tower Maintenance

Due to its lack of fan, motor, and drive assembly, routine maintenance for the induction draft or Venturi cooling tower is somewhat simpler than for a mechanical draft cooling tower. Basically, except for the fan elements, routine maintenance should include the same elements and schedule outlined in the section "Mechanical Maintenance."

There are several special maintenance aspects of this type of tower that should be addressed on a *monthly* basis as follows:

*Strainers.* There are two strainers that must be removed and cleaned by high-pressure water spray. The *sump strainers* are flat screens that require the basin to be drained before removal. A *final strainer* is located in a vertical cylindrical enclosure on the inlet side of the tower. This strainer is removed by removing the bolted cover plate.

*Spray nozzles.* Inspect the spray nozzles while the tower is in operation. If cleaning is necessary (indicated by poor or erratic spray), it can usually be accomplished while the tower is in operation by scrubbing the nozzle with a nonmetallic brush and using a nonmetallic pick. If this cleaning is unsatisfactory, shut down the tower, remove the nozzle, and clean thoroughly in the shop.

*Inlet louvers.* The inlet louvers are much more closely spaced than on other types of towers and must be kept clean. They are removable to facilitate cleaning (and to provide access to the spray nozzles).

## Heat Exchanger Maintenance

When a waterside economizer cycle is utilized, the heat exchangers applied must also be maintained on a regular basis as follows:

*Plate-and-frame heat exchanger.* When the economizer cycle is in use, perform the following routine maintenance:

a. At the beginning of the mechanical cooling season, when the economizer cycle is no longer used, dismantle and clean the heat transfer plates. High-pressure water can be used, avoiding the gaskets, along with brushing with a fiber bristle brush or a brush of the same alloy as the plates. Replace gaskets as necessary.
b. At least one manufacturer recommends that during the summer months when the heat exchanger is not in use, the plate pack should be loosened until all gasket compression is removed.
c. Make periodic inspection of control valves to ensure proper operation.

*Shell-and-tube heat exchanger.* When the economizer cycle is in use, perform the following routine maintenance:

a. Monthly, flush the shell-and-tube bundle and inspect the interior and exterior condition of the tubes. (Tubes can be cleaned with a 10% muriatic acid solution allowed to stand for 2 h.) Replace gaskets when replacing heads.

b. Make periodic inspection of control valves to ensure proper operation.

## TOWER PERFORMANCE TROUBLESHOOTING

Selected correctly, installed correctly, and properly maintained, an HVAC cooling tower should perform as required for many years. However, the tower may not perform as anticipated due to numerous problems that must be investigated and evaluated in the field.

Typical HVAC cooling tower performance problems fall into three major categories:

1. Selection problems
2. Installation problems
3. Maintenance problems

### Selection Problems

Performance problems related to the tower selection refers to inadequate tower performance because one or more of the tower performance requirements were incorrectly or inadequately specified.

From Chapter 9, two factors were defined as establishing the HVAC cooling tower performance:

1. The cooling load imposed on the tower, which is defined by the condenser water flow rate and the selected range
2. The approach, which is dictated by the design outdoor wet bulb temperature and the required condenser water supply temperature

Of these variables, the most common errors are made in defining the cooling load and in selecting the design outdoor wet bulb temperature. If either or both of these variables are underestimated, the cooling tower selected will simply be too small.

Selection errors are usually identified by performance problems when at or near design conditions and the design condenser water supply temperature cannot be achieved. Measuring the outdoor wet bulb temperature under these conditions will demonstrate whether the tower selection was based on a too low design wet bulb temperature. The imposed cooling load can be computed based on the design flow rate and the actual entering and leaving condenser water temperature and this value compared to the design requirement.

Unfortunately, there is relatively little that can be done to correct selection problems short of adding additional cooling tower capacity since the tower characteristic is too small for the performance required.

## Installation Problems

There are a host of potential installation problems that can adversely affect cooling tower performance:

1. Excess recirculation may exist because the tower is not installed with sufficient clearance around it for free airflow. This condition can be determined by measuring the ambient air wet bulb temperature and the wet bulb temperature entering the tower. An increase of more than about 1°F in the entering wet bulb over the ambient wet bulb indicates that tower discharge is reentering the tower.
2. There is inadequate airflow through the tower because the fan speed is not correctly adjusted. This problem can be solved by adjustment of the fan drive for the proper fan speed.
3. There is inadequate water flow through the tower because the pump is incorrectly sized or installed incorrectly. Correcting the pump usually corrects the flow problem.
4. The make-up water system is inadequately sized and the tower runs dry. The make-up water supply must be sized in accordance with the section "Make-Up Water Piping" in Chapter 11.
5. Bypass piping is incorrectly installed and the chiller will not start on cold mornings. This problem is corrected by eliminating the potential air gap in the pump suction as discussed in the section "Start/Stop Control" in Chapter 12.
6. With two-speed fan systems, time delay is not included in the control scheme and the drive is damaged. See the section "Fan Speed Control" in Chapter 12.
7. For multicell tower installations, the equalizer line may be inadequately sized if one basin floods while another runs dry. This problem is usually corrected by installing an equalizer line sized in accordance with Table 11.3.
8. As obvious as it may sound, some cooling towers do not get installed with the basin level. Thus, the basin operating depth varies from one end to the other (or even worse, from corner to corner). Obviously, basin level control under these conditions can be difficult and the only real solution is to disconnect all of the piping, level the tower, and reinstall the piping.
9. As discussed in Chapter 11, if any of the condenser water supply piping between the tower basin and the pump suction is installed at an elevation higher than the tower operating level, the pump will lose prime due to air entrainment. The only fix for this problem is to lower the piping or raise the tower.

## Maintenance Problems

Proper water treatment and regular mechanical maintenance of the HVAC cooling tower are required to keep it performing as needed. (See the section "Cooling Tower Maintenance.")

## ENHANCING TOWER PERFORMANCE

When the load on a cooling system increases, the load imposed on the condenser water increases in direct proportion. However, if the load increases to a point where the cooling tower capacity is exceeded, the tower must be modified, if possible, to enhance its performance.

Increased load imposed on an HVAC cooling tower is represented by increased condenser water flow rate and/or increased range. Increased tower performance is achieved by increasing the airflow through the tower and/or increasing the tower heat exchange (fill) area. Thus, the first step in the analysis is to determine a new tower characteristic and from that determine the required additional heat transfer area and airflow rate required. The manufacturer can then determine if increasing airflow and amount of fill is possible. If not, the next step is to add a new cooling tower to operate in parallel with the existing tower.

Induced draft towers have, usually, more potential for increasing airflow than do forced draft towers. The induced draft fan speed and/or blade pitch may be changed to produce more airflow. The fan wheel can be replaced with one or more blades. As a last resort, the entire fan assembly can be replaced with a larger one. *In each of these cases, the fan motor size will increase as the cube of the increased airflow and must be replaced, also requiring wiring changes to support the increased motor horsepower.*

Fill can be replaced with a different type that is more efficient. Bar splash fill can be replaced with formed PVC or fiberglass splash fill that is 15–20% more effective. It may even be possible to replace splash fill with film fill, improving heat transfer by as much as 25%. Existing film fill can be replaced with more efficient film fill, but at the expense of increased air pressure drop and fouling potential.

The tower configuration itself can be changed to improve performance. For example, a forced draft tower can be reconfigured as an induced draft tower or the small fans of a forced draft tower can be replaced with ducted air delivered from much larger fans. Usually, however, this type of modification, expected in very large towers, is not cost effective and an inadequate tower should simply be replaced with a new one.

## COOLING TOWERS IN FREEZING CLIMATES

### WINTER TOWER OPERATION

Winter tower operation may be dictated by the need to either provide cooling when systems have no economizer cycle or serve in a waterside economizer application. Typically, the freezing problem is not related to the main condenser water flow stream, or even the water in the tower basin, but to ice formation from stray water droplets and/or saturated air impinging on cold surfaces.

*Since forced draft cooling towers are far more susceptible to icing problems and potential mechanical damage, induced draft towers are usually used where winter operation is required and, therefore, this discussion is limited to that tower configuration.*

Ice, in small quantities, will often form on tower louvers, structure, and the entering edge of the tower fill, and is rarely a cause for concern. However, if the ice thickness grows, it can extend across and into the fill, effectively blocking airflow and, by its weight, cause structural damage to the tower. Thus, it is necessary for tower operators to monitor ice formation and to take the following routine steps to control it.

1. Never operate the cooling tower without a heat load. Without a heat load being imposed, the water temperature will continue to fall until (a) the ambient wet bulb temperature is reached, or (b) ice forms, whichever occurs first.
2. Do not modulate water flow through the tower. Reducing the water flow will provide low flow regions in the fill that will promote icing. If flow control to prevent the water temperature from going too low is required, use the bypass valve as a two-position device rather than a modulating device—send all of the water through the tower, then bypass all of the water.
3. Control airflow through the tower to control condenser water supply temperature above freezing. As discussed in Chapter 12, tower capacity control is routinely done by controlling the amount of air through the tower. Two-speed or pony motor arrangements are much better than single-speed fan cycling and variable speed control is even better for winter operation.
4. Deice the tower by airflow reversal. If, despite best efforts to prevent it, a significant amount of ice does build up on the tower, deicing can be accomplished by reversing the fan rotation and, thus, the airflow direction through the tower. Warm air is circulated back through the fill and louvers to melt the ice that has formed on these surfaces. Fan reversal can be accomplished by installing a "forward-off-reverse" switch in the fan circuit or, for three-phase motors, simply switching two of the motor wires manually. To prevent damage to the fan and tower, fan reversal should follow this procedure:
   a. Turn the fan off for 2–3 min to let the fan come to a complete stop.
   b. Restart and run the fan in reverse for no more than 15–20 min at one time.
   c. If the tower has a two-speed or pony motor, run the fan in reverse at low speed.
   d. If the tower consists of multiple cells, deice each cell individually while the adjacent cells are shut down.
   e. After reversal, turn the fan off for 5–10 min before restarting in the correct direction.
   f. Check fan and drive for proper operation, vibration, noise, and so on.
5. *Never use antifreeze solutions in condenser water.* The use of antifreeze chemicals in the water changes the evaporation rate of the water and reduces the efficiency of the tower and, on evaporation, pollutes the area around the tower.

6. During cold weather, monitor towers carefully. Never allow a tower to operate unattended.

## Winter Tower Shutdown

If the tower is not required to operate during the winter, it can be shut down for the heating season. To protect the tower from freezing and to minimize adverse effects to the system, the following guidelines are recommended for winter shutdown:

1. The tower wet deck and basin must be drained to prevent freezing.
2. Condenser water piping should be drained *only* to where it enters the building or is otherwise protected from freezing. Studies have shown

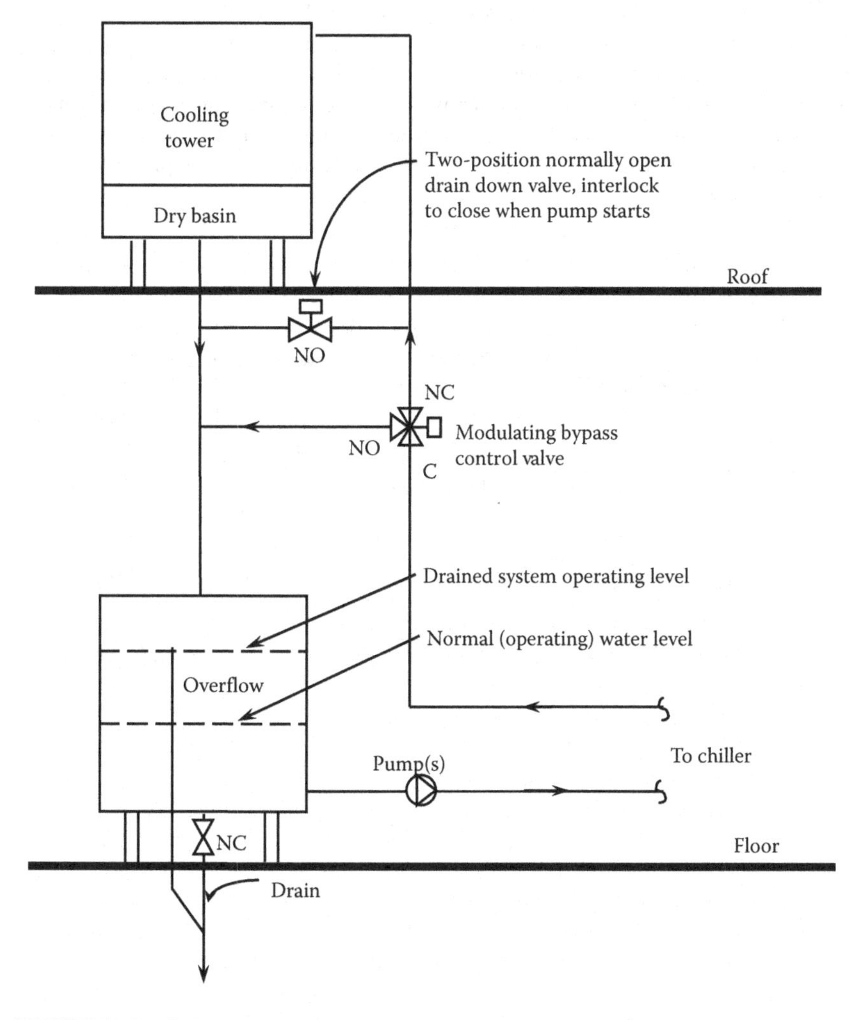


**FIGURE 15.2** Indoor sump schematic.

that piping systems that are drained tend to have significantly increased corrosion rates due to exposure to air and the resulting oxidation. Thus, drain the minimum amount of piping needed to prevent freezing and keep the rest of the system flooded.

3. Drain exposed makeup water piping. (This piping is usually copper or plastic and corrosion is not a problem.)

An alternative to the manual shutdown and draining outlined above is to design the tower system with a remote sump within the building so that the tower operates effectively "dry." When the tower is shut down, all water drains to the indoor sump and no additional freeze protection is required. A typical tower and remote sump arrangement is illustrated in Figure 15.2.

Corrosion may occur and fouling and/or scale may form in the shutdown condenser water system. Corrosion shows up during the following cooling season when iron rust and scale breaks loose and begins to plug wet deck nozzles, strainers, and/or condenser tubes. Scale forms when wetted pipe dries and begins to rust. Other fouling is formed by the dissolved solids left behind as the water evaporates. To mitigate these problems, the following shutdown procedures are recommended:

1. As the cooling season ends, decrease the cycles of concentration to ~5–6 to reduce the level of dissolved solids in the water.
2. Before shutdown, add a nonoxidizing biocide to maximize biological "kill," since stagnant water left in the system will enhance the environment for biological growth.
3. Add additional dispersant to loosen and penetrate existing fouling.
4. To minimize corrosion during the shutdown period, circulate a film-type corrosion inhibitor for 3–5 days before the system is shut down and drained, but only after sufficient time as elapsed for the dispersant to have been effective.
5. All water treatment systems should be shut down, cleaned, and flushed with fresh water in preparation for the next cooling system. Conductivity and pH probes should be removed, cleaned, and stored in a cool, dry area.

# 16 Buying a Cooling Tower

## DEFINING TOWER PERFORMANCE REQUIREMENTS

The first step in purchasing a new or replacement HVAC cooling tower is to define the performance or capacity requirements for it. To do this, the following four parameters must be specified:

1. Condenser water flow rate (gpm)
2. Entering (return) condenser water temperature (°F)
3. Leaving (supply) condenser water temperature (°F)
4. Entering (ambient) wet bulb temperature (°F)

The difference between the entering and leaving condenser water temperatures defines the required range, while the difference between the leaving condenser water temperature and the entering wet bulb temperature defines the required approach. With range and flow rate, the total Btu/h of heat rejection capacity is defined. With these data, any manufacturer can select a cooling tower.

The next step is to define one or more tower configurations that would be acceptable as follows:

| | |
|---|---|
| Draft/fan | A—Forced/propeller fan |
| | B—Forced/centrifugal fan |
| | C—Induced/propeller fan, gear-drive |
| | D—Induced/propeller fan, belt-drive |
| | E—Induced/propeller fan, direct-drive |
| Flow arrangement | 1—Crossflow |
| | 2—Counterflow |
| Assembly | Factory-assembled |
| | Field-erected |
| Construction | Wood |
| | Galvanized steel |
| | Galvanized steel with stainless-steel basin (and wet deck, for a crossflow tower) |
| | All stainless steel |
| | Fiberglass reinforced plastic (or equivalent) |
| | Concrete/masonry |

Table 16.1 summarizes the typical types of construction available for various draft and flow arrangements. As discussed in earlier chapters, wood and concrete/masonry towers are used infrequently in HVAC applications unless the tower requirement is very large (>2000 tons) or there is a specific aesthetic element that must be addressed.

**TABLE 16.1**
**Cooling Tower Construction Options**

| | Typical Construction Availability for Tower Type | | | |
|---|---|---|---|---|
| **Construction** | **A1/B1** | **C1/D1/E1** | **A2/B2** | **C2/D2/E2** |
| Wood | | × | | |
| Galvanized steel | × | × | | × |
| Galvanized steel with stainless-steel basin (and wet deck, as applicable) | × | × | | × |
| All stainless steel | × | × | | × |
| Fiberglass-reinforced plastic | | × | × | × |
| Concrete/masonry | | × | × | × |

*When one of the acceptable alternatives is a galvanized tower, it is recommended that the basin and, for crossflow towers, the wet deck be constructed of stainless steel.* This option is offered by all of the tower manufacturers and will increase tower costs by 20–40%. But, as shown in Table 16.2, typical cooling tower service life is extended by at least 30+%, making it an excellent investment.

Factory-assembled towers are much easier and faster to install and tend to be less expensive. Thus, field-erected towers are rarely used for HVAC applications except, again, for very large loads (2000+ tons).

## CTI RATINGS AND PERFORMANCE GUARANTEES

Except in very special circumstances or for large field-erected towers, HVAC cooling towers should be specified to have their performance certified by the CTI in accordance with CTI Standard 201. This certification amounts to a third-party guarantee that the tower will perform as advertised, provided the tower is installed correctly and the condenser water is "clean."

**TABLE 16.2**
**Typical Service Life**

| Equipment | Life (Years) |
|---|---|
| Pump, base-mounted | 20 |
| Pump, line-mounted | 10–15 |
| Cooling tower, galvanized steel | 15–20 |
| Cooling tower, FRP | 20 |
| Cooling tower, galvanized steel w/stainless steel basin and wet deck | 25 |
| Cooling tower, wood | 20–25 |
| Cooling tower, all stainless steel or masonry | 35+ |

**TABLE 16.3**
**CTI Certified Cooling Towers (January 2011)**

**Advance GRP Cooling Towers Pvt. Ltd. & Advance Cooling Towers Pvt. Ltd.**
- Advance 2020 – Series A Line

**Aggreko Cooling Tower Services**
- AG Cooling Tower Line

**Amcot Cooling Tower Corporation**
- LC Line

**American Cooling Tower, Inc.**
- ACF Series Line

**AONE E&C Corporation, Ltd.**
- ACT-R and ACT-RU Series Line

**Baltimore Aircoil Company, Inc.**
- ACT Series Line
- FXT Line
- FXV Closed Circuit Cooling Tower Line
- Series V Closed Circuit Cooling Tower Line
- Series V Open Cooling Tower Line
- Series 1500 Line
- Series 3000A, C & D Line
- PT2 Series Line
- PCT Series Line

**The Cooling Tower Company, L. C.**
- Series TCI Line

**Delta Cooling Towers, Inc.**
- TM Series

**Evapco, Inc.**
- AT Line
- ATW Line of Closed Circuit Coolers
- ESWA Line of Closed Circuit Coolers
- L Series Line
- L Series Line of Closed Circuit Coolers
- PMWQ Series Line of Closed Circuit Coolers
- PMTQ Series Line of Counter-flow, Forced-draft Cooling Towers

**Fabrica Mexicana De Torres, S.A.**

**Reymsa Cooling Towers**
- HFC Series of CTI Certified Closed Circuit Fluid Coolers
- HR Line
- LSFG Line
- SLSFG Line

**HVAC/R International, Inc.**
- Therflow TFW Series Line
- Therflow TFC Series Line of Closed Circuit Cooling Towers

**KIMCO (Kyung In Machinery Company, Ltd.)**
- CKL Line of Closed Circuit Cooling Towers
- EnduraCool Line
- ECO-DYNA COOL Line

*continued*

**TABLE 16.3 (continued)**
**CTI Certified Cooling Towers (January 2011)**

**King Sun Industry Company, Ltd.**
- HKB Line
- HKD Line
- KC Line of CTI Certified Closed-Circuit Cooling Towers

**Liang Chi Industry Company, Ltd.**
- LC Line
- C-LC Line
- V-LC Line of Counter-flow, Forced-draft Cooling Towers
- U-LC Line of Cross-flow, Induced-draft, Low-noise Cooling Towers

**Marley (SPX Cooling Technologies)**
- Aquatower Series Line
- AV Series Line
- MCW Series
- MD Series
- MHF Series of Closed-Circuit Fluid Coolers
- NC Class Line
- Quadraflow Line

**Mesan Cooling Tower, Ltd.**
- MCR Series Line
- MXR-KM Series Line
- MCR-KM Series Cooling Tower Line
- MCX Series Cooling Tower Line

**Nihon Spindle Manufacturing Company, Ltd.**
- CTA-KX Series

**Polacel, b. v.**
- CR Series Line
- XR Series Line

**Protec Cooling Towers, Inc.**
- FRS Series Line
- FWS Series Line

**RSD Cooling Towers**
- RSS Series Line

**Ryowo (Holding) Company, Ltd.**
- FRS Series Line
- FWS Series Line
- FXS Series Line
- FCS Series Line

**Sinro Air-Conditioning (Fogang) Co., Ltd.**
- Sinro SC-H Series Line

**Ta Shin F. R. P. Company, Ltd.**
- TSS Series Line

**Tower Tech, Inc.**
- TTXL Line
- TTRT Cooling Tower Line

**TABLE 16.3 (continued)**
**CTI Certified Cooling Towers (January 2011)**

**The Trane Company**
Series Quiet (TQ) Line
**Waltco System Limited**
WGI Series Line
**Zhejiang Jinling Refrigeration Engineering Company, Ltd.**
JNT Series Line
JNC Series Line of Closed Circuit Cooling Towers

CTI certification eliminates the need and expense of field testing a cooling tower to confirm that it performs as specified. Table 16.3 lists the towers that are CTI certified. However, new towers are added to CTI's list from time to time, and so the list should be checked (http://www.cti.org) to confirm any proposed cooling tower's certification.

## ECONOMIC EVALUATION OF ALTERNATIVE COOLING TOWER SYSTEMS

Life cycle cost analysis is the best way to compare alternative cooling tower selections. It provides for a consistent method of analyzing the economic aspects of each tower and allows realistic comparison so that the most cost-effective, for example, least life cycle cost, tower can be selected. The process requires only that each potential new tower be evaluated using the same criteria. The result is an "apples-to-apples" comparison, not "fruit salad."

The computations to determine life cycle cost utilizing the *total owning and operating cost* methodology are simple. However, the methodology's accuracy depends wholly on the accuracy of the data utilized. Anyone can calculate "garbage" life cycle costs simply because they use data and/or assumptions that are "garbage." Two different individuals, faced with the same evaluation, may compute wildly different life cycle costs because they use significantly different data and/or assumptions in their computations.

The following subsections define the basic elements that make up the life cycle cost.

### FIRST COSTS

The initial capital costs associated with each potential cooling tower are all of the costs that would be incurred in the design and construction of that tower and the associated condenser water pump. Equipment costs can be obtained directly from the prospective equipment vendors. Installation cost estimates can be obtained from local contractors or, lacking that, from cost data published by R. S. Means

Co., Inc. (Construction Plaza, 63 Smiths Lane, Kingston, MA 02364-0800, (871) 585-7880 or (800) 334-3509).

The construction cost estimate must include the following, in addition to the cost of the tower itself:

1. Tower dunnage and grillage
2. Rigging
3. Demolition
4. Electrical power
5. Controls
6. Contractor overhead (insurance, bonds, taxes, and general office operations, special conditions), typically 15–20%
7. Contractor profit, typically 5–20%

Other costs that may be included in the capital requirement are design fees, which may increase or decrease as a function of the selected alternative; special consultants' fees; testing fees; and so on. Also, condenser water, makeup water, and/or drainage piping may change configuration and cost between alternative towers.

Unless the cost of the tower is being met from operating revenues, at least a portion of the capital expense will be met with borrowed funds. The use of this money has a cost in the form of the applied *interest rate*, and information about the amount of borrowed funds, the applied interest rate, and the period of the loan must be determined for the analysis. Then, the total capital cost, including both principal and interest, can be computed using the relationships from Chapter 8.

## Annual Recurring Costs

Once the tower system is placed into operation, two annual recurring costs must be met each year of its economic life: energy costs and maintenance costs.

The economic life for an alternative is the time frame within which it provides a positive benefit to the owner. Thus, when it costs more to operate and maintain a piece of equipment than it would to replace it, the economic life has ended. The economic life or "service life" is the period over which the equipment is expected to last physically. The typical economic life for the cooling towers and pumps is provided in Table 16.2.

The computation of annual energy cost requires that two quantities be known: (1) the amount of electrical energy consumed by the tower and associated condenser water pump(s), and (2) the unit cost or rate schedule for that energy. The second quantity is relatively easy to determine by contacting the utilities serving the site or, for some campus facilities, obtaining the cost for steam, power, chilled water, and so on that may be furnished from a central source.

To accurately evaluate and compare alternative cooling towers, it is necessary to also consider the condenser water pump(s). Since different towers may impose different pressure requirements on the pump(s), this energy and cost component must be incorporated into the analysis.

To determine the total energy use by the condenser water pump and cooling tower, the first step is to develop a "load profile" for the condenser water system in terms of system load (tons) as a function of outdoor temperature, usually for each 5°F "bin" of outdoor temperature. At each bin, the power input to the condenser water pump and cooling tower must be determined in terms of kW and then multiplied by the hours of occurrence of that bin to yield energy consumption in terms of kWh. Adding the energy consumption in all of the bins yields the total annual energy consumption. Figure 16.1 provides a form that can be used for this calculation.

With this form, for single-speed motors, the "percent load" is 100% any time the fan is on. For two-speed or pony motor arrangements, the full-speed load is 100%, while the half-speed load is ~30%. When a VFD is used, the percent load can be estimated from Figure 12.6.

| Outdoor temp range (°F) | (A) Annual hours (Note 1) | (B) Cooling load (tons) | Tower fan | | CDW pump | | (G) Total kW (Note 3) | (H) Total kWh (Note 4) |
|---|---|---|---|---|---|---|---|---|
| | | | (C) Percent load (Note 2) | (D) kW | (E) Percent load (Note 2) | (F) kW | | |
| 95–99 | | | | | | | | |
| 90–94 | | | | | | | | |
| 85–89 | | | | | | | | |
| 80–84 | | | | | | | | |
| 75–79 | | | | | | | | |
| 70–74 | | | | | | | | |
| 65–69 | | | | | | | | |
| 60–64 | | | | | | | | |
| 55–59 | | | | | | | | |
| 50–54 | | | | | | | | |
| 45–49 | | | | | | | | |
| 40–44 | | | | | | | | |
| 35–39 | | | | | | | | |
| 30–34 | | | | | | | | |
| 25–29 | | | | | | | | |
| 20–24 | | | | | | | | |
| 14–19 | | | | | | | | |
| 10–14 | | | | | | | | |
| 5–9 | | | | | | | | |
| 0–4 | | | | | | | | |

Notes:

1. These data are available from *Engineering Weather Data*, published by the Department of Defense as Air Force AFM 88-29, Army TM 5-785, and Navy NAVFAC P-89.
2. Percent of fan or pump motor load to satisfy imposed cooling load.
3. (D) + (F)
4. (G) × (A)

**FIGURE 16.1** Condenser water system energy consumption calculation form.

Cooling system load profiles fall into three basic forms:

1. Where airside economizer systems are used or the system is located in a cold climate and is not operated in winter, the load can be prorated between the design summer temperature and about 55°F. In each of these bins, since some cooling is required, the condenser water pump must run and, therefore, its load is 100% in each bin. The tower fan, with cycling capacity control, will operate a percentage of the time essentially equal to the percent load imposed on it.
2. When the cooling system operates on a year-round basis, the cooling load will be prorated to all bins of temperature.
3. When a waterside economizer is in use, the winter use of the condenser water pump and cooling tower must be estimated. Again, if there is any cooling load imposed in a bin, the condenser water pump power requirement is 100% in that bin. The tower fan use can be prorated in direct proportion to the imposed cooling load in winter.

The energy cost, then, is computed by multiplying the electrical energy consumption by the unit cost for electricity.

Annual recurring maintenance cost is a very difficult element to estimate. Lacking other information, the annual routine maintenance cost associated with pumps and cooling towers can be estimated as a percentage of the initial equipment cost as follows:

| | |
|---|---|
| Galvanized steel or wood cooling tower | 3% |
| Galvanized steel tower with stainless wet deck and basin or fiberglass-reinforced plastic (FRP) tower | 2% |
| Stainless-steel or masonry tower | 1% |

These costs do not include the cost of makeup water or the water treatment program. Since the amount of makeup water and water treatment chemical consumption is a function of (a) cooling tower loads, (b) the water treatment program required, and (c) the water flow rate, it is essentially independent of the tower itself. Therefore, this element can be ignored when comparing alternative cooling towers *unless* there is a significant difference in drift losses between alternative towers. (*The one exception to this is the additional water treatment required for wooden towers and this additional cost should be included in the analysis.*)

## Nonrecurring Repair and Replacement Costs

Nonrecurring costs represent repair and/or replacement costs that occur at intervals longer than 1 year. For example, a galvanized wet deck and basin in a crossflow induced draft tower may require significant repair (or even replacement) after 10 years of service. These costs must be determined and the year of their occurrence estimated.

### Total Owning and Operating Cost Comparison

The total owning and operating cost for a cooling tower and its associated condenser water pump, over the system economic life, can be computed as follows:

$$\text{Life cycle cost} = C + (\text{sum of repair and replacement costs}) + [(\text{economic life}) \times (\text{annual energy cost} + \text{annual maintenance cost})]$$

where $C$ is the total capital cost.

## PROCUREMENT SPECIFICATIONS

A recommended specification for induced draft cooling towers is provided as Appendix B4. Since the size range, performance characteristics, and, most importantly, energy consumption by induced draft towers are much better than for forced draft towers, the use of forced draft tower is rarely recommended. These specifications are based on factory-assembled galvanized steel towers with stainless-steel basins and, for crossflow towers, stainless-steel wet decks. Each specification includes both crossflow and counterflow configurations and, therefore, must be edited carefully if only one of these configurations is acceptable. Drives may be gear type, belt type, or direct-drive variable speed.

Appendix B5 contains a recommended specification for closed-circuit coolers.

## WATER TREATMENT PROGRAM CONTRACTING

The following are the recommended minimum standards that should be applied for the selection of a water treatment service company:

1. To be considered, a water treatment service company should be an established company with full-service capabilities. The company should have been in business for at least 5 years, have corporate staff with sufficient expertise and experience to competently address all aspects of water-related issues, and be capable of providing a reasonable list of references for whom they have provided service for a minimum of 2 years. *Call and verify the listed references.*
2. The local service personnel should have sufficient expertise and experience to competently address all aspects of water-related issues. After all, it is the local service technician who will actually provide the day-to-day service, not a "water expert" at the corporate office. Ask for a detailed resume of the service technician. Further, check into the company's service technician turnover rate. If the local technician is constantly changing, the quality of service provided by this company will be uneven.
3. The proposed approach to water treatment should incorporate proven technology. This does not mean totally excluding "new" technologies,

products, or methods, but it does mean that the water treatment service company must demonstrate that their proposed technology has been successfully applied at other locations with similar water treatment conditions and needs. Ask for a list of these locations and call them.

4. Ensure the program performance and its cost. Even with the most reputable water service companies, there are many instances where the water treatment program performance and/or cost did not meet promises. Therefore, develop detailed and rigid performance and cost standards and include them in the contract with the service company, including penalties if the standards are not met.

As a basis of comparing alternative water treatment programs by different service companies, prepare a detailed "request for proposals" (RFP) that defines requirements and standards and send it to several service companies who potentially meet the standards outlined above. Thus, there will be several different proposals that will have a common basis for comparison. For larger systems, it may be necessary to retain an independent consultant to develop the RFP and/or evaluate the vendor proposals. Costs incurred for the consultant will be more than offset by the proper selection of a cost-effective water treatment program.

Initial water service treatment contracts should be for a 2-year period. This gives the service company ample opportunity to address all of the problems that may be in the system and meet the cost and performance goals that have been established. After this initial period, contracts should be for 1 year.

# 17 *In Situ* Tower Performance Testing

For large towers or for towers with special requirements that are not CTI certified, *in situ* testing is the only way to guarantee that the tower will perform as required. This is an expensive and a time-consuming process that should be undertaken only after due consideration.

## WHY *IN SITU* TESTING?

Thermal performance testing of an operating cooling tower is a complicated and expensive undertaking. These tests are normally conducted for one of the two purposes:

1. Acceptance tests may be required to demonstrate that the installed cooling tower meets the performance standards that were specified. For the majority of HVAC applications, tower performance for many brands and types of cooling towers are certified by CTI and it more cost effective to simply select a certified tower and, thus, avoid the cost of acceptance testing entirely. In some cases, such as for very large towers or field-erected towers, an acceptance test may be necessary.
2. Performance testing to evaluate a cooling tower's performance relative to correcting problems and/or changing the tower's thermal requirements. *In situ* performance testing yields actual tower performance data that can be used to indicate required tower modifications or improvements.

## TESTING CRITERIA AND METHODS

The field testing of cooling tower performance must be done in accordance with CTI Standard ATC-105 or Standard PTC-23 issued by ASME. *Testing is typically done by third-party agencies and it is recommended that the selected agency be an agency that is licensed by CTI.*

Commonly, thermal performance tests are referred to as either Class A or B tests. The Class A test is one conducted using mercury-in-glass thermometers and grade-level psychrometers, while the Class B test uses a data acquisition system with psychrometers arranged in an array over the entire air inlet face of the tower. This does not mean that a data acquisition system cannot be used in conjunction with grade-level psychrometers. It is common to call the air temperature measurement device used in cooling tower tests a psychrometer. This is technically incorrect as a psychrometer measures the wet bulb and dry bulb temperatures while those instruments widely used today measure only a single temperature. Depending on how the "cooling tower" psychrometers are assembled, they may be used to measure either a wet or a dry bulb temperature.

It is also very important to recognize the difference between an ambient and entering wet bulb test. Both ASME and CTI recommend that towers be sized and tested based on entering wet bulb temperatures. This consideration can affect the size of the tower selected and the results of thermal tests. Ambient wet bulb is defined as the temperature of the air mass entering the tower, excluding recirculation. For an ambient WB test, at least three wet bulb instruments are located 50–100 ft upwind of the tower.

It is also necessary to measure (or otherwise to account for) the temperature and quantity of any other air streams (interference) entering the tower, other than its own recirculation. This interference can come from any other source, including other cooling towers. This can be very difficult, if not impossible to do accurately. The entering wet bulb temperature attempts to measure the average temperature of all the air entering the tower regardless of its source. While this is easier than trying to separate the influence of several air masses, it still requires careful analysis by the test staff to ensure that the number of instruments and their locations are adequate.

Most testing today is the Class B test, using data acquisition systems to measure temperatures. With this test method, the first step is to inspect the tower to ensure that it is ready for the test and identify points of measurement for the various parameters. While test agencies may consider condition of the tower, it is not their obligation to clean, balance, or otherwise adjust the tower for the test—this is the owner's or installer's responsibility.

In the case of an acceptance test, the installer and/or manufacturer will normally be much more thorough in this area to ensure the tower's full potential is measured.

Once all parties are satisfied that the tower is ready for testing, instruments are deployed as illustrated in Figure 17.1 and the testing begins. For typical large HVAC cooling towers, the testing process requires 1–2 days. Weather and operating conditions can sometimes increase this time, however.

To begin the testing process, the test technicians begins taking data. Usually, the thermal data loggers are started and monitored for a brief period. If any problems with instrumentation or conditions are noted, efforts will be made to correct them. Once any corrections are completed, the test begins and must last at least 1 h after steady-state conditions have been achieved.

During the test, the technicians will monitor the system temperatures and measure the water flow rate and fan power. The two test standards offer recommendations on deviation from design conditions for the test parameters, as shown in Table 17.1.

While it is preferable to comply with all these limitations, it is not always possible. CTI-licensed testing agencies report deviations from recommended parameters in 70–75% of all tests. Recognizing this, the standards allow for deviation *provided all parties agree.* If at any time during the process, it is determined that a parameter

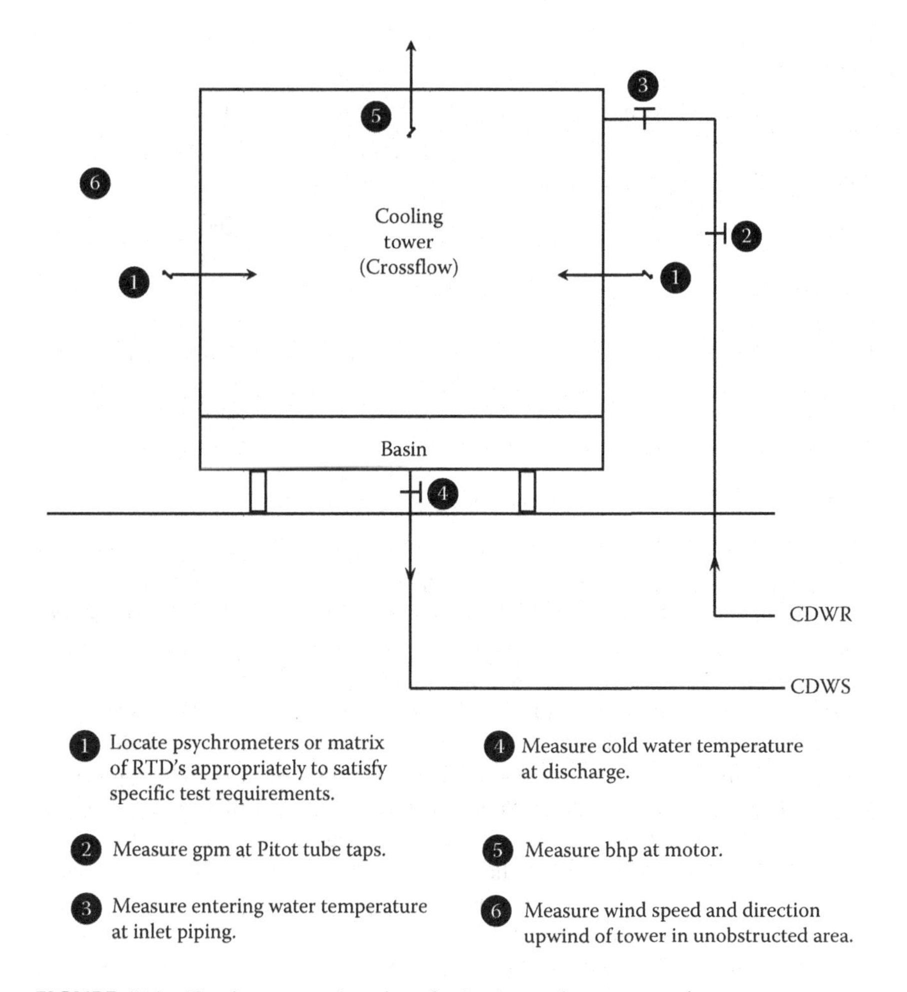


**FIGURE 17.1** Test instrument locations for *in situ* performance testing.

is outside the recommended limitations, all parties must review the situation and reach a unanimous solution. This can result in data being discarded and restarts required.

Two parameters, the limits on oil, tar, or fatty substances and the total dissolved solids in the condenser water, are not routinely checked during a tower test. However, if the tower fails the test, and any party thinks that these agents are present and could have contributed to the failure, these parameters are measured.

To measure the water flow rate, a Pitot tube traverse of the piping carrying water to the tower is the preferred method. A wattmeter is used to measure fan input power on mechanical draft tower systems up to 600 V. Temperatures are measured using thermometers, RTDs, or thermistors.

The test parameters must be measured at regular intervals during the test, as shown in Table 17.2.

**TABLE 17.1**
**Recommended Maximum Parameter Variations from Design**

| Parameter | CTI Standard | ASME Standard |
|---|---|---|
| Condenser water flow rate | ±10% | ±10% |
| Design heat load | Not specified | ±20% |
| Range | ±20% | ±20% |
| Fan motor power | ±10% | ±15% |
| WB temperature | ±15°F | +5°F/−15°F |
| Wind velocity (maximum) | 10 mph with 1 min gusts not to exceed 15 mph | 10 mph with 1 min gusts not to exceed 15 mph |
| Barometric pressure | ±1.0″ Hg | Not specified |
| Oil, tar, or fatty substances in water | 10 ppm | 10 ppm |
| TDS in water | The greater of 5000 ppm or 1.1 times design concentration | 10% above design |

In addition, any other factor affecting the tower's operation or the data taken must be recorded. These other factors may include pump discharge pressure, make-up flow and temperature, blowdown flow and temperature, auxiliary streams entering the collection basin, and so on.

The hot-water temperature is normally taken in the wet deck or in a tap in the piping carrying water to the tower. The cold-water temperature is normally taken at taps on the discharge side of the pumps. In most cases, installations have pressure gauges at this location and these gauges can be replaced with flowing wells for temperature measurement. If this is not possible, separate taps are required.

The CTI and ASME standards have defined the instrumentation and procedures very clearly. Unfortunately, the many installation variations and test circumstances provide multiple obstacles. This can create serious problems for the technicians and increased uncertainty of results. The straightforward process alone does not protect against completely meaningless results. For this reason, the CTI carefully tests those individuals licensed by CTI to lead tests and inspects and approves their test equipment. Manufacturers also have highly skilled and

**TABLE 17.2**
**Tower Test Parameter Recording Requirements**

| Parameter | Test Frequency (Readings/h) |
|---|---|
| Water flow rate | 3 |
| Hot-water temperature | 12 |
| Cold-water temperature | 12 |
| Wet bulb temperature | 36 per station |
| Fan power | 4 (or monitor continuously) |
| Wind speed | 12 |

trained staff to participate in the testing process, particularly for an acceptance test, to help ensure that the products are properly evaluated.

Evaluation of the data collected during the test must follow the requirements outlined by the test standard that is being used, either CTI or ASME.

## TOWER INSTALLATION REQUIREMENTS FOR TESTING

For a thermal performance test, especially an acceptance test, there are certain site requirements that need to be met by the customer or his representative. The following list includes the common considerations for a standard tower installation operating on a closed-loop system:

1. Pitot tube taps must be installed in the pipe(s) delivering water to the cooling tower.
2. Sensor taps must be installed for measurement of water temperatures. Hot-water temperature can normally be measured in the distribution basin of crossflow towers. Often the Pitot taps can also be a measurement point for hot-water temperature. If site-specific circumstances make neither of these options acceptable, special taps will be required. The cold-water temperature is normally measured at the discharge of the circulating water pumps. The most common location is at the pressure gauge tap present on most systems. If this is not available or applicable on a system, special taps or another solution must be identified. Measurement in a tower basin is not acceptable. Measurement of cold-water temperature in a flume or channel can sometimes be accomplished with acceptable accuracy, but specifics should be reviewed.
3. At the time of the test, safe access to any elevated points of measurement must be provided. All access must conform to safe work practices, OSHA requirements, and any local requirements.
4. Power for test instruments (typically 120/1/60) must be available adjacent to the tower. On very large towers multiple sources around the tower may be necessary.

The tower must be prepared for testing in the following ways before the test technicians arrive:

1. The tower must be clean. The wet deck must not have damaged, missing, or plugged nozzles or orifices and must be balanced to their design allows. The air inlet should be cleared of any blockage. If the tower has louvers, they should be in the normal design position, if adjustable. The eliminators should be free of foreign matter. Fan discharge should be clear and unobstructed.
2. Water flow and heat load to the tower, or representative cells, should be as close to design as the system will permit. If the test standard recommended limitations cannot be met, all parties should review the situation to agree on the deviation or delay/cancel the test.

3. Any water bypass should be closed and inspected to ensure that there is no leakage.
4. Any source of air leakage such as access doors, mechanical equipment supports, or holes in the casing or fan cylinders must be closed/blocked.
5. All fans must be operating at full speed and cannot cycle during the test period. In the case of tower fans operating with VFDs, they should be placed in bypass mode.
6. The owner or his representative should designate a coordinator qualified to integrate the testing activity and the normal process operation being served by the cooling tower.
7. The owner or his representative should have an electrician or qualified operator available to assist in the measurement of fan power.

All parties to the test must be advised in advance of any special safety issues required at the site.

# Appendix A

# *Design Ambient Wet Bulb Temperatures (Recommended for Cooling Tower Selection)*

| State | City | WB Temperature (°F) |
|---|---|---|
| Alabama | Birmingham | 79.5 |
| | Mobile | 81.1 |
| Alaska | Anchorage | 61.5 |
| Arizona | Flagstaff | 62.1 |
| | Phoenix | 77.1 |
| | Tucson | 73.5 |
| Arkansas | Little Rock | 82.1 |
| California | Los Angeles | 71.2 |
| | Sacramento | 72.2 |
| | San Francisco | 66.4 |
| Colorado | Denver | 61.9 |
| Florida | All cities | 82.0 |
| Georgia | Atlanta | 78.2 |
| | Augusta | 80.4 |
| | Savannah | 81.2 |
| Illinois | Chicago | 79.0 |
| | Springfield | 80.4 |
| Indiana | Indianapolis | 79.2 |
| | Terre Haute | 80.6 |
| Iowa | Des Moines | 79.0 |
| Kansas | Dodge City | 75.0 |
| | Topeka | 79.9 |
| Kentucky | Lexington | 78.3 |
| | Louisville | 79.4 |
| Louisiana | Baton Rouge | 81.3 |
| | New Orleans | 82.4 |
| Maine | Portland | 74.9 |
| Maryland | Baltimore | 79.1 |
| Massachusetts | Boston | 77.2 |
| Michigan | Detroit | 77.9 |

*continued*

| State | City | WB Temperature (°F) |
|---|---|---|
| Minnesota | Minneapolis/St. Paul | 77.9 |
| Mississippi | Biloxi | 84.5 |
| | Jackson | 80.8 |
| Missouri | Columbia | 80.1 |
| | Kansas City | 80.5 |
| | St. Louis | 80.4 |
| Montana | Butte | 61.7 |
| | Missoula | 66.4 |
| Nebraska | Omaha | 80.0 |
| Nevada | Las Vegas | 73.6 |
| | Reno | 65.0 |
| New Hampshire | Concord | 75.7 |
| New Jersey | Atlantic City | 78.9 |
| | Newark | 78.7 |
| New Mexico | Albuquerque | 66.3 |
| | Roswell | 71.6 |
| New York | Albany | 76.8 |
| | Buffalo | 75.8 |
| | New York City (JFK Airport) | 75.9 |
| North Carolina | Asheville | 75.2 |
| | Charlotte | 78.1 |
| | Fayetteville | 80.3 |
| | Greensboro | 77.9 |
| | Raleigh | 79.3 |
| | Wilmington | 81.3 |
| North Dakota | Bismarck | 75.3 |
| Ohio | Akron | 76.4 |
| | Cincinnati | 78.9 |
| | Columbus | 77.7 |
| Oklahoma | Oklahoma City | 78.7 |
| Oregon | Eugene/Portland | 70.4 |
| Pennsylvania | Philadelphia | 79.3 |
| | Pittsburgh | 76.2 |
| Rhode Island | Providence | 77.5 |
| South Carolina | Charleston | 81.5 |
| | Columbia | 79.4 |
| | Greenville | 78.0 |
| South Dakota | Rapid City | 71.9 |
| | Sioux Falls | 78.4 |
| Tennessee | Chattanooga | 78.8 |
| | Knoxville | 78.1 |
| | Memphis | 81.2 |
| | Nashville | 79.2 |

| State | City | WB Temperature (°F) |
|---|---|---|
| Texas | Abilene | 76.4 |
| | Amarillo | 72.1 |
| | Austin | 80.1 |
| | Dallas/Fort Worth | 79.6 |
| | El Paso | 71.3 |
| | Houston | 81.5 |
| | Laredo | 80.0 |
| | San Antonio | 79.0 |
| Utah | Salt Lake City | 68.0 |
| Vermont | Burlington | 75.2 |
| Virginia | Norfolk | 80.9 |
| | Richmond | 80.0 |
| | Roanoke | 76.4 |
| Washington | District of Columbia | 79.6 |
| Washington | Seattle | 67.5 |
| | Yakima | 69.4 |
| West Virginia | Charleston | 79.7 |
| Wisconsin | Madison | 78.7 |
| | Milwaukee | 78.0 |
| Wyoming | Cheyenne | 64.0 |

*Source:* Adapted from the Appendix to Chapter 14, *2009 ASHRAE Handbook—Fundamentals* (2009).

*Note:* 0.4% Evaporation WB Temperature, Plus 1°F.

# Appendix B1

# *Centrifugal Compressor Water Chillers*

1
2
3 **PART 1: GENERAL**
4
5 **SUBMITTALS**
6
7 General: Submittals shall demonstrate compliance with technical requirements by refering to each
8 subsection of this specification. Where a submitted item does not *comply fully* with each and
9 every requirement of the specifications, the submittal shall clearly indicate such deviations.
10 Identification requirements for noncomplying features of items are very specific.
11
12 Manufacturer's Data: Submit manufacturer's product data, including rated capacities, weights
13 (shipping, installed, and operating), furnished specialties and accessories, and installation and
14 start-up instructions.
15
16
17 **QUALITY ASSURANCE**
18
19 ARI Compliance
20
21 Fabricate, test, and certify chillers to comply with ARI Standard 550/590-98 "Water-
22 Chilling Packages Using the Vapor Compression Cycle."
23
24 Rate water-cooled chiller sound power levels according to ARI 575 "Method of

Measuring Machinery Sound within an Equipment Space."

ASHRAE Compliance

Fabricate and install rotary water chillers to comply with ASHRAE 15 "Safety Code for Mechanical Refrigeration."

Comply with ASHRAE 147 for refrigerant leaks, recovery, handling, and storage requirements.

UL Compliance: Fabricate rotary water chillers to comply with UL 465 "Central Cooling Air Conditioners."

ASME Compliance: Fabricate water chillers to comply with ASME Boiler and Pressure Vessel Code, Section VIII, Division 1. Include ASME U-stamp and nameplate certifying compliance.

Warranty: The contractor shall provide a 5-year parts and labor manufacturer's warranty on each refrigeration compressor, including refrigerant. This warranty shall be directly from the manufacturer to the owner and shall be in addition to guarantees and warranties required under the general conditions to the construction contract. This warranty shall provide for repair or replacement of the covered compressor that becomes inoperative as a result of defects in materials or workmanship within 5 years after the date of substantial completion.

Acoustic Criteria: Noise levels due to chillers shall result in maximum sound levels in occupied

1 spaces conforming to the following room criteria (RC):
2
3

| Maximum RC | Environment | Typical Occupancy |
|---|---|---|
| 25 | Extremely quiet environment, suppressed speech is audible, suitable for acute pickup of all sounds | Broadcast or recording studios, concert halls, music rooms, bedrooms, special classrooms for the very young or hearing impaired, etc. |
| 30 | Very quiet, suitable for large conferences; telephone use satisfactory; sleeping unimpaired | Residences, hotel or hospital sleeping rooms, theaters, auditoriums, libraries, executive offices, directors' rooms, large conference rooms, etc. |
| 35 | Quiet, suitable for conference at 15 ft. table; normal voice 10-30 ft; telephone use satisfactory | Private offices, school cafeterias, court rooms, churches, small conference rooms, etc. |
| 40 | Satisfactory for conferences at 6-8 ft. table; normal voice 6-12 ft.; telephone use satisfactory | General (open) offices, school corridors, laboratories, restaurants, etc. |
| 45 | Satisfactory for conferences at 4-5 ft. table; normal voice 3-6 ft., raised voice 6-12 ft; telephone use is somewhat difficult | Retail stores, cafeterias and fast food dining, lobbies or public areas, etc. |
| 50 | Unsatisfactory for conferences; | Workshops, machine rooms, |

1

| Maximum RC | Environment | Typical Occupancy |
|---|---|---|
| | normal voice 1-2 ft; raised voice 3-6 ft; telephone use is difficult | industrial process areas, etc. |

2 For classrooms for primary education, colleges and universities, training facilities, etc., HVAC
3 systems and their components shall be selected/installed to comply with ANSI Standard S12.60.
4
5 Outdoor HVAC equipment shall be selected, located, and oriented to limit the "Equivalent Sound
6 Level" ($LA_{eq}$,T) A-scale sound pressure level (dBA), averaged over $T = 16$ h, to 50 dBA or less
7 at the site boundary or at a distance of 100 ft, whichever is closer to the equipment.
8
9
10 **GUARANTEE PERIOD SERVICES**
11
12 Maintenance of Chiller: The contractor shall inspect and maintain chillers and chiller components,
13 including manufacturer-furnished controls, during the guarantee period. The contractor shall then
14 furnish a report describing the status of the equipment, problem areas (if any) noticed during
15 service work, and description of the corrective actions taken. The report shall clearly certify that
16 all chiller controls are functioning correctly.
17
18 Service Documentation: A copy of the service report associated with each routine service visit or
19 owner-initiated service call shall be provided to the owner within 10 days after the date of each
20 service call.
21

## PART 2: PRODUCTS

### WATER-COOLED CHILLERS

Description: Packaged, factory—assembled, hermetic or open-drive type high or medium pressure water chillers utilizing refrigerant R-134A, consisting of centrifugal single- or multi-stage electric-drive compressor(s), compressor motor(s), motor starter(s), evaporator, condenser, controls and panels, including gages and indicating lights, auxiliary components, and accessories.

**//SELECT ONE OR THE OTHER OF THE TWO FOLLOWING COMPRESSOR SECTIONS. OIL-LUBRICATED COMPRESSORS ARE CONVENTIONAL, WHILE ONLY MCQUAY AND JCI/YORK CURRENTLY OFFER OIL-FREE MAGNETIC BEARING SYSTEMS.//**

Oil-Lubricated Centrifugal Compressor(s) (150–1600 Tons):

Casing: Cast iron, precision ground with gasket sealed casing joints.

Shaft and Impeller Assembly: Carbon or forged steel shaft with cast high-strength aluminum alloy impellers, designed and assembled for no critical speeds within operating

1 range; statically and dynamically balanced. Shaft main bearings shall be roller bearings in
2 accordance with ABMA 9 with L10 life rating of at least 200,000 h.
3
4 Drive Assembly: Helical gear transmission integral with compressor and lubricated
5 through compressor lubrication system. For chillers with open drives, provide flexible disk
6 coupling with all-metal construction and no wearing parts or need for lubrication.
7
8 Oil Lubrication System: Consisting of pump, filtration, heater (as applicable), cooler,
9 factory-wired power connection, and controls.
10
11 Provide lubrication to bearings, gears, and other rotating surfaces at all operating,
12 startup, coastdown, and standby conditions, including power failure.
13
14 Oil shall be compatible with refrigerant and chiller components in contact with the
15 refrigerant.
16
17 Provide dual oil filers, one redundant, shall be the easily replaceable cartridge type,
18 minimum 0.5-μm efficiency, with means of positive isolation while servicing.
19
20 Oil cooler shall be refrigerant- or water-cooled.
21
22 Include manufacturer's standard method to remove refrigerant from oil.
23
24 Provide factory-installed and pressure-tested oil piping with isolation valves and

accessories.

Provide positive visual indication of oil level.

On two-compressor chillers, provide redundant oil pump.

Compressor Motor(s)

Motor and Accessories (Hermetic Type): Hermetically sealed, continuous duty, single speed, squirrel cage, induction type, refrigerant cooled and lubricated; full-load operation of the motor shall not exceed nameplate rating; rotor shaft shall be heat-treated carbon steel and designed such that the first critical speed is well above the operating speed. Provide for removal of the stator for service or replacement without breaking the main refrigerant piping connections. Motor bearings shall be oil-lubricated by refrigerant-entrained oil.

Motor and Accessories (Open-Drive Type): Two-pole, continuous duty, squirrel cage induction type, and with open drip-proof enclosure. Motor full-load amperes at design conditions shall not exceed motor nameplate FLA. Motor shall be designed for use with the type starter specified. Motor shall be factory mounted. Provide internal electrical heater, internally powered from the chiller power supply.

Oil-Free Magnetic Bearing Centrifugal Compressor(s) (150–500 Tons):

Casing: Cast iron, precision ground with gasket-sealed casing joints.

Shaft and Impeller Assembly: Carbon or forged steel shaft with single- or multistage cast high–strength aluminum alloy impeller(s), designed and assembled for no critical speeds within operating range; statically and dynamically balanced.

Compressor Motor(s): Hermetically sealed, continuous duty, single speed, squirrel cage, induction type, refrigerant cooled; full-load operation of the motor shall not exceed nameplate rating; rotor shaft shall be heat-treated carbon steel and designed such that the first critical speed is well above the operating speed.

The motor shaft shall be supported on active magnetic radial and thrust bearings. Magnetic bearing control shall be equipped with automatic vibration reduction and balancing systems. Magnetic bearing system shall remain active during compressor coast-down. Provide secondary bearings as back-up to the magnetic bearing system for emergency "touch down" situations.

Provide for removal of the stator for service or replacement without breaking the main refrigerant piping connections.

Selection to Avoid Surge: Select compressors/impellors/controls to avoid surge conditions at any point on the performance curve between minimum load and maximum load. To avoid the potential for surge, one or more of the following elements may be provided:

Select chiller for 10–15% over-capacity, resulting in a larger impellor and wider separation between operating conditions and surge conditions.

Provide internal hot-gas bypass to maintain higher refrigerant flow rate through the compressor.

Provide "anti-surge" controls. However, such controls shall not result in loss of capacity control in response to the imposed cooling load.

<u>Capacity Control</u>: Provide variable guide vanes and/or a variable frequency drive to provide stable operation without surge, cavitation, or vibration from 100% to 15% of full-load capacity. Capacity control shall be based on maintaining chilled water supply temperature at setpoint, +0.25°F.

Provide hot-gas bypass piping and control valve integrated with capacity control sequence to maintain chiller operation down to 0.5°F differential chilled water temperature.

<u>Compressor Motor Starter(s)</u>: Provide unit-mounted or free-standing factory-wired starters. All power and control conductors between starters and other chiller components, whether unit mounted or free standing, shall be sized, furnished, and installed, complete and ready for operation.

Starter shall incorporate a single point for chiller power connection. All chiller heaters, pumps, control circuits, and other miscellaneous items and accessories shall be supplied directly from the starter without dependence on external power from other sources. Where ancillary items require a different supply voltage from the chiller motor, a suitable transformer(s) and appropriate wiring shall be provided as a part of the chiller starter to supply these items.

Starter shall be reduced voltage, closed transition wye–delta, electronic, or auto-transformer type as indicated on the drawings. Starter shall be non-reversing, enclosed in a suitable NEMA-1 enclosure, and shall have as a minimum the following features:

*Nonfused disconnecting* means to disconnect all power sources from the chiller and related accessories.

*Electronic internal self-contained monitor system* to protect against three-phase and single–phase overloads during running and starting, loss of phase, phase reversal, phase unbalance voltage or current, overvoltage, and undervoltage.

A total run-time meter mounted on the starter door.

An electronic multifunction metering module with display mounted on the starter door providing the following instantaneous, demand, and energy information:

Instantaneous: Digital readings of line and phase voltages, line and phase currents, total watts, total VARS, total kVA, power factor, and line frequency.

Demand: Digital readings of line and phase currents, and total watts.

1
2 Energy: Digital readings of accumulated watthours and accumulated VARhours.
3
4 Variable Frequency Controller: If required, provide variable frequency controller (VFC) to
5 control compressor speed. VFC shall be factory mounted and factory wired as an integral part of
6 the chiller.
7
8 General: VFCs shall be integrated gate bipolar transistor PWM type in accordance with
9 NEMA ICS 2; listed and labeled as a complete unit and arranged to provide variable speed
10 of a inverter duty induction motor specified in Section 230513 by adjusting output voltage
11 and frequency.
12
13 Design and Rating: Match load type such as fans, blowers, and pumps, and type of
14 connection used between motor and load such as direct or through a power transmission
15 connection.
16
17 Output Rating: Three-phase; 6–66 Hz, with torque constant as speed changes.
18
19 Unit Operating Requirements:
20
21 Input AC voltage tolerance: 380–500 V, ±10 %
22
23 Input frequency tolerance: 50/60 Hz, ±6%
24

Ambient Temperature: 0–40 C.

Humidity: <90% (noncondensing)

Altitude: 3300 ft.

Minimum efficiency: 96% at 60-Hz, full load

Minimum displacement primary-side power factor: 96%

Overload capability: 1.1 times the base load current for 60 s; 2.0 times the base load current for 3 s

Starting torque: 100% of rated torque or as indicated

Speed regulation: ±1% over an 11:1 speed range

User Adjustability Capabilities:

Minimum speed: 5–25% of maximum rpm

Maximum speed: 80–100% of maximum rpm

Acceleration: 2 to a maximum of 22 s

Deceleration: 2 to a maximum of 22 s

Current limit: 50 to a maximum of 110% of maximum rating

Self-Protection and Reliability Features

Input transient protection by means of surge suppressors

Snubber networks to protect against malfunction due to system voltage transients

Under- and overvoltage trips; inverter overtemperature, overload, and overcurrent trips

Motor overload relay: Adjustable and capable of NEMA 250, Class 20 performance. Include notch filter to prevent operation of the controller–motor–load combination at a natural frequency of the combination.

Instantaneous line-to-line and line-to-ground overcurrent trips

Loss-of-phase protection

Reverse-phase protection

1 Short-circuit protection
2
3 Motor overtemperature fault
4
5 Automatic Reset and Restart: VFC shall attempt three restarts after controller fault or on
6 return of power after an interruption and before shutting down for manual reset or fault
7 correction.
8
9 Power-Interruption Protection: VFC shall prevent motor from re-energizing after a power
10 interruption until motor has stopped.
11
12 Torque Boost: VFC shall automatically vary starting and continuous torque to at least 1.5
13 times the minimum torque to ensure high starting torque and increased torque at slow
14 speeds.
15
16 Motor Temperature Compensation at Slow Speeds: Adjustable current fall-back based on
17 output frequency for temperature protection of self-cooled fan-ventilated motors at slow
18 speeds.
19
20 Input Line Conditioning: Provide line reactor.
21
22 VFC Output Filtering: For distance greater than 50 ft between motor and VFC, provide
23 DV/DT filter for insulation protection and reduced bearing current problems.
24

1 Panel-Mounted Operator Station: Provide operator panel-mounted flush in controller door
2 and connected to indicate the following controller parameters:
3
4 Output frequency (Hz)
5
6 Motor speed (rpm)
7
8 Motor status (running, stop, fault)
9
10 Motor current (A)
11
12 Motor torque (%)
13
14 Fault or alarming status (code)
15
16 PID feedback signal (%)
17
18 DC-link voltage (VDC)
19
20 Set-point frequency (Hz)
21
22 Motor output voltage (V)
23
24 Manual Bypass: Arrange magnetic contactor to safely transfer motor between controller

1 output and bypass controller circuit when motor is at zero speed. Controller-off-bypass
2 selector switch sets mode, and indicator lights give indication of mode selected. The unit
3 shall be capable of stable operation (starting, stopping, and running), with the motor
4 completely disconnected from the controller (no load).
5
6 Isolating Switch: Non-load-break switch arranged to service/isolate VFC and permit safe
7 troubleshooting and testing, both energized and de-energized, while the motor is operating
8 in bypass mode.
9
10 Integral Disconnecting Means: Provide NEMA–1 input disconnect switch interlocked with
11 the enclosure door with a lockable handle.
12
13
14 Evaporator and Condenser:
15
16 Shell and Water Boxes: Fabricated from welded carbon steel plate. Provide 150-psig
17 maximum working pressure water boxes and nozzle connections.
18
19 Provide vents, drains, and covers in water boxes to permit tube cleaning within the
20 space shown on the drawings.
21
22 Provide suitable tappings in the water boxes and nozzles for control sensors,
23 gages, and thermometers.
24

Provide with standard water boxes for chillers of less than 500-tons capacity and marine water boxes for chillers of 500 tons or greater capacity. Piping connections shall be grooved mechanical or flanged and piping shall be arranged so that heat exchanger heads can be easily removed for maintenance with minimal removal of piping required.

Tube Sheets: Fabricated of carbon steel sheets welded to the shell and drilled for tubes. Include intermediate tube support sheets as required to prevent tube vibration.

Tubes: Individually replaceable, finned, seamless copper tubes; removable from either end of the heat exchanger without affecting strength and durability of the tube sheets and without causing leakage in adjacent tubes. Expand ends of tubes in tube sheets and intermediate tube support sheets for tight fit to prevent vibration of tubes. Provide suitable baffles or distributing plates in condenser tubes to evenly distribute refrigerant discharge gas on heat transfer tubes.

Pressure Limiting and Pressure Relief Devices: Manufacturer's standard complying with ASHRAE 15. Provide refrigerant charging and transfer connections, pressure relief device on the evaporator to prevent excessive pressure in refrigerant side, and means to sense refrigerant pressure or temperature. Pressure relief valve shall close once pressure is reduced below setpoint pressure.

Pump-Out System: Provide factory-installed positive shutoff, manual isolation valves in the compressor discharge line to the condenser and the refrigerant liquid line leaving the condenser to

1 allow for isolation and storage of full refrigerant charge in the chiller condenser shell. In addition,
2 provide isolation valve on suction side of compressor from evaporator to allow for isolation and
3 storage of full refrigerant charge in the chiller evaporator shell.
4
5 Unit Controller: Manufacturer-furnished, factory-installed unit controls, shall interface seamlessly
6 with the facility direct digital control (DDC) system, as follows:
7
8 **// SELECT TYPE OF DDC ARCHITECTURE/COMMUNICATION REQUIRED FOR**
9 **CHILLER UNIT CONTROLLER//**
10
11 BACnet Compliance: Unit controllers shall use ANSI/ASHRAE Standard 135 protocol
12 and communicate using ISO 8802-3 (Ethernet) datalink/physical layer protocol. Comply
13 with ANSI/ASHRAE Standard 135 for all hardware and software.
14
15 LonWorks Compliance: Unit controllers shall use LonTalk protocol and communicate
16 using EIA/CEA 709.1 datalink/physical layer protocol. Comply with LONMARK
17 Interoperability Association "Interoperability Guidelines" for all control hardware and
18 software. Utilize published functional profiles for all product network message and
19 configuration parameters.
20
21 Gateways: **The use of gateways is prohibited.**
22
23 Coordination: The contractor shall be responsible for coordination between the controls
24 system sub-contractor and the unit controller manufacturer/vendor to ensure that required

control interface is implemented, resulting in full interoperability between unit controls and the DDC system.

Refrigerant Flow Control Devices: Provide refrigerant flow control devices between evaporator and condensers (and elsewhere as required) to regulate refrigerant flow at volume and pressure required to maintain evaporator liquid refrigerant at a level sufficient to keep cooler heat transfer tubes adequately wetted through a full range of chiller operation.

Design Devices to Permit Chiller Operation: This is done at scheduled conditions, and to allow condenser entering water temperature to decrease to minimum permissible temperature or 1°F above return chilled water temperature.

Safety Controls

Design cutouts to operate independently and factory wire to control panel.

Design controls to stop compressor motor in event of low refrigerant pressure or temperature in evaporator, high condenser pressure, high compressor discharge temperature, low evaporator leaving water temperature ("freezestat"), high motor temperature, high bearing temperature, low oil pressure, high oil temperature, compressor motor overcurrent or over voltage, loss of phase, or power interruption.

1 Design each cutout to require manual restarting of compressor in the event of
2 safety failure but allowing automatic restart after power interruption.
3
4 Provide anti recycle timer for limiting compressor motor restarts at scheduled time
5 intervals.
6
7 Operational Controls: Provide controls to ensure that compressor will start only under
8 unloaded condition.
9
10 Provide Sequencing Controls: This is to ensure lubrication of compressor motor bearings
11 and seals (if any). Sequence as follows:
12
13 Run lubrication system oil pump so that compressor motor bearing is lubricated
14 before start-up.
15
16 Start compressor motor.
17
18 Provide lubrication during coast-down after compressor motor shutdown.
19
20 Design Controls to Automatically Restart Compressor: This is done after power failure
21 interruptions, provided minimum time between starts has been complied with.
22

1 Manual Control and Display Panel: Factory mounted and factory wired. Provide gages or
2 meters to indicate low refrigerant pressure in evaporator, high condenser pressure, and
3 low oil pressure.
4
5 Provide operator manual input functions designed to permit indicated operations
6 including the following:
7
8 Manual and automatic operation of oil pump
9
10 Manual and automatic operation of oil separator heater
11
12 Provide pilot lights or visual display for indicated operations and cutouts including
13 the following:
14
15 Oil pump operation
16
17 Low chilled water temperature cutout
18
19 Low evaporator refrigerant pressure or temperature cutout
20
21 High condenser pressure cutout
22
23 High motor winding temperature cutout
24

Low oil pressure cutout

Motor overload cutout

Insulation: Insulate evaporators and all other cold surfaces to prevent condensation, with ambient humidity of 75% and dry bulb temperature of 90°F, no air movement, with flexible elastomeric, closed cell, thermal insulation in accordance with ASTM C534, Type II for sheet materials, black in color, rated for temperatures from –40°F to 220°F. Insulation shall be AR/Armaflex or equivalent.

## PART 3: EXECUTION

### INSTALLATION OF CHILLERS

Install chillers in accordance with the manufacturer's written instructions. Install units plumb and level, firmly anchored in locations indicated; maintain the manufacturer's recommended clearances.

Provide relief piping from indoor water-cooled chiller refrigerant pressure relief rupture disk to atmosphere; size piping as recommended by chiller manufacturer, and terminate with gooseneck facing down.

1 **// INCLUDE FACTORY PERFORMANCE TEST <u>ONLY</u> WHEN REQUIRED.//**
2
3 **<u>FACTORY PERFORMANCE TEST</u>**
4
5 The manufacturer shall conduct witnessed performance acceptance test for each typical machine
6 to verify the performance submitted. The test will be in accordance with ARI 550/590-98 with the
7 following exceptions:
8
9 Capacity will have no tolerances. Capacity must meet or better required tonnage.
10
11 Heat balance must be verified at 3% or below during test conditions.
12
13 Test shall be conducted at an approved ARI-certified test facility of the manufacturer's factory.
14 Instrumentation used for testing must be calibrated within 6 months of the test date and traceable
15 to the National Bureau of Standards (NBS). All documentation verifying NBS traceability shall be
16 included in a bound folder for presentation to the owner's representative. The Owner's
17 representative may elect to contact ARI for verification of performance and test conditions.
18
19 The chiller shall be tested with water temperature and adjustment cooler/condenser per standard
20 ARI 550-90 to simulate specified fouling versus no fouling during test. Verification of this
21 procedure will require inside surface area and number of tubes per vessel. This information is to
22 be submitted prior to the test for formula verification of fouling per ARI 550-90 if so requested by
23 engineer.
24

1 Performance test shall be a four-point test, 25%, 50%, 75%, and 100% load.
2
3 The chiller will be accepted if the test indicates that all conditions of the performance test meet the
4 specified conditions. If the performance test results are outside of specified tolerances, the
5 manufacturer will be allowed to make repairs and retest, and the manufacturer will assume all
6 expenses incurred for the owner's representatives to witness the retest.
7
8 If after the second test the machines do not meet the test requirements, an energy cost penalty, of
9 $4000 per kW above test requirements, will be deducted from the manufacturer's contract. In the
10 case of an ARI certified performance chiller, ARI will be notified of any test failures. If there is
11 more than one typical machine, the energy penalty will be the penalty of the tested machine times
12 the number of typical machines on the project. The total energy penalty will be the sum of all
13 machines on the project whether tested or typical.
14
15
16 **DEMONSTRATION**
17
18 Start-up chillers, in accordance with the manufacturer's start-up instructions, and provide the service of
19 a factory-authorized service representative to provide start-up service and to demonstrate and train the
20 owner's maintenance personnel as specified below. Evacuate, dehydrate, and charge with specified
21 refrigerant, and leak test in accordance with manufacturer's instructions.
22

Perform lubrication service, including filling of reservoirs, and confirming that lubricant is of quantity and type recommended by the manufacturer. Test controls and demonstrate compliance with requirements. Replace damaged, or malfunctioning, controls and equipment and retest.

Do not place chillers in sustained operation prior to initial balancing of mechanical systems that interface with the chillers.

## OWNER INSTRUCTION AND TRAINING

Provide services of the manufacturer's technical representative to instruct owner's personnel in operation and maintenance of chillers. Review with the owner's personnel the data contained in the manufacturer's operating and maintenance manual(s).

# Appendix B2

## *Scroll Compressor Water Chillers*

1
2
3 **PART 1: GENERAL**
4
5 **SUBMITTALS**
6
7 General: Submittals shall demonstrate compliance with technical requirements by reference to
8 each subsection of this specification. Where a submitted item does not *comply fully* with each and
9 every requirement of the specifications, the submittal shall clearly indicate such deviations.
10 Identification requirements for non-complying features of items are very specific.
11
12 Manufacturer's Data: Submit the manufacturer's product data, including rated capacities, weights
13 (shipping, installed, and operating), furnished specialties and accessories, and installation and
14 start-up instructions.
15
16
17 **QUALITY ASSURANCE**
18
19 ARI Compliance
20
21 Fabricate, test, and certify chillers to comply with ARI Standard 550/590 "Water-Chilling
22 Packages Using the Vapor Compression Cycle."
23

1 Rate air-cooled chillers sound power levels according to ARI 270 "Sound Rating of
2 Outdoor Unitary Equipment" or ARI 370 "Sound Rating of Large Outdoor Refrigerating
3 and Air-Conditioning Equipment," as applicable.
4
5 Rate water-cooled chillers sound power levels according to ARI 575 "Method of
6 Measuring Machinery Sound within an Equipment Space."
7
8 ASHRAE Compliance
9
10 Fabricate and install rotary water chillers to comply with ASHRAE 15 "Safety Code for
11 Mechanical Refrigeration."
12
13 Comply with ASHRAE 147 for refrigerant leaks, recovery, handling, and storage
14 requirements.
15
16 UL Compliance: Fabricate water chillers to comply with UL 465 "Central Cooling Air
17 Conditioners."
18
19 ASME Compliance: Fabricate water chillers to comply with ASME Boiler and Pressure Vessel
20 Code, Section VIII, Division 1. Include ASME U-stamp and nameplate certifying compliance.
21
22 Warranty: The contractor shall provide a 5-year parts and labor manufacturer's warranty on each
23 refrigeration compressor, including refrigerant. This warranty shall be directly from the
24 manufacturer to the owner and shall be in addition to guarantees and warranties required under

1 the general conditions to the construction contract. This warranty shall provide for repair or
2 replacement of the covered compressor that becomes inoperative as a result of defects in materials
3 or workmanship within 5 years after the date of substantial completion.
4
5 Acoustic Criteria: Noise levels due to chillers shall result in maximum sound levels in occupied
6 spaces conforming to the following room criteria (RC):
7
8

| Maximum RC | Environment | Typical Occupancy |
|---|---|---|
| 25 | Extremely quiet environment, suppressed speech is audible, suitable for acute pickup of all sounds | Broadcast or recording studios, concert halls, music rooms, bedrooms, special classrooms for the very young or hearing impaired, etc. |
| 30 | Very quiet, suitable for large conferences; telephone use satisfactory; sleeping unimpaired | Residences, hotel or hospital sleeping rooms, theaters, auditoriums, libraries, executive offices, directors' rooms, large conference rooms, etc. |
| 35 | Quiet, suitable for conference at 15 ft table; normal voice 10–30 ft; telephone use satisfactory | Private offices, school cafeterias, court rooms, churches, small conference rooms, etc. |
| 40 | Satisfactory for conferences at 6–8 ft table; normal voice 6–12 ft; telephone use satisfactory | General (open) offices, school corridors, laboratories, restaurants, etc. |

1

| Maximum RC | Environment | Typical Occupancy |
|---|---|---|
| 45 | Satisfactory for conferences at 4–5 ft table; normal voice 3–6 ft, raised voice 6–12 ft; telephone use is somewhat difficult. | Retail stores, cafeterias and fast food dining, lobbies or public areas, etc. |
| 50 | Unsatisfactory for conferences; normal voice 1–2 ft, raised voice 3–6 ft.; telephone use is difficult | Workshops, machine rooms, industrial process areas, etc. |

2 For classrooms for primary education, colleges and universities, training facilities, etc., HVAC
3 systems and their components shall be selected/installed to comply with ANSI Standard S12.60.
4
5 Outdoor HVAC equipment shall be selected, located, and oriented to limit the "Equivalent Sound
6 Level" ($LA_{eq}$,T) A-scale sound pressure level (dBA), averaged over $T$ = 16 h, to 50 dBA or less
7 at the site boundary or at a distance of 100 ft, whichever is closer to the equipment.
8
9
10 **GUARANTEE PERIOD SERVICES**
11
12 Maintenance of Chiller: The contractor shall inspect and maintain, as required, chillers and chiller
13 components, including manufacturer-furnished controls, during the guarantee period. The
14 contractor shall then furnish a report describing the status of the equipment, problem areas (if any)
15 noticed during service work, and description of the corrective actions taken. The report shall
16 clearly certify that all chiller controls are functioning correctly.

1
2 Service Documentation: A copy of the service report associated with each routine service visit or
3 owner-initiated service call shall be provided to the owner within 10 days after the date of each
4 service call.
5
6
7
8 **PART 2: PRODUCTS**
9
10
11 **WATER-COOLED SCROLL COMPRESSOR WATER CHILLERS**
12
13 Description: Factory-assembled and run-tested water chiller complete with compressor(s),
14 compressor motors and motor controllers, evaporator, condenser where indicated, electrical
15 power, controls, and indicated accessories.
16
17 Refrigerant: R-407C, R-410A, or R-134A. **//EDIT REFRIGERANT TYPE AS REQUIRED//**
18
19 Compressors
20
21 Provide multiple positive-displacement direct drive scroll compressors with hermetically
22 sealed casing. At least two compressors are required and the maximum single compressor
23 capacity shall not exceed 40 tons.
24

1 Each compressor shall be piped as a separate and independent refrigerant circuit so that in
2 the event of failure of any one compressor, the remaining compressors shall continue to
3 operate. Each circuit shall include a thermal-expansion valve, refrigerant charging
4 connections, a hot-gas muffler, compressor suction and discharge shutoff valves, a liquid-
5 line shutoff valve, a replaceable-core filter-dryer, a sight glass with moisture indicator, a
6 liquid-line solenoid valve, and an insulated suction line.
7
8 Provide each compressor with crankcase oil heater, and suction strainer.
9
10 Compressor motors shall be high-torque, two-pole induction type with inherent thermal-
11 overload protection on each phase, hermetically sealed and cooled by refrigerant suction
12 gas.
13
14 Provide full voltage, across-the-line starters in accordance with NEMA ICS 2, Class A,
15 full voltage, non-reversing. for each compressor.
16
17 Mount individual compressors on vibration isolators.
18
19 Oil Lubrication System: Automatic pump with strainer, sight glass, filling connection, filter with
20 magnetic plug, and initial oil charge.
21
22 Refrigerant Isolation: Factory-install positive shutoff isolation valves in the compressor discharge
23 line and the refrigerant liquid-line to allow the isolation and storage of the refrigerant charge in the
24 chiller condenser.

1
2 Evaporator: Brazed-plate or shell-and-tube design as follows.
3
4 Shell-and-Tube Evaporator:
5
6 Description: Direct-expansion, shell-and-tube design with fluid flowing through the
7 shell and refrigerant flowing through the tubes within the shell.
8
9 Code compliance: Tested and stamped according to ASME Boiler and Pressure
10 Vessel Code.
11
12 Shell material: Carbon steel.
13
14 Shell heads: Removable carbon-steel heads with multipass baffles designed to
15 ensure positive oil return and located at each end of the tube bundle.
16
17 Shell nozzles: Fluid nozzles located along the side of the shell and terminated with
18 mechanical-coupling end connections for connection to field piping.
19
20 Tube construction: Individually replaceable copper tubes with enhanced fin design,
21 expanded into tube sheets.
22
23 Provide standard water boxes with grooved mechanical or flanged piping
24 connections and piping shall be arranged so that heat exchanger heads can be easily

1 removed for maintenance.
2
3 Brazed Plate Evaporator
4
5 Direct-expansion, single-pass, brazed-plate design.
6
7 Type 316 stainless-steel construction.
8
9 Code compliance: Tested and stamped according to ASME Boiler and Pressure
10 Vessel Code.
11
12 Fluid nozzles: Terminate with mechanical-coupling end connections for connection
13 to field piping.
14
15 Condenser
16
17 Description: Shell-and-tube design with refrigerant flowing through the shell and fluid
18 flowing through the tubes within the shell.
19
20 Provides positive sub-cooling of liquid refrigerant.
21
22 Code compliance: Tested and stamped according to ASME Boiler and Pressure Vessel
23 Code.
24

Shell material: Carbon steel.

Water boxes: Provide standard water boxes with grooved mechanical or flanged piping connections and piping shall be arranged so that heat exchanger heads can be easily removed for maintenance.

Tube construction: Individually replaceable copper tubes with enhanced fin design, expanded into tube sheets.

Provide each condenser with a pressure relief device, purge cock, and liquid-line shutoff valve.

Capacity Control: On–off compressor staging/cycling. Provide hot-gas bypass piping and control valve integrated with capacity control sequence to maintain chiller operation down to 0.5°F differential chilled water temperature.

Controls: Manufacturer-furnished, factory-installed unit controls, shall interface seamlessly with the facility direct digital control (DDC) system, as follows:

**// SELECT TYPE OF DDC ARCHITECTURE/COMMUNICATION REQUIRED FOR CHILLER UNIT CONTROLLER//**

BACnet Compliance: Unit controllers shall use ANSI/ASHRAE Standard 135 protocol and communicate using ISO 8802-3 (Ethernet) datalink/physical layer protocol. Comply with ANSI/ASHRAE Standard 135 for all hardware and software.

LonWorks Compliance: Unit controllers shall use LonTalk protocol and communicate using EIA/CEA 709.1 datalink/physical layer protocol. Comply with LonMark Interoperability Association "Interoperability Guidelines" for all control hardware and software. Utilize published functional profiles for all product network message and configuration parameters.

Gateways: **The use of gateways is prohibited**.

Coordination: The Contractor shall be responsible for coordination between the controls system subcontractor and the unit controller manufacturer/vendor to ensure that required control interface is implemented, resulting in full interoperability between unit controls and the DDC system.

Manual-Reset Safety Controls: The following conditions shall shut down water chiller and require manual reset:

Low evaporator pressure or high condenser pressure

Low chilled water temperature

Refrigerant high pressure

1 High or low oil pressure
2
3 High oil temperature
4
5 Loss of chilled water flow
6
7 Loss of condenser-water flow
8
9 Control device failure
10
11 Insulation: Insulate evaporators and all other cold surfaces to prevent condensation, with ambient
12 humidity of 75% and dry bulb temperature of 90°F, no air movement.
13
14
15
16 **AIR-COOLED SCROLL COMPRESSOR WATER CHILLERS**
17
18 Description: Factory-assembled and run-tested water chiller complete with base and frame,
19 condenser casing, compressors, compressor motors and motor controllers, evaporator, condenser
20 coils, condenser fans and motors, electrical power, controls, and accessories.
21
22 Cabinet
23
24 Base: Galvanized-steel base extending the perimeter of water chiller. Secure frame,

1 compressors, and evaporator to base to provide a single-piece unit.
2
3 Frame: Rigid galvanized-steel frame secured to base and designed to support cabinet,
4 condenser, control panel, and other chiller components not directly supported from base.
5
6 Casing: Galvanized steel.
7
8 Finish: Coat base, frame, and casing with a corrosion-resistant coating capable of
9 withstanding a 500-h salt-spray test according to ASTM B 117.
10
11 Refrigerant: R-407C, R-410A, or R-134A. **//EDIT REFRIGERANT TYPE AS REQUIRED//**
12
13 Compressors
14
15 Provide multiple positive-displacement direct-drive scroll compressors with hermetically
16 sealed casing. At least two compressors are required and the maximum single compressor
17 capacity shall not exceed 40 tons.
18
19 Each compressor shall be piped as a separate and independent refrigerant circuit so that in
20 the event of failure of any one compressor, the remaining compressors shall continue to
21 operate. Each circuit shall include a thermal-expansion valve, refrigerant charging
22 connections, a hot-gas muffler, compressor suction and discharge shutoff valves, a liquid-
23 line shutoff valve, a replaceable-core filter-dryer, a sight glass with moisture indicator, a
24 liquid-line solenoid valve, and an insulated suction line.

1
2 Provide each compressor with crankcase oil heater, and suction strainer.
3
4 Compressor motors shall be high-torque, two-pole induction type with inherent thermal-
5 overload protection on each phase, hermetically sealed and cooled by refrigerant suction
6 gas.
7
8 Provide full voltage, across-the-line starters in accordance with NEMA ICS 2, Class A,
9 full voltage, nonreversing for each compressor.
10
11 Mount individual compressors on vibration isolators.
12
13 Capacity Control: On–off compressor staging/cycling. Provide hot-gas bypass piping and control
14 valve integrated with capacity control sequence to maintain chiller operation down to 0.5°F
15 differential chilled water temperature.
16
17 Oil Lubrication System: Automatic pump with strainer, sight glass, filling connection, filter with
18 magnetic plug, and initial oil charge.
19
20 Refrigerant Isolation: Factory-install positive shutoff isolation valves in the compressor discharge
21 line and the refrigerant liquid-line to allow the isolation and storage of the refrigerant charge in the
22 chiller condenser.
23
24 Evaporator: Brazed-plate or shell-and-tube design as follows.

1
2 Shell-and-Tube Evaporator
3
4 Description: Direct-expansion, shell-and-tube design with fluid flowing through the
5 shell and refrigerant flowing through the tubes within the shell.
6
7 Code compliance: Tested and stamped according to ASME Boiler and Pressure
8 Vessel Code.
9
10 Shell material: Carbon steel.
11
12 Shell heads: Removable carbon-steel heads with multipass baffles designed to
13 ensure positive oil return and located at each end of the tube bundle.
14
15 Shell nozzles: Fluid nozzles located along the side of the shell and terminated with
16 mechanical-coupling end connections for connection to field piping.
17
18 Tube construction: Individually replaceable copper tubes with enhanced fin design,
19 expanded into tube sheets.
20
21 Provide standard water boxes with grooved mechanical or flanged piping
22 connections and piping shall be arranged so that heat exchanger heads can be easily
23 removed for maintenance.
24

1 Brazed Plate Evaporator
2
3 Direct-expansion, single-pass, brazed-plate design.
4
5 Type 316 stainless-steel construction.
6
7 Code compliance: Tested and stamped according to ASME Boiler and Pressure
8 Vessel Code.
9
10 Fluid nozzles: Terminate with mechanical-coupling end connections for connection
11 to field piping.
12
13 Heater: Factory-installed and factory-wired electric heater with integral controls designed
14 to protect the evaporator to minus 20°F.
15
16 Air-Cooled Condenser: Plate-fin coil with integral sub-cooling on each circuit, rated at 450 psig.
17 Construct coils of copper tubes mechanically bonded to aluminum fins, factory-coated with
18 epoxy-phenolic after fabrication.
19
20 Condenser fans: Direct-drive propeller type with statically and dynamically balanced fan
21 blades, arranged for vertical air discharge.
22
23 Fan motors: Totally enclosed nonventilating (TENV) or totally enclosed air over (TEAO)
24 enclosure, with permanently lubricated bearings, and having built-in overcurrent- and

1 thermal-overload protection.
2
3 Provide condenser coils with louvers, baffles, or hoods to protect against hail damage.
4
5 Unit Controller: Manufacturer-furnished, factory-installed unit controls, shall interface seamlessly
6 with the facility DDC system, as follows:
7
8 **// SELECT TYPE OF DDC ARCHITECTURE/COMMUNICATION REQUIRED FOR**
9 **CHILLER UNIT CONTROLLER//**
10
11 BACnet Compliance: Unit controllers shall use ANSI/ASHRAE Standard 135 protocol
12 and communicate using ISO 8802-3 (Ethernet) datalink/physical layer protocol. Comply
13 with ANSI/ASHRAE Standard 135 for all hardware and software.
14
15 LonWorks Compliance: Unit controllers shall use LonTalk protocol and communicate
16 using EIA/CEA 709.1 datalink/physical layer protocol. Comply with LONMARK
17 Interoperability Association "Interoperability Guidelines" for all control hardware and
18 software. Utilize published functional profiles for all product network message and
19 configuration parameters.
20
21 Gateways: **The use of gateways is prohibited.**
22
23 Coordination: The contractor shall be responsible for coordination between the controls system
24 sub-contractor and the unit controller manufacturer/vendor to ensure that required control

interface is implemented, resulting in full interoperability between unit controls and the DDC system.

Manual-Reset Safety Controls: The following conditions shall shut down water chiller and require manual reset:

Low evaporator pressure or high condenser pressure

Low chilled water temperature

Refrigerant high pressure

High or low oil pressure

High oil temperature

Loss of chilled water flow

Control device failure

Insulation: Insulate evaporators and all other cold surfaces to prevent condensation, with ambient humidity of 75% and dry bulb temperature of 90°F, no air movement, with flexible elastomeric, closed cell, thermal insulation in accordance with ASTM C534, Type II for sheet materials, black in color, rated for temperatures from –40°F to 220°F. Insulation shall be

1 AR/Armaflex or equivalent.
2
3
4 **PART 3: EXECUTION**
5
6 **INSTALLATION OF CHILLERS**
7
8 Install chillers in accordance with the manufacturer's written instructions. Install units plumb and level,
9 firmly anchored in locations indicated; maintain the manufacturer's recommended clearances.
10
11 Provide relief piping from indoor water-cooled chiller refrigerant pressure relief rupture disk to
12 atmosphere; size piping as recommended by the chiller manufacturer, and terminate with
13 gooseneck facing down.
14
15
16 **DEMONSTRATION**
17
18 Start-up chillers, in accordance with the manufacturer's start-up instructions, and provide the service of
19 a factory-authorized service representative to provide start-up service and to demonstrate and train the
20 owner's maintenance personnel as specified below. Evacuate, dehydrate, and charge with specified
21 refrigerant, and leak test in accordance with the manufacturer's instructions.
22

Perform lubrication service, including filling of reservoirs, and confirming that lubricant is of quantity and type recommended by the manufacturer. Test controls and demonstrate compliance with requirements. Replace damaged, or malfunctioning, controls and equipment and retest.

Do not place chillers in sustained operation prior to initial balancing of mechanical systems that interface with the chillers.

## OWNER INSTRUCTION AND TRAINING

Provide services of the manufacturer's technical representative to instruct the owner's personnel in operation and maintenance of chillers. Review with the owner's personnel the data contained in the manufacturer's operating and maintenance manual(s).

# Appendix B3

# *Rotary Screw Compressor Water Chillers*

1
2
3 **PART 1: GENERAL**
4
5
6 **SUBMITTALS**
7
8 General: Submittals shall demonstrate compliance with technical requirements by reference to
9 each subsection of this specification. Where a submitted item does not *comply fully* with each and
10 every requirement of the specifications, the submittal shall clearly indicate such deviations.
11 Identification requirements for noncomplying features of items are very specific.
12
13 Manufacturer's Data: Submit the manufacturer's product data, including rated capacities, weights
14 (shipping, installed, and operating), furnished specialties and accessories, and installation and
15 start-up instructions.
16
17
18 **QUALITY ASSURANCE**
19
20 ARI Compliance:
21
22 Fabricate, test, and certify chillers to comply with ARI Standard 550/590 "Water-Chilling
23 Packages Using the Vapor Compression Cycle."
24

1 Rate air-cooled chillers sound power levels according to ARI 270 "Sound Rating of
2 Outdoor Unitary Equipment" or ARI 370 "Sound Rating of Large Outdoor Refrigerating
3 and Air-Conditioning Equipment," as applicable.
4
5 Rate water-cooled chillers sound power levels according to ARI 575 "Method of
6 Measuring Machinery Sound within an Equipment Space."
7
8 ASHRAE Compliance
9
10 Fabricate and install rotary water chillers to comply with ASHRAE 15 "Safety Code for
11 Mechanical Refrigeration."
12
13 Comply with ASHRAE 147 for refrigerant leaks, recovery, handling, and storage
14 requirements.
15
16 UL Compliance: Fabricate rotary water chillers to comply with UL 465 "Central Cooling Air
17 Conditioners."
18
19 ASME Compliance: Fabricate and stamp rotary water chillers to comply with ASME Boiler and
20 Pressure Vessel Code, Section VIII, Division 1.
21
22 Warranty: The contractor shall provide a 5-year parts and labor manufacturer's warranty on each
23 refrigeration compressor, including refrigerant. This warranty shall be directly from the
24 manufacturer to the owner and shall be in addition to guarantees and warranties required under

1 the general conditions to the construction contract. This warranty shall provide for repair or
2 replacement of the covered compressor that becomes inoperative as a result of defects in materials
3 or workmanship within 5 years after the date of substantial completion.
4
5 **Acoustic Criteria:** Noise levels due to chillers shall result in maximum sound levels in occupied
6 spaces conforming to the following room criteria (RC):
7

| Maximum RC | Environment | Typical Occupancy |
|---|---|---|
| 25 | Extremely quiet environment, suppressed speech is audible, suitable for acute pickup of all sounds | Broadcast or recording studios, concert halls, music rooms, bedrooms, special classrooms for the very young or hearing impaired, etc. |
| 30 | Very quiet, suitable for large conferences; telephone use satisfactory; sleeping unimpaired | Residences, hotel or hospital sleeping rooms, theaters, auditoriums, libraries, executive offices, directors' rooms, large conference rooms, etc. |
| 35 | Quiet, suitable for conference at 15 ft table; normal voice 10–30 ft; telephone use satisfactory | Private offices, school cafeterias, court rooms, churches, small conference rooms, etc. |
| 40 | Satisfactory for conferences at 6–8 ft table; normal voice 6–12 ft; telephone use satisfactory | General (open) offices, school corridors, laboratories, restaurants, etc. |
| 45 | Satisfactory for conferences at 4–5 ft | Retail stores, cafeterias and fast food |

1

| Maximum RC | Environment | Typical Occupancy |
|---|---|---|
| | table; normal voice 3–6 ft, raised voice 6–12 ft; telephone use is somewhat difficult | dining, lobbies or public areas, etc. |
| 50 | Unsatisfactory for conferences; normal voice 1–2 ft, raised voice 3-6 ft; telephone use is difficult. | Workshops, machine rooms, industrial process areas, etc. |

2 For classrooms for primary education, colleges and universities, training facilities, etc., HVAC
3 systems and their components shall be selected/installed to comply with ANSI Standard S12.60.
4
5 Outdoor HVAC equipment shall be selected, located, and oriented to limit the "Equivalent Sound
6 Level" ($LA_{eq}$,T) A-scale sound pressure level (dBA), averaged over $T = 16$ h, to 50 dBA or less
7 at the site boundary or at a distance of 100 ft, whichever is closer to the equipment.
8
9
10 **GUARANTEE PERIOD SERVICES**
11
12 Maintenance of Chiller: The contractor shall inspect and maintain chillers and chiller components,
13 including manufacturer-furnished controls, during the guarantee period. The contractor shall then
14 furnish a report describing the status of the equipment, problem areas (if any) noticed during
15 service work, and description of the corrective actions taken. The report shall clearly certify that
16 all chiller controls are functioning correctly.
17

1 Service Documentation: A copy of the service report associated with each routine service visit or
2 owner-initiated service call shall be provided to the owner within 10 days after the date of each
3 service call.
4
5
6 **PART 2: PRODUCTS**
7
8
9 **WATER-COOLED CHILLERS**
10
11 Description: Packaged, factory-assembled, hermetic or open-drive-type high- or medium-pressure
12 water chillers consisting of single or multiple rotary screw compressor(s), compressor motor(s),
13 motor starter(s), evaporator, condenser, controls and panels including gages and indicating lights,
14 auxiliary components, and accessories.
15
16 Refrigerants: Chiller shall utilize R-407C, R-410A, or R-134A. **//EDIT REFRIGERANTS AS**
17 **REQUIRED//**
18
19 Rotary Screw Compressor(s)
20
21 General: Compressors shall be rotary twin screw type operating at not more than 3600
22 rpm. Compressors shall be designed to provide access to compressor components for
23 inspection, maintenance, and repair.
24

1 Construction
2
3 Casing shall be cast iron with gasket-sealed casing joints, machined to provide
4 minimum, uniform clearance for the rotors.
5
6 Rotors shall be forged steel with asymmetric profiles.
7
8 Provide sleeve or roller bearings in accordance with ABMA 9 or ABMA 11 to
9 adequately contain both radial and axial loadings and maintain accurate rotor
10 positioning at all pressure ratios, with L10 life rating of at least 200,000 h.
11
12 Provide check valve on compressor discharge to prevent rotor backspin.
13
14 Open-Drive Shaft: For chillers with open drives, provide flexible disk with all-metal
15 construction and no wearing parts or need for lubrication. Provide spring-loaded,
16 precision carbon ring, high-temperature elastomer "O" ring static shaft seal, and stress-
17 relieved, precision-lapped collars. The entire shaft seal cavity shall be at low pressure, and
18 shall be vented to the oil drain from the compressor.
19
20 Lubrication System: Provide oil separator on discharge of the compressor to provide
21 minimum 99% separation prior to refrigerant entering condenser. Main oil reservoir shall
22 be located so as to provide gravity lubrication of rotor and rotor bearings during start-up
23 and coast-down; system differential pressure shall provide proper oil flow during normal
24 operation. Provide crankcase oil heater as required.

1
2 Capacity Control: Achieve by use of compressor staging, for multi-compressor units, and
3 individual compressor slide valves, which shall provide fully modulating capacity control
4 from 100% to 10% of full load. Slide valves shall be actuated by oil pressure, controlled
5 by external solenoid valves, via the chiller integral controls.
6
7 Provide hot-gas bypass piping and control valve integrated with capacity control sequence
8 to maintain chiller operation down to 0.5°F differential chilled water temperature.
9
10 Compressor Motor(s)
11
12 Motor and Accessories (Hermetic Type): Hermetically sealed, continuous duty, single
13 speed, squirrel cage, induction type; full-load operation of the motor shall not exceed
14 nameplate rating; rotor shaft shall be heat-treated carbon steel and designed such that the
15 first critical speed is well above the operating speed. Provide for removal of the stator for
16 service or replacement without breaking the main refrigerant piping connections.
17
18 Motor and Accessories (Open-Drive Type): Two-pole, continuous duty, squirrel cage
19 induction type, and shall have an open drip-proof enclosure. Motor full-load amperes at
20 design conditions shall not exceed motor nameplate (FLA). Motor shall be designed for
21 use with the type starter specified. Motor shall be factory mounted and directly supported
22 by the compressor in providing controlled alignment.
23

Compressor Motor Starter(s): Provide unit-mounted or free-standing factory-wired starters. All power and control conductors between starters and other chiller components, whether unit mounted or free standing, shall be sized, furnished, and installed, complete and ready for operation.

Starter shall incorporate a single point for chiller power connection. All chiller heaters, pumps, control circuits, and other miscellaneous items and accessories shall be supplied directly from the starter without dependence on external power from other sources. Where ancillary items require a different supply voltage from the chiller motor, a suitable transformer(s) and appropriate wiring shall be provided as a part of the chiller starter to supply these items.

Starter shall be reduced voltage, closed transition wye–delta, electronic, or auto-transformer type as indicated on the drawings. Starter shall be non-reversing, enclosed in a suitable NEMA-1 enclosure, and shall have as a minimum the following features:

> Nonfused disconnecting means to disconnect all power sources from the chiller and related accessories.
>
> Electronic internal self-contained monitor system to protect against three-phase and single-phase overloads during running and starting, loss of phase, phase reversal, phase unbalance voltage or current, overvoltage, and undervoltage.
>
> A total run-time meter mounted on the starter door.

1 An electronic mulitfunciton metering module with display mounted on the starter door
2 providing the following instantaneous, demand, and energy information:
3
4 Digital readings of line and phase voltages, line and phase currents, total watts,
5 total VARS, total kVA, power factor, and line frequency.
6
7 Digital readings of line and phase currents, and total watts.
8
9 Digital readings of accumulated watthours and accumulated VARhours.
10
11 Evaporator and Condenser
12
13 Shell and Water Boxes: Fabricated from welded carbon steel plate. Provide 150 psig
14 maximum working pressure water boxes and nozzle connections.
15
16 Provide vents, drains, and covers in water boxes to permit tube cleaning within the
17 space shown on the drawings.
18
19 Provide suitable tappings in the water boxes and nozzles for control sensors,
20 gages, and thermometers.
21
22 Provide standard water boxes with grooved mechanical or flanged piping
23 connections and piping shall be arranged so that heat exchanger heads can be easily
24 removed for maintenance.

Water Heads: Fabricated steel water heads with integral water connections.

Tube Sheets: Fabricated of carbon steel sheets welded to the shell and drilled for tubes. Include intermediate tube support sheets as required to prevent tube vibration.

Tubes: Individually replaceable, finned, seamless copper tubes; removable from either end of the heat exchanger without affecting strength and durability of the tube sheets and without causing leakage in adjacent tubes. Expand ends of tubes in tube sheets and intermediate tube support sheets for tight fit to prevent vibration of tubes. Provide suitable baffles or distributing plates in condenser tubes to evenly distribute refrigerant discharge gas on heat transfer tubes.

Pressure Limiting and Pressure Relief Devices: Manufacturer's standard complying with ASHRAE 15. Provide refrigerant charging and transfer connections, pressure relief device on the evaporator to prevent excessive pressure in refrigerant side, and means to sense refrigerant pressure or temperature. Pressure relief valve shall close once pressure is reduced below setpoint pressure.

Pump-Out System: Include compressor and drive, piping, wiring, motor starter, and manual isolation valves so that the entire refrigerant charge can be contained and isolated in either the condenser or evaporator, as required.

Unit Controller: Manufacturer-furnished, factory-installed unit controls shall interface seamlessly with the facility direct digital control (DDC) system, as follows:

**// SELECT TYPE OF DDC ARCHITECTURE/COMMUNICATION REQUIRED FOR CHILLER UNIT CONTROLLER//**

BACnet Compliance: Unit controllers shall use ANSI/ASHRAE Standard 135 protocol and communicate using ISO 8802-3 (Ethernet) datalink/physical layer protocol. Comply with ANSI/ASHRAE Standard 135 for all hardware and software.

LonWorks Compliance: Unit controllers shall use LonTalk protocol and communicate using EIA/CEA 709.1 datalink/physical layer protocol. Comply with LONMARK Interoperability Association "Interoperability Guidelines" for all control hardware and software. Utilize published functional profiles for all product network message and configuration parameters.

Gateways: **The use of gateways is prohibited.**

Coordination: The Contractor shall be responsible for coordination between the control system subcontractor and the unit controller manufacturer/vendor to ensure that required control interface is implemented, resulting in full interoperability between unit controls and the DDC system.

Refrigerant Flow Control Devices: Provide refrigerant flow control devices between evaporator and condensers (and elsewhere as required) to regulate refrigerant flow at volume and pressure required to maintain evaporator liquid refrigerant at a level sufficient to keep cooler heat transfer tubes adequately wetted through a full range of chiller operation.

Design Devices to Permit Chiller Operation: This is done at scheduled conditions, and to allow condenser entering water temperature to decrease to minimum permissible temperature or 1°F above return chilled water temperature.

Safety Controls

Design cutouts to operate independently and factory wire to control panel.

Design controls to stop the compressor motor in the event of low refrigerant pressure or temperature in evaporator, high condenser pressure, high compressor discharge temperature, low evaporator leaving water temperature ("freezestat"), high motor temperature, high bearing temperature, low oil pressure, high oil temperature, compressor motor overcurrent or over voltage, loss of phase, or power interruption.

Design each cutout to require manual restarting of compressor in the event of safety failure but allowing automatic restart after power interruption.

Provide an anti recycle timer for limiting compressor motor restarts at scheduled time intervals.

Operational Controls: Provide controls to ensure that compressor will start only under unloaded condition.

Provide Sequencing Controls: This is done to ensure lubrication of compressor motor bearings and seals (if any). Sequence as follows:

Run lubrication system oil pump so that compressor motor bearing is lubricated before start-up.

Start compressor motor.

Provide lubrication during coast-down after compressor motor shutdown.

Design Controls to Automatically Restart Compressor: This is done after power failure interruptions, provided minimum time between starts has been complied with.

Manual Control and Display Panel: Factory–mounted and factory wired. Provide gages or meters to indicate low refrigerant pressure in evaporator, high condenser pressure, and low oil pressure.

1 Provide operator manual input functions designed to permit indicated operations
2 including the following:
3
4 Manual and automatic operation of oil pump
5
6 Manual and automatic operation of oil separator heater
7
8 Provide pilot lights or visual display for indicated operations and cutouts including
9 the following:
10
11 Oil pump operation
12
13 Low chilled water temperature cutout
14
15 Low evaporator refrigerant pressure or temperature cutout
16
17 High condenser pressure cutout
18
19 High motor winding temperature cutout
20
21 Low oil pressure cutout
22
23 Motor overload cutout
24

1 Insulation: Insulate evaporators and all other cold surfaces to prevent condensation, with ambient
2 humidity of 75% and dry–bulb temperature of 90°F, no air movement, with flexible elastomeric,
3 closed cell, thermal insulation in accordance with ASTM C534, Type II for sheet materials, black
4 in color, rated for temperatures from –40°F to 220°F. Insulation shall be AR/Armaflex or
5 equivalent.
6
7
8 **AIR-COOLED CHILLERS**
9
10 General: Provide factory—assembled and tested multi compressor outdoor air—cooled liquid
11 chillers as indicated, consisting of at least two compressors, evaporator, condensers, thermal
12 expansion valves, and control panels. Provide capacity and electrical characteristics as scheduled.
13
14 Refrigerant: Chiller shall utilize R-407C, R-410A, or R-134A. **//EDIT REFRIGERANTS AS**
15 **REQUIRED//**
16
17 Housing: Provide the manufacturer's standard equipment housing construction, corrosion
18 protection coating, and exterior finish. Provide removable panels and/or access doors for
19 inspection and access to internal parts and components.
20
21 Rotary Screw Compressors:
22
23 General: Provide multiple positive-displacement direct drive rotary twin screw
24 compressors with hermetically sealed casing, operating at not more than 3600 rpm.

Maximum single compressor capacity shall not exceed 50° tons.

Each compressor shall be piped as a separate and independent refrigerant circuit so that in the event of failure of any one compressor, the remaining compressors shall continue to operate.

Each circuit shall include a thermal-expansion valve, refrigerant charging connections, a hot-gas muffler, compressor suction and discharge shutoff valves, a liquid-line shutoff valve, a replaceable-core filter-dryer, a sight glass with moisture indicator, a liquid-line solenoid valve, and an insulated suction line.

Provide each compressor with crankcase oil heater and suction strainer.

Compressor motors shall be high-torque, two-pole induction type with inherent thermal-overload protection on each phase, hermetically sealed and cooled by the refrigerant suction gas.

Provide full voltage, across-the-line starters in accordance with NEMA ICS 2, Class A, full voltage, nonreversing for each compressor.

Mount individual compressors on vibration isolators.

Construction

1 Casing shall be cast iron with gasket-sealed casing joints, machined to provide
2 minimum, uniform clearance for the rotors.
3
4 Rotors shall be forged steel with asymmetric profiles.
5
6 Provide sleeve or roller bearings in accordance with ABMA 9 or ABMA 11 to
7 adequately contain both radial and axial loadings and maintain accurate rotor
8 positioning at all pressure ratios, with L10 life rating of at least 200,000 h.
9
10 Provide check valve on compressor discharge to prevent rotor backspin.
11
12 Compressors shall be designed to provide access to compressor components for
13 inspection, maintenance, and repair.
14
15 Open-Drive Shaft: For chillers with open drives, provide flexible disk with all-metal
16 construction and no wearing parts or need for lubrication. Provide a spring-loaded,
17 precision carbon ring, high-temperature elastomer "O" ring static shaft seal, and stress-
18 relieved, precision-lapped collars. The entire shaft seal cavity shall be at low pressure, and
19 shall be vented to the oil drain from the compressor.
20
21 Lubrication System: Provide oil separator on discharge of compressor to provide
22 minimum 99% separation prior to refrigerant entering condenser. Main oil reservoir shall
23 be located so as to provide gravity lubrication of rotor and rotor bearings during start-up

and coast-down; system differential pressure shall provide proper oil flow during normal operation. Provide crankcase oil heater as required.

Capacity Control: Achieve by the use of compressor staging, for multi compressor units, and individual compressor slide valves, which shall provide fully modulating capacity control from 100% to 10% of full load. Slide valves shall be actuated by oil pressure, controlled by external solenoid valves, via the chiller integral controls.

Lubrication System: Provide oil separator on discharge of compressor to provide minimum 99% separation prior to refrigerant entering condenser. Main oil reservoir shall be located so as to provide gravity lubrication of rotor and rotor bearings during start-up and coast-down; system differential pressure shall provide proper oil flow during normal operation.

Capacity Control: Achieve by the use of compressor staging and compressor slide valves, which shall provide fully modulating capacity control from 100% to 10% of full load. The slide valve shall be actuated by oil pressure, controlled by external solenoid valves via the chiller integral controls.

Evaporator:

Shell and Water Boxes: Fabricated from welded carbon steel plate. Provide 150 psig maximum working pressure water boxes and nozzle connections. Provide vents, drains, and covers in water boxes to permit tube cleaning within the space shown on the drawings. Provide suitable tappings in the water boxes and nozzles for control sensors, gages, and thermometers.

Water Heads: Provide standard water boxes with grooved mechanical or flanged piping connections and piping shall be arranged so that heat exchanger heads can be easily removed for maintenance.

Tube Sheets: Fabricated of carbon steel sheets welded to the shell and drilled for tubes. Include intermediate tube support sheets as required to prevent tube vibration.

Tubes: Individually replaceable, finned, seamless copper tubes; removable from either end of the heat exchanger without affecting strength and durability of the tube sheets and without causing leakage in adjacent tubes. Expand the ends of tubes in tube sheets and intermediate tube support sheets for tight fit to prevent vibration of tubes. Provide suitable baffles or distributing plates in condenser tubes to evenly distribute refrigerant discharge gas on heat transfer tubes.

Pressure Limiting and Pressure Relief Devices: Manufacturer's standard complying with ASHRAE 15. Provide refrigerant charging and transfer connections, pressure relief device on the evaporator to prevent excessive pressure in refrigerant side, and means to sense refrigerant pressure or temperature. Pressure relief valve shall close once pressure is reduced below setpoint pressure.

Heater: Provide electrical resistance heater to protect evaporator to protect against

1 freezing at –20°F ambient at no-flow condition.
2
3 Air-Cooled Condenser: Plate-fin coil with integral sub-cooling on each circuit, rated at 450 psig.
4 Construct coils of copper tubes mechanically bonded to aluminum fins, factory-coated with
5 epoxy-phenolic after fabrication.
6
7 Condenser fans: Direct-drive propeller type with statically and dynamically balanced fan
8 blades, arranged for vertical air discharge.
9
10 Fan motors: Totally enclosed nonventilating (TENV) or totally enclosed air over (TEAO)
11 enclosure, with permanently lubricated bearings, and having built-in overcurrent- and
12 thermal-overload protection.
13
14 Provide condenser coils with louvers, baffles, or hoods to protect against hail damage.
15
16 Pump-out System: Include compressor and drive, piping, wiring, motor starter, and manual
17 isolation valves so that entire refrigerant charge can be contained and isolated in the evaporator.
18
19 Unit Controller: Manufacturer-furnished, factory-installed unit controls, shall interface seamlessly
20 with the facility DDC system, as follows:
21
22 **// SELECT TYPE OF DDC ARCHITECTURE/COMMUNICATION REQUIRED FOR**
23 **CHILLER UNIT CONTROLLER//**
24 BACnet Compliance: Unit controllers shall use ANSI/ASHRAE Standard 135 protocol

and communicate using ISO 8802-3 (Ethernet) datalink/physical layer protocol. Comply with ANSI/ASHRAE Standard 135 for all hardware and software.

LonWorks Compliance: Unit controllers shall use LonTalk protocol and communicate using EIA/CEA 709.1 datalink/physical layer protocol. Comply with LonMark Interoperability Association "Interoperability Guidelines" for all control hardware and software. Utilize published functional profiles for all product network message and configuration parameters.

Gateways: **The use of gateways is prohibited.**

Coordination: The contractor shall be responsible for coordination between the controls system sub-contractor and the unit controller manufacturer/vendor to ensure that required control interface is implemented, resulting in full interoperability between unit controls and the DDC system.

Refrigerant Flow Control Devices: Provide refrigerant flow control devices between evaporator and condensers (and elsewhere as required) to regulate refrigerant flow at volume and pressure required to maintain evaporator liquid refrigerant at a level sufficient to keep cooler heat transfer tubes adequately wetted through a full range of chiller operation.

Design Devices to Permit Chiller Operation: This is done at scheduled conditions, and to allow condenser entering water temperature to decrease to minimum permissible temperature or 1°F above return chilled water temperature.

Safety Controls:

Design cutouts to operate independently and factory wire to control panel.

Design controls to stop compressor motor in event of low refrigerant pressure or temperature in evaporator, high condenser pressure, high compressor discharge temperature, low evaporator leaving water temperature ("freezestat"), high motor temperature, high bearing temperature, low oil pressure, high oil temperature, compressor motor overcurrent or over voltage, loss of phase, or power interruption.

Design each cutout to require manual restarting of compressor in the event of safety failure but allowing automatic restart after power interruption.

Provide an anti recycle timer for limiting compressor motor restarts at scheduled time intervals.

Operational Controls: Provide controls to ensure that compressor will start only under unloaded condition.

Provide Sequencing Controls: This is to ensure lubrication of compressor motor bearings and seals (if any). Sequence as follows:

1 Run lubrication system oil pump so that compressor motor bearing is lubricated
2 before start-up.
3
4 Start compressor motor.
5
6 Provide lubrication during coast-down after compressor motor shutdown.
7
8 Design Controls to Automatically Restart Compressor: This is done after power failure
9 interruptions, provided minimum time between starts has been complied with.
10
11 Manual Control and Display Panel: Factory mounted and factory wired. Provide gages or
12 meters to indicate low refrigerant pressure in evaporator, high condenser pressure, and
13 low oil pressure.
14
15 Provide operator manual input functions designed to permit indicated operations
16 including the following:
17
18 Manual and automatic operation of oil pump
19
20 Manual and automatic operation of oil separator heater
21
22 Provide pilot lights or visual display for indicated operations and cutouts including
23 the following:
24

1 Oil pump operation
2
3 Low chilled water temperature cutout
4
5 Low evaporator refrigerant pressure or temperature cutout
6
7 High condenser pressure cutout
8
9 High motor winding temperature cutout
10
11 Low oil pressure cutout
12
13 Motor overload cutout
14
15 Insulation: Insulate evaporators and all other cold surfaces to prevent condensation, with ambient
16 humidity of 75% and dry bulb temperature of 90°F, no air movement, with flexible elastomeric,
17 closed cell, thermal insulation in accordance with ASTM C534, Type II for sheet materials, black
18 in color, rated for temperatures from –40°F to 220°F. Insulation shall be AR/Armaflex or
19 equivalent.
20
21
22
23 **PART 3: EXECUTION**
24

**INSTALLATION OF CHILLERS**

Install chillers in accordance with the manufacturer's written instructions. Install units plumb and level, firmly anchored in locations indicated; maintain the manufacturer's recommended clearances.

Provide relief piping from indoor water-cooled chiller refrigerant pressure relief rupture disk to atmosphere; size piping as recommended by the chiller manufacturer, and terminate with gooseneck facing down.

**// INCLUDE FACTORY PERFORMANCE TEST ONLY WHEN REQUIRED.//**

**FACTORY PERFORMANCE TEST**

The manufacturer shall conduct witnessed performance acceptance test for each typical machine to verify the performance submitted. The test will be in accordance with ARI 550/590-98 with the following exceptions:

Capacity will have no tolerances. Capacity must meet or better required tonnage.

Heat balance must be verified at 3% or below during test conditions.

The test shall be conducted at an approved ARI certified test facility of the manufacturer's factory. Instrumentation used for testing must be calibrated within 6 months of the test date and traceable to the National Bureau of Standards. All documentation verifying NBS traceability shall be included in a bound folder for presentation to the owner's representative. The owner's representative may elect to contact ARI for verification of performance and test conditions.

Chiller shall be tested with water temperature and adjustment cooler/condenser per standard ARI 550-90 to simulate specified fouling versus no fouling during test. Verification of this procedure will require inside surface area and number of tubes per vessel. This information is to be submitted prior to the test for formula verification of fouling per ARI 550-90 if requested by the engineer.

Performance test shall be a four point test, 25%, 50%, 75%, and 100% load.

The chiller will be accepted if the test indicates that all conditions of the performance test meet the specified conditions. If the performance test results are outside of specified tolerances, the manufacturer will be allowed to make repairs and retest, and the manufacturer will assume all expenses incurred for the owner's representatives to witness the retest.

If, after the second test, the machines do not meet the test requirements, an energy cost penalty of $4,000 per kW above test requirements will be deducted from the manufacturer's contract. In the case of an ARI certified performance chiller, ARI will be notified of any test failures. If there is more than one typical machine, the energy penalty will be the penalty of the tested machine times the number of typical machines on the project. The total energy penalty will be the sum of all machines on the project whether tested or typical.

1 **DEMONSTRATION**
2
3 Start-up chillers, in accordance with the manufacturer's start-up instructions, and provide the service of
4 a factory-authorized service representative to provide start-up service and to demonstrate and train the
5 owner's maintenance personnel as specified below. Evacuate, dehydrate, and charge with specified
6 refrigerant, and leak test in accordance with the manufacturer's instructions.
7
8 Perform lubrication service, including filling of reservoirs, and confirming that the lubricant is of
9 quantity and type recommended by the manufacturer. Test controls and demonstrate compliance
10 with requirements. Replace damaged, or malfunctioning, controls and equipment and retest.
11
12
13 **OWNER INSTRUCTION AND TRAINING**
14
15 Provide services of manufacturer's technical representative to instruct the owner's personnel in
16 operation and maintenance of chillers. Review with the owner's personnel the data contained in
17 the manufacturer's operating and maintenance manual(s).
18
19

# Appendix B4

## *Induced Draft Cooling Towers*

1
2
3
4 **PART 1 - GENERAL**
5
6
7 **QUALITY ASSURANCE**
8
9 Cooling Technology Institute (CTI) Compliance:
10
11 Rate cooling tower capacity based on performance tests in accordance with CTI Standard
12 201.
13
14 Rate tower sound power levels in accordance with CTI Code ATC-128.
15
16 Manufacturer Guarantee: Provide written manufacturer guarantee that the tower will perform in
17 complete compliance with performance requirements, provided the tower is installed in
18 accordance with the manufacturer's written instructions. This guarantee shall include the
19 manufacturer providing in-situ testing of the tower performance in accordance with CTI Standard
20 ATC-105, by a testing agency licensed by CTI, if required by the owner.
21
22 **Acoustic Criteria:** Outdoor HVAC equipment shall be selected, located, and oriented to limit the
23 "Equivalent Sound Level" ($LA_{eq}$,T) A-scale sound pressure level (dBA), averaged over $T$ = 16 h,
24 to 50 dBA or less at the site boundary or at a distance of 100 ft, whichever is closer to the

equipment.

## SUBMITTALS

General: Submittals shall demonstrate compliance with technical requirements by reference to each subsection of this specification. Where a submitted item does not *comply fully* with each and every requirement of the specifications, the submittal shall clearly indicate such deviations. Identification requirements for non-complying features of items are very specific.

Manufacturer's Data: Submit manufacturer's technical product data, including rated capacities, pressure drop, fan performance data, weights (shipping, installed, and operating), and installation and start-up instructions.

## PART 2: PRODUCTS

Description: Provide crossflow- or counterflow-induced draft cooling tower(s) as indicated on the drawings. Cooling tower shall be a complete factory-assembled, totally noncombustible unit.

Structure: Structure shall be steel with G235 galvanized finish coating. Assemble structural system, including basin and casings, by providing bolted connections with fasteners having equal or better corrosion resistance than the materials fastened. Seal joints to make a watertight enclosure.

Certify tower structure for wind loading caused by 100 mph wind from any direction.

Provide rigging supports on structure for final rigging.

Casing: Provide fiberglass-reinforced polyester or minimum 22 ga. steel panels with G235 galvanized finish coating and fabricated to make tower watertight.

Basin:

Basin shall be constructed of one of the following materials, designed and fabricated to contain and support the water and ensure water tightness:

Fiberglass-reinforced polyester

Stainless steel, Type 306, minimum 20 ga.

Provide integral collecting basin and sump with side outlet and with connections for drain, overflow and water makeup. Drain connection and basin operating level shall be designed to prevent vortexing and air entrainment.

1 Wetted-Surface (Film) Fill: Provide vertical sheets of PVC or CPVC plastic having flame spread
2 rating of 5 in accordance with ASTM E84 and fabricated into wave-formed configurations
3 installed by the manufacturer to assure maximum wetted heat transfer surface.
4
5 Drift Eliminators: Provide UV-inhibited PVC or CPVC plastic having flame spread rating of 5
6 in accordance with ASTM E84, which is fabricated into configuration to limit drift loss to a
7 maximum of 0.1% of the condenser water flow rate. Drift eliminators may be integral with
8 the fill.
9
10 Louvers: Provide fiberglass-reinforced plastic or polyvinyl chloride plastic of sufficient thickness
11 and rigidity to prevent visible sagging or fluttering. Plastic louvers shall be UV-inhibited.
12
13 Wet Deck Water Distribution
14
15 Crossflow Tower: Provide stainless-steel open basin with gravity-flow plastic metering
16 orifices providing even distribution of water over fill. Provide flow control valves for
17 balancing flow to each distribution basin where multiple inlets are required. Provide
18 removable fiberglass-reinforced polyester panels to prevent debris from entering the wet
19 deck and to inhibit algae growth by eliminating sunlight.
20
21 Counterflow Tower: Provide spray distribution system to ensure even distribution of
22 water over wetted-surface fill.
23
24 Schedule 40 PVC pipe header and removable schedule 40 PVC pipe branches.

1
2 Provide removable plastic, brass, or ceramic nozzles with a maximum 12 psig inlet
3 pressure required.
4
5 Inlet Screens: Provide stainless-steel mesh, mounted in removable frames by the manufacturer.
6
7 **//SELECT ONE OF THE FOLLOWING TYPES OF BASIN HEATERS//**
8
9 Basin Heaters: Provide steam injection heaters, with two-position steam control valve and
10 thermostat set at 40°F.
11
12 Provide basin heaters sized by the manufacturer to maintain basin water at 40°F at ambient
13 temperature of 0°F and wind velocity of 15 mph.
14
15 Basin Heater: Penberthy Model NWK, minimum 20 psig steam pressure rating.
16
17 Basin Heaters: Provide electric immersion heaters including thermostat, low-water cutout
18 contactor, and transformers, in weatherproof enclosure.
19
20 Provide basin heaters sized by the manufacturer to maintain basin water at 40°F at ambient
21 temperature of 0°F and wind velocity of 15 mph.
22
23 Basin heater: Screw-plug type, 2½ inch pipe thread, electric immersion heater. Heaters
24 specified for three-phase service shall be provided with necessary contactors. The heater

1 element shall be copper sheathed, brazed to brass screw plugs. Multiple heaters shall be
2 provided, wired in parallel, as required to meet indicated capacity. Heater shall be
3 equivalent to Chromolox Type ARMT.
4
5 **//SELECT ONE OF THE FOLLOWING METHODS OF BASIN WATER LEVEL**
6 **CONTROL//**
7
8 Float Valve Water Level Control: Provide float-activated makeup water valve. Float shall be fully
9 adjustable as to depth control and angle to provide tight shutoff. Set-screw adjustments are not
10 acceptable.
11
12 Electronic Water-Level Control: Provide electric water-level control, package including
13 conductance-actuated level controller and slow closing solenoid makeup valve. Level control shall
14 provide for independently adjustable high operating level to close makeup water valve and low
15 operating level to open makeup water valve. Additional alarm contacts for high and low water
16 levels shall be provided.
17
18 Fan and Drive: Provide propeller fan with cast-aluminum adjustable pitch blades driven by one of
19 the following methods. Fan shall be statically and dynamically balanced and hub shall be keyed to
20 ground and polished steel shaft:
21
22 Gear Drive: Provide right-angle, industrial duty, oil-lubricated geared speed reducer with
23 shafts to connect the motor and fan. Gear drive shall be designed for cooling tower
24 applications and comply with CTI 111. Include low-oil-level warning switch that shall, on

1 reaching a low-oil-level set point recommended by cooling tower manufacturer, signal an
2 alarm and shutdown the tower or cell affected.
3
4 Belt Drive: Provide cast aluminum fan and motor sheaves with one-piece multi-groove
5 neoprene/polyester drive belt. Provide belt guard in accordance with OSHA requirements.
6
7 Direct Drive: Provide Baldor-Reliance Model RPM-AC cooling tower direct-drive motor
8 with Model VS1CTD variable speed control system. Fan shaft shall be directly coupled to
9 motor shaft in a vertical position.
10
11 Fan Bearings:. Bearings shall be grease-lubricated ball or roller bearings with an L10-rated life in
12 accordance with ABMA 9 or ABMA 11 of at least 100,000 h at maximum operating speed and
13 horsepower.
14 Polyphase Motors:
15
16 Provide totally enclosed motors. Motors located in the tower air stream may be "air over"
17 type (TEAO), while motors located out of the tower air steam shall be "fan-cooled" type
18 (TEFC).
19
20 Multispeed motors with 2:1 speed ratio shall be consequent pole, single winding type.
21 Multispeed motors with other than 2:1 speed ratio shall have separate winding for each
22 speed. Rotor shall be random-wound, squirrel cage type.
23
24 Motor bearings shall be regreasable, shielded, antifriction ball bearings suitable for radial

and thrust loadings.

Motor frames shall be cast iron for motors 10 hp and larger; rolled steel for motors smaller than 10 hp.

Motors used with reduced-voltage and multi speed controllers: Match wiring connection requirements for controller with required motor leads. Provide terminals in motor terminal box, suited to the control method.

Motors used with variable frequency controllers: Windings shall be copper magnet wire with moisture-resistant insulation varnish, designed and tested to resist transient spikes, high frequencies, and short time rise pulses produced by pulse-wide modulated inverters.

Vibration Cutout Switch: Provide switch to provide alarm and deenergize fan motor(s) if excessive vibration occurs due to fan imbalance.

Service Access:

Crossflow Tower: Provide minimum 24″ wide by 30″ high access doors on both end walls to provide access to basin, fan plenum, motor, drive, and fill. Provide minimum 18″ wide galvanized steel grated walkway across the tower above the basin overflow level. If the tower is installed so that the bottom of the access door is more than 5′–0″ above the roof or grade, provide an access door platform on each side with ladder rigidly attached to the tower.

Top of Tower Handrail and Ladder: Provide top of tower galvanized steel structural tubing handrail, complete with knee rail and toeboard, constructed in accordance with OSHA guidelines. Provide ladder from roof or grade to top of tower. Provide ladder safety cage if the top of tower is 20′ 0″ or greater above grade or roof.

Apply phosphatized pretreatment on any zinc-coated surfaces that have not been mill-phosphatized or polymer-coated. Apply gasoline-soluble rust-preventive compound on ferrous parts that cannot be galvanized, including shafts and machined parts.

## PART 3: EXECUTION

### INSTALLATION

Install cooling towers where indicated, in accordance with equipment manufacturer's written instructions and with recognized industry practices, to ensure that cooling towers comply with requirements and serve intended purposes.

Install cooling tower on structural supports as indicated on the drawings. Anchor cooling tower to stand with removable fasteners. Tower shall be installed level. Where multiple tower cells are indicated, the entire tower assembly shall be leveled end to end within 0.25″.

1 Provide fan motor disconnect switch in NEMA 3R enclosure rigidly attached to the cooling tower
2 and located immediately adjacent to the primary access door. Provide service clearance required
3 by NFPA 70.
4
5 Provide equalizer piping between multiple tower or cell basins that are part of a common
6 condenser water system, whether indicated on the drawings or not. Equalizer shall be full size of
7 cooling tower connection and include a shutoff valve.
8
9
10 **DEMONSTRATION**
11
12 Operate the condenser water system to maintain setpoint temperature under its automatic controls
13 for at least 24 h. All aspects of condenser water pump, cooling tower, controls, and water
14 treatment shall be checked and evaluated frequently during this period and any unusual noises or
15 vibration, erratic performance or operation, or other problems shall be repaired before being
16 placed into fulltime service.
17
18
19
20 **OWNER INSTRUCTION AND TRAINING**
21
22 Provide services of the manufacturer's technical representative to instruct owner's personnel in
23 operation and maintenance of chillers. Review with the owner's personnel the data contained in
24 the manufacturer's operating and maintenance manual(s).

# Appendix B5

## *Closed-Circuit Liquid Coolers*

**PART 1: GENERAL**

**QUALITY ASSURANCE**

Cooling Technology Institute (CTI) Compliance

Rate closed-circuit cooler capacity based on performance tests in accordance with CTI Standard 201.

Rate cooler sound power levels in accordance with CTI Code ATC-128.

Manufacturer Guarantee: Provide written manufacturer guarantee that the tower will perform in complete compliance with performance requirements, provided the tower is installed in accordance with the manufacturer's written instructions. This guarantee shall include the manufacturer providing *in situ* testing of the tower performance in accordance with CTI Standard ATC-105, by a testing agency licensed by CTI, if required by the owner.

Acoustic Criteria: Outdoor HVAC equipment shall be selected, located, and oriented to limit the "Equivalent Sound Level" ($LA_{eq}$,T) A-scale sound pressure level (dBA), averaged over $T = 16$ h, to 50 dBA or less at the site boundary or at a distance of 100 ft, whichever is closer to the equipment.

**SUBMITTALS**

General: Submittals shall demonstrate compliance with technical requirements by reference to each subsection of this specification. Where a submitted item does not *comply fully* with each and every requirement of the specifications, the submittal shall clearly indicate such deviations. Identification requirements for non-complying features of items are very specific.

Manufacturer's Data: Submit manufacturer's technical product data, including rated capacities, pressure drop, fan and coil performance data, weights (shipping, installed, and operating), installation and start-up instructions, and rating curves with selected points clearly indicated.

## PART 2: PRODUCTS

Description: Provide induced draft evaporative liquid cooler as indicated on the drawings. Cooler shall be a complete factory-assembled, totally noncombustible unit.

Structure: Structure shall be steel with G235-galvanized finish coating. Assemble structural system, including basin and casings, by providing bolted connections with fasteners having equal

1 or better corrosion resistance than the materials fastened. Seal joints to make watertight
2 enclosure.
3
4 Certify cooler structure for wind loading caused by 100 mph wind from any direction.
5
6 Provide rigging supports on structure for final rigging.
7
8 Casing: Provide fiberglass-reinforced polyester or minimum 22 ga. steel panels with G235-
9 galvanized finish coating and fabricated to make cooler watertight.
10
11 Basin:
12
13 Basin shall be constructed of one of the following materials, designed and fabricated to
14 contain and support the water and ensure water tightness:
15
16 Fiberglass-reinforced polyester
17
18 Stainless steel, Type 306, minimum 20 ga.
19
20 Provide integral collecting basin and sump with side outlet and with connections for drain,
21 overflow and water makeup. Drain connection and basin operating level shall be designed
22 to prevent vortexing and air entrainment.
23

Coil: Coil(s) shall be G235 steel tubes rated for 350 psig and shall be designed for low pressure drop and with slope for free drainage of fluid. After fabrication, factory coat coils with epoxy-phenolic coating.

Drift Eliminators: Provide UV-inhibited PVC or CPVC plastic having flame spread rating of 5 in accordance with ASTM E84, which is fabricated into configuration to limit drift loss to a maximum of 0.1% of the spray water flow-rate.

Inlet Louvers: Provide fiberglass-reinforced plastic or polyvinyl chloride plastic of sufficient thickness and rigidity to prevent visible sagging or fluttering. Plastic louvers shall be UV inhibited.

Inlet Screens: Provide stainless-steel mesh, mounted in removable frames.

Water Distribution:

Water Spray: Provide PVC or CPVC main header and lateral branch piping designed for even distribution over heat-exchanger coil throughout the flow range without the need for balancing valves and for connecting individual, removable, non-clogging spray nozzles.

Spray Pump: Close-coupled, end-suction, single-stage, bronze-fitted centrifugal pump with mechanical seal, suction strainer, and flow balancing valve, all suitable for outdoor service. Pump shall be mounted on the cooler, factory-installed, piped, and wired to disconnect switch.

1 **//SELECT ONE OF THE FOLLOWING TYPES OF BASIN HEATERS//**
2
3 Basin Heaters: Provide steam injection heaters, with two-position steam control valve and
4 thermostat set at 40°F.
5
6 Provide basin heaters sized by the manufacturer to maintain basin water at 40°F at
7 ambient temperature of 0°F and wind velocity of 15 mph.
8
9 Basin heater: Penberthy Model NWK, minimum 20 psig steam pressure rating.
10
11 Basin Heaters: Provide electric immersion heaters, including thermostat, low-water cutout
12 contactor, and transformers, in a weatherproof enclosure.
13
14 Provide basin heaters sized by manufacturer to maintain basin water at 40°F at
15 ambient temperature of 0°F and wind velocity of 15 mph.
16
17 Basin Heater: Screw-plug type, 2½ inch pipe thread, electric immersion heater.
18 Heaters specified for three-phase service shall be provided with necessary
19 contactors. Heater element shall be copper sheathed, brazed to brass screw plugs.
20 Multiple heaters shall be provided, wired in parallel, as required to meet indicated
21 capacity. The heater shall be equivalent to Chromolox Type ARMT.
22
23 **//SELECT ONE OF THE FOLLOWING METHODS OF BASIN WATER LEVEL**
24 **CONTROL//**

1
2 Float Valve Water-Level Control: Provide float-activated makeup water valve. Float shall
3 be fully adjustable as to depth control and angle to provide tight shutoff. Set-screw
4 adjustments are not acceptable.
5
6 Electronic Water-Level Control: Provide electric water level control package, including
7 conductance-actuated level controller and slow closing solenoid makeup valve. Level
8 control shall provide for independently adjustable high operating level to close makeup
9 water valve and low operating level to open makeup water valve. Additional alarm
10 contacts for high and low water levels shall be provided.
11
12 Fan and Drive: Provide propeller fan with cast-aluminum adjustable pitch blades, direct drive or
13 with belt drive as follows.
14
15 Belt Drive: Provide cast-aluminum fan and motor sheaves with one-piece multi groove
16 neoprene/polyester drive belt. Provide belt guard in accordance with OSHA requirements.
17
18 Direct Drive: Provide Baldor-Reliance Model RPM-AC cooling tower direct-drive motor
19 with Model VS1CTD variable speed control system. Fan shaft shall be directly coupled to
20 motor shaft in a vertical position.
21
22

1 Fan Bearings:. Bearings shall be grease-lubricated ball or roller bearings with an L10-rated life in
2 accordance with ABMA 9 or ABMA 11 of at least 100,000 h at maximum operating speed and
3 horsepower.
4
5 Polyphase Motors
6
7 Provide totally enclosed motors. Motors located in the tower air stream may be "air over"
8 type (TEAO), while motors located out of the tower air steam shall be "fan cooled" type
9 (TEFC).
10
11 Multispeed motors with 2:1 speed ratio shall be consequent pole, single-winding type.
12 Multispeed motors with other than 2:1 speed ratio shall have separate winding for each
13 speed. Rotor shall be random-wound, squirrel cage type.
14
15 Motor bearings shall be regreasable, shielded, antifriction ball bearings suitable for radial
16 and thrust loadings.
17
18 Motor frames shall be cast iron for motors 10 hp and larger; rolled steel for motors smaller
19 than 10 hp.
20
21 Motors used with reduced-voltage and multispeed controllers: Match wiring connection
22 requirements for controller with required motor leads. Provide terminals in motor terminal
23 box, suited to control method.
24

Motors used with variable frequency controllers: Windings shall be copper magnet wire with moisture-resistant insulation varnish, designed and tested to resist transient spikes, high frequencies, and short time rise pulses produced by pulse-wide modulated inverters.

Vibration Cutout Switch: Provide switch to provide alarm and deenergize fan motor(s) if excessive vibration occurs due to fan imbalance.

Capacity-Control Dampers: Dampers shall be G235-galvanized steel with linkages, electric operator, controller, limit switches, transformer, and weatherproof enclosure. Dampers shall be interlocked with fan(s) to close when the fan is deenergized.

Service Access

Cooler Access: Provide minimum 24″-wide by 24″-high access doors on both end walls to provide access to basin and cooling coil. If the cooler is installed so that the bottom of the access door is more than 5′ 0″ above the roof or grade, provide an access door and platform on each side with ladder rigidly attached to the cooler.

Top of Cooler Handrail and Ladder: Provide top of cooler galvanized steel structural tubing handrail, complete with knee rail and toeboard, constructed in accordance with OSHA guidelines. Provide a ladder from the roof or grade to the top of the cooler. Provide a ladder safety cage if the top of the cooler is 20′ 0″ or greater above the grade or roof.

1 Apply phosphatized pretreatment on any zinc-coated surfaces that have not been mill
2 phosphatized or polymer coated. Apply gasoline-soluble rust-preventive compound on ferrous
3 parts that cannot be galvanized, including shafts and machined parts.
4
5
6
7 **PART 3: EXECUTION**
8
9
10
11 **INSTALLATION**
12
13 Install coolers where indicated, in accordance with the equipment manufacturer's written
14 instructions and with recognized industry practices, to ensure that cooling coolers comply with
15 requirements and serve intended purposes.
16
17 Install cooler on structural supports as indicated on the drawings. Anchor cooling cooler to stand
18 with removable fasteners. Cooler shall be installed level. Where multiple cooler cells are indicated,
19 the entire cooler assembly shall be leveled end to end within 0.25″.
20
21 Provide factory-mounted nonfused disconnect switches in NEMA 3R enclosures for fan(s) and
22 spray pump, rigidly attached to the cooler and located immediately adjacent to the primary access
23 door. Provide service clearance required by NFPA 70.
24

## DEMONSTRATION

Operate the cooling water system to maintain setpoint temperature under its automatic controls for at least 24 h. All aspects of cooler, controls, and water treatment shall be checked and evaluated frequently during this period and any unusual noises or vibration, erratic performance or operation, or other problems shall be repaired before being placed into fulltime service.

## OWNER INSTRUCTION AND TRAINING

Provide services of the manufacturer's technical representative to instruct the owner's personnel in operation and maintenance of chillers. Review with the owner's personnel the data contained in the manufacturer's operating and maintenance manual(s).

# Appendix C

# *References and Resources*

## WATER CHILLERS

### Water Chiller General References

ASHRAE, *ASHRAE Handbook, Refrigeration,* Chapters 41 and 43, 1998.
ASHRAE, *ASHRAE Handbook, Heating, Ventilating, and Air-Conditionings Systems and Equipment,* Chapter 42, 2008.
ANSI/ASHRAE Standard 15-2010, *Safety Standard for Refrigeration Systems.*
McQuay International, Application Guide AG 31-002, McQuay Air-Conditioning, *Centrifugal Chiller Fundamentals,* 2000.

### Refrigerants

ANSI/ASHRAE Standard 34-2010. *Designation and Safety Classification of Refrigerants.*
ASHRAE, Ozone-Depleting Substances: Position Paper, *American Society of Heating, Refrigerating, and Air-Conditioning Engineers*, 2001.
W. Goetzler, J. Burgos, and T. Sutherland, Ultra-Low GWP Refrigerants, *ASHRAE Journal*, 34–43, 2010.
New HFO-Based Option, *ASHRAE Journal*, 16–20, 2010.
G.J. Williams, The greenhouse-gas impact of various chiller technologies, *HPAC Engineering*, 36–44, 2010.

### Chiller Performance Factors

New ARI Rating Allows More Accurate Chiller-Energy Specification, *HVAC&R Engineering Update*, York International Corporation, 1998.
G. Nowakowski and R. Busby, Advances in Natural Gas Cooling, *ASHRAE Journal*, 47–52, 2001.
System Part Load Value: A Case for Chiller System Optimization, *Carrier Synopsis*, (Carrier), 3(3).
White Paper on ARI Standard 550/590-98, American Refrigeration Institute, 1998.

### Chiller Configuration and Control

Asymmetry as a Basis of Design, *Engineers Newsletter,* The Trane Company, 28(4), 1999.
J.W. Benson, Designing Chilled Water Systems for the 21st Century, The Trane Company, *Semiconductor FabTech*, 9th Edition.
M.J. Bitondo and M.J. Tozzi, Chiller Plant Control: Multiple Chiller Controls, Carrier Corporation, 1999.

A. Burd and G. Burd, Primary/Secondary-Loop vs. Primary-Loop-Only Systems, Advanced Research Technology LLC, *HPAC Engineering*, 36–45, 2010.
W.J. Coad, A Fundamental Perspective on Chilled Water Systems, *HPAC Interactive*, 1998.
*HVAC&R, Chiller-Plant Design in a Deregulated Electric Environment*, York International Corporation (Update).
*McQuay International*, Application Guide AG 31-003, McQuay Air-Conditioning, *Chiller Plant Design*, 2001.
Off-Design Chiller Performance, *Engineers Newsletter*, The Trane Company, 25(5), 1996.
Variable-Primary-Flow Systems, *Engineers Newsletter*, The Trane Company, 28(3), 1999.

## Cooling Thermal Energy Storage

W.P. Bahnfleth, Cool Thermal Storage: Is It Still Cool?, *Heating/Piping/AirConditioning*, 49–54, 2002.
Hot Technology for Large Cooling Systems, Penn State College of Engineering (posted online in 2002).
HVAC: Cool Thermal Storage, Los Angeles Department of Water and Power Energy Advisor (Available at http://www.ladwp.com).
H. Paksoy, O. Andersson, H. Evliya, and S. Abaci, Aquifer Thermal Energy Storage System for Cooling and Heating of Cukurova University Balcali Hospital, (Cukurova University, Adana, Turkey).
M. Ravikumar and P.S.S. Srinivasan, Phase Change Material as a Thermal Energy Storage Material for Cooling of Building, *Journal of Theoretical and Applied Information Technology*, 503–511, 2008.
Real-Time Pricing and Thermal Energy Storage, *EPRI HVAC&R Center Quarterly Newsletter*, 2000.
A. Sharma, V.V. Tyagi, C.R. Chen, and D. Buddhi, Review on Thermal Energy Storage with Phase Change Materials and Applications, *Renewable and Sustainable Energy Reviews*, 13, 318–345, 2009.
B.M. Silvetti, The Application of Thermal Storage in an Unregulated Power Marketplace, (Conference Paper, Association of Energy Engineers).
U.S. Department of Energy, Thermal Energy Storage for Space Cooling, Federal Energy Management Program, 2000. (http://www1.eere.energy.gov/femp/pdfs/FTA_coolstorage.pdf).
D.R. Wulfinghoff, Measure 2.11.1 Install Cooling Thermal Storage, *Energy Efficiency Manual*, 1999.

## Noise and Vibration

C.M. Harris, *Noise Control in Buildings, McGraw-Hill, Inc., 1994.*
M.E. Schaffer, *ASHRAE, A Practical Guide to Noise and Vibration Control for HVAC Systems*, 1991.

## Chiller Installation, Operation, and Maintenance

H. Crowther and P. Eng, Installing Absorption Chillers, *ASHRAE Journal*, 41–42, 2000.
C.B. Dorgan and C.E. Dorgan, ASHRAE's New Chiller Heat Recovery Application Guide, *ASHRAE Transaction*, 106, Pt. 1, 2000.
C. Julian, A Screw Compressor and Chiller Service Plan, *Contracting Business Interactive*, 2002.

J.Y. Kao, *Evaluation of GSA Maintenance Practices of Large Centrifugal Chillers and Review of GSA Refrigerant Management Practices*, National Institute of Standards and Technology, NISTIR Publication, 5336, 1994.
H.W. Stanford III, *Effective Building Maintenance*, Fairmont Press, 2010.

## COOLING TOWERS

### Cooling Tower General References

ASHRAE, *ASHRAE Handbook, Heating, Ventilating, and Air-Conditionings Systems and Equipment,* Chapter 39, 2008.
R. Burger, *Cooling Tower Technology: Maintenance, Upgrading, and Rebuilding* (3rd Edition), The Fairmont Press, 1994.
N.P. Cheremisinoff and P.N. Cheremisinoff, *Cooling Towers: Selection, Design, and Practice*, SciTech Publishers, Inc., 1989.
Cooling-Tower Motors Save Energy, Reduce Maintenance for University, Baldor Electric Co. (HPAC Engineering, November 2009).
D. Davis, Cooling Tower Doctor, 1999.
J.C. Hensley, ed., *Cooling Tower Fundamentals* (2nd Edition), The Marley Cooling Tower Company, 1985.
G.B. Hill et. al., *Cooling Towers: Principles and Practice* (3rd Edition), Butterworth-Heinemann, *1990.*
J. Katzel, Trends in Cooling Towers, *Plant Engineering*, 2000.
I.F. Kuharic, Psychrometrics and the Psychrometer, Cooling Technology Institute, Paper No. TP81-13, 1981.
T. Pannkoke, Cooling Tower Basics, *Heating/Piping/Air Conditioning*, 137–155, 1996.
RPM ACTM Cooling Tower Direct Drive Motor & VS1CTD Variable Speed Control System, Baldor Electric Company, Publication FL476, 2009.
D. Sellers, Commissioning Cooling Towers, *HPAC Engineering*, 2002.
J.L. Willa, Evolution of the Cooling Tower, Cooling Technology Institute, Paper No. TP91-01, 1991.

### Tower Placement and Layout

Cooling Tower Layout, Electronic Library Documents Nos. 1071, 1074, and 1103, Baltimore Aircoil Company, Baltimore, MD, 1999.
Equipment Layout Manual, Bulletin 311G, EVAPCO, Inc., Westminster, MD, 1999.

### HVAC Systems References

H.W. Stanford III, *Analysis and Design of Heating, Ventilating, and Air-Conditioning Systems*, 2nd Edition, Prentice-Hall, 1988, 2000, unpublished.

### Noise and Vibration

B. Gerglund et al. Eds., *Guidelines for Community Noise*, World Health Organization, 1999.
C.M. Harris, *Noise Control in Buildings*, McGraw-Hill, Inc., 1994.
M.E. Schaffer, ASHRAE, *A Practical Guide to Noise and Vibration Control for HVAC Systems*, 1991.

## Controls

Butterfly Valves, Section Vb2, *Engineering Data Book*, Johnson Controls, Inc., 1990.
*Engineering Manual of Automatic Control for Commercial Buildings* (I-P Edition), Honeywell, Inc., 1997, Publication LC No. 97-072971.

## Water Treatment

B. Smithee, Cooling Tower Maintenance Improves with Ozone, *Facilities Engineering Journal*, 2000.
Fundamentals of Cooling Water Treatment, Annual Conference 2000 Panel Discussion, Cooling Technology Institute, *CTI Journal*, 22(1), 56–74.
*Industrial Water Conditioning* (9th Edition) (Betz Dearborn, 1991).
T.P. Ruisinger, Ozonation in Cooling Water Systems, *Plant Engineering*, 1996.
A. Meitz, Water Treatment for Cooling Towers, *Heating/Piping/Air Conditioning*,125–134, 1999.

## Fire Protection

NFPA 214, *Standard on Water-Cooling Towers*, National Fire Protection Association, Boston.

## Plume Control

P. Lindahl and K. Mortensen, Plume Abatement—The Next Generation, *CTI Journal*, 31(2), 8–23.

### *Legionella*

ASHRAE Guideline 12-2000, *Minimizing the Risk of Legionellosis Associated with Building Water Systems.*
T. Bugler, B. Fields, and R.D. Miller, Cooling Towers, Drift, and Legionellosis, *CTI Journal*, 31(1), 30–47.
A.J. Cooper, H.R. Barnes, and E.R. Myers. Assessing Risk of Legionella, *ASHRAE Journal,* 22–27, 2004.
CTI Guideline WTB-148, *Legionellosis, Guideline: Best Practices for Control of Legionella,* 2008.
*H. Finkelstein, Deadly Legionnaires Disease and HVAC Systems*, The National Resource Center, Washington, DC, 1998.
ISO Standard 11731, *Water Quality—Detection and Enumeration of Legionella,* 1998.
J. Springston, Legionella Lives, CIH, CSP, *Engineered Systems*, 80–90, 2000.
Standard Culture Method, Centers for Disease Control and Prevention, Atlanta, GA.

# Index

## B



## C

## D

## E



## F



## G

## M



## N

## S

## T



## U



## V

Printed in the United States
By Bookmasters